R Graphics

Second Edition

Chapman & Hall/CRC
The R Series

Series Editors

John M. Chambers
Department of Statistics
Stanford University
Stanford, California, USA

Torsten Hothorn
Institut für Statistik
Ludwig-Maximilians-Universität
München, Germany

Duncan Temple Lang
Department of Statistics
University of California, Davis
Davis, California, USA

Hadley Wickham
Department of Statistics
Rice University
Houston, Texas, USA

Aims and Scope

This book series reflects the recent rapid growth in the development and application of R, the programming language and software environment for statistical computing and graphics. R is now widely used in academic research, education, and industry. It is constantly growing, with new versions of the core software released regularly and more than 2,600 packages available. It is difficult for the documentation to keep pace with the expansion of the software, and this vital book series provides a forum for the publication of books covering many aspects of the development and application of R.

The scope of the series is wide, covering three main threads:
- Applications of R to specific disciplines such as biology, epidemiology, genetics, engineering, finance, and the social sciences.
- Using R for the study of topics of statistical methodology, such as linear and mixed modeling, time series, Bayesian methods, and missing data.
- The development of R, including programming, building packages, and graphics.

The books will appeal to programmers and developers of R software, as well as applied statisticians and data analysts in many fields. The books will feature detailed worked examples and R code fully integrated into the text, ensuring their usefulness to researchers, practitioners and students.

R Graphics

Second Edition

Paul Murrell

The University of Auckland

New Zealand

CRC Press
Taylor & Francis Group
Boca Raton   London   New York

CRC Press is an imprint of the
Taylor & Francis Group   an **informa** business

A CHAPMAN & HALL BOOK

CRC Press
Taylor & Francis Group
6000 Broken Sound Parkway NW, Suite 300
Boca Raton, FL 33487-2742

Printed in the United States of America on acid-free paper
Version Date: 20110504

International Standard Book Number: 978-1-4398-3176-2 (Hardback)

Visit the Taylor & Francis Web site at
http://www.taylorandfrancis.com

and the CRC Press Web site at
http://www.crcpress.com

Contents

List of Figures   xiii

List of Tables   xix

Preface   xxi

1 An Introduction to R Graphics   **1**
  1.1   R graphics examples 3
      1.1.1   Standard plots 3
      1.1.2   Trellis plots 4
      1.1.3   The Grammar of Graphics 5
      1.1.4   Specialized plots 5
      1.1.5   General graphical scenes 6
  1.2   The organization of R graphics 18
      1.2.1   Types of graphics functions 20
      1.2.2   Traditional graphics versus **grid** graphics 21

I   TRADITIONAL GRAPHICS   23

2 Simple Usage of Traditional Graphics   **25**
  2.1   The traditional graphics model 26
  2.2   The `plot()` function 26
  2.3   Plots of a single variable 30
  2.4   Plots of two variables 34
  2.5   Plots of many variables 35
  2.6   Arguments to graphics functions 38
      2.6.1   Standard arguments to graphics functions 41
  2.7   Specialized plots 42
  2.8   Interactive graphics 44

3 Customizing Traditional Graphics   **47**
  3.1   The traditional graphics model in more detail 48
      3.1.1   Plotting regions 48
      3.1.2   The traditional graphics state 51
  3.2   Controlling the appearance of plots 57
      3.2.1   Colors 58

3.2.2   Lines 59

3.2.3   Text 60

3.2.4   Data symbols 64

3.2.5   Axes 65

3.2.6   Plotting regions 69

3.2.7   Clipping 71

3.2.8   Moving to a new plot 72

3.3   Arranging multiple plots 72

3.3.1   Using the traditional graphics state 72

3.3.2   Layouts 73

3.3.3   The split-screen approach 76

3.4   Annotating plots 78

3.4.1   Annotating the plot region 78

3.4.2   Annotating the margins 86

3.4.3   Legends 89

3.4.4   Axes 90

3.4.5   Coordinate systems 94

3.4.6   Special cases 102

3.5   Creating new plots 107

3.5.1   A simple plot from scratch 108

3.5.2   A more complex plot from scratch 108

3.5.3   Writing traditional graphics functions 111

II   GRID GRAPHICS   117

4   Trellis Graphics: The lattice Package   119

4.1   The **lattice** graphics model 120

4.1.1   Why another graphics system? 123

4.2   **lattice** plot types 123

4.3   The `formula` argument and multipanel conditioning 126

4.4   The `group` argument and legends 129

4.5   The `layout` argument and arranging plots 129

4.6   The `scales` argument and labeling axes 132

4.7   The `panel` argument and annotating plots 135

4.7.1   Adding output to a **lattice** plot 137

4.8   `par.settings` and graphical parameters 139

4.9   Extending **lattice** plots 142

4.9.1   The **latticeExtra** package 142

5   The Grammar of Graphics: The ggplot2 Package   145

5.1   Quick plots 146

5.2   The **ggplot2** graphics model 148

5.2.1   Why another graphics system? 149

5.2.2   An example data set 149

5.3	Data	150
5.4	Geoms and aesthetics	150
5.5	Scales	154
5.6	Statistical transformations	155
5.7	The **group** aesthetic	159
5.8	Position adjustments	161
5.9	Coordinate transformations	163
5.10	Facets	164
5.11	Themes	166
5.12	Annotating	169
5.13	Extending **ggplot2**	170

6 The grid Graphics Model   **173**

6.1	A brief overview of **grid** graphics	174
	6.1.1 A simple example	175
6.2	Graphical primitives	178
	6.2.1 Standard arguments	184
	6.2.2 Clipping	185
6.3	Coordinate systems	185
	6.3.1 Conversion functions	188
	6.3.2 Complex units	189
6.4	Controlling the appearance of output	192
	6.4.1 Specifying graphical parameter settings	195
	6.4.2 Vectorized graphical parameter settings	196
6.5	Viewports	199
	6.5.1 Pushing, popping, and navigating between viewports	200
	6.5.2 Clipping to viewports	205
	6.5.3 Viewport lists, stacks, and trees	206
	6.5.4 Viewports as arguments to graphical primitives	210
	6.5.5 Graphical parameter settings in viewports	211
	6.5.6 Layouts	212
6.6	Missing values and non-finite values	216
6.7	Interactive graphics	217
6.8	Customizing **lattice** plots	219
	6.8.1 Adding **grid** output to **lattice** output	219
	6.8.2 Adding **lattice** output to **grid** output	220
6.9	Customizing **ggplot2** output	222
	6.9.1 Adding **grid** output to **ggplot2** output	222
	6.9.2 Adding **ggplot2** output to **grid** output	223

7 The grid Graphics Object Model   **227**

7.1	Working with graphical output	228
	7.1.1 Standard functions and arguments	229
7.2	Grob lists, trees, and paths	232
	7.2.1 Graphical parameter settings in gTrees	233

	7.2.2	Viewports as components of gTrees	234
	7.2.3	Searching for grobs	235
7.3		Working with graphical objects off-screen	236
	7.3.1	Capturing output	238
7.4		Placing and packing grobs in frames	239
	7.4.1	Placing and packing off-screen	242
7.5		Other details about grobs	242
	7.5.1	Calculating the sizes of grobs	242
	7.5.2	Calculating the positions of grobs	245
	7.5.3	Editing graphical context	247
7.6		Saving and loading **grid** graphics	249
7.7		Working with **lattice** grobs	250
7.8		Working with **ggplot2** grobs	251

8 Developing New Graphics Functions and Objects   **255**
8.1		An example	256
	8.1.1	Modularity	257
8.2		Simple graphics functions	257
	8.2.1	Embedding graphical output	259
	8.2.2	Facilitating annotation	261
	8.2.3	Editing output	263
	8.2.4	Absolute versus relative sizes	264
8.3		Graphical objects	265
	8.3.1	Overview of creating a new graphical class	265
	8.3.2	Defining a new graphical class	266
	8.3.3	Validating grobs	268
	8.3.4	Drawing grobs	270
	8.3.5	Editing grobs	275
	8.3.6	Querying grobs	279
	8.3.7	Pre-drawing and post-drawing	280
	8.3.8	Summary of graphical object methods	282
	8.3.9	Completing the example	283
	8.3.10	Reusing graphical elements	284
	8.3.11	Other details	285
8.4		Debugging **grid**	296

III   THE GRAPHICS ENGINE   **303**

9 Graphics Formats   **305**
9.1		Graphics devices	306
9.2		Graphical output formats	307
	9.2.1	Vector formats	307
	9.2.2	Raster formats	312
9.3		Including R graphics in other documents	314

9.3.1 LaTeX 315
9.3.2 "Productivity" software 315
9.3.3 Web pages 315
9.4 Device-specific features 316
9.5 Multiple pages of output 317
9.6 Display lists 317
9.7 Extension packages 318

10 Graphical Parameters   **321**
10.1 Colors 321
10.1.1 Semitransparent colors 323
10.1.2 Converting colors 324
10.1.3 Color sets 324
10.1.4 Device Dependency of Color Specifications 326
10.2 Line styles 327
10.2.1 Line widths 327
10.2.2 Line types 327
10.2.3 Line ends and joins 327
10.3 Data symbols 330
10.4 Fonts 331
10.4.1 Font family 331
10.4.2 Font face 331
10.4.3 Multi-line text 333
10.4.4 Locales 333
10.5 Mathematical formulae 333

IV GRAPHICS PACKAGES   **337**

11 Graphics Extensions   **339**
11.1 Tricks with text 339
11.1.1 Drawing formatted text on a plot 340
11.1.2 Avoiding text overlaps 341
11.2 Peculiar primitives 347
11.2.1 Confidence bars 350
11.3 Calculations on colors 351
11.3.1 The **colorspace** package 351
11.3.2 The **RColorBrewer** package 352
11.3.3 The **munsell** package 353
11.3.4 The **dichromat** package 354
11.4 Custom coordinates 354
11.4.1 Converting between traditional coordinate systems . . 354
11.4.2 Subplots 355
11.5 Atypical axes 357

12 Plot Extensions   **361**
12.1   Venn diagrams 362
12.2   Chernoff faces 363
12.3   Ternary plots 365
     12.3.1   Soil texture diagrams 368
12.4   Polar plots 368
     12.4.1   Wind roses 375
12.5   Hexagonal binning 378

13 Graphics for Categorical Data   **381**
13.1   The **vcd** package 381
13.2   XMM-Newton 382
13.3   Plots of categorical data 383
13.4   Categorical data on the y-axis 383
13.5   Visualizing contingency tables 385
13.6   Categorical plot matrices 388
13.7   Multipanel categorical plots 392
13.8   Customizing categorical plots 393
13.9   The **vcdExtra** package 395

14 Maps   **397**
14.1   Map data 399
     14.1.1   The **maps** package 399
     14.1.2   Shapefiles 400
14.2   Map annotation 401
14.3   Complex polygons 405
14.4   Map projections 406
14.5   Raster maps 409
14.6   Other packages 411

15 Node-and-edge Graphs   **413**
15.1   Creating graphs 414
     15.1.1   The **graph** package 414
15.2   Graph layout and rendering 417
     15.2.1   The **Rgraphviz** package 417
     15.2.2   Graph attributes 418
     15.2.3   Customization 420
     15.2.4   Output formats 420
     15.2.5   Hypergraphs 422
15.3   Other packages 423
     15.3.1   The **igraph** package 424
     15.3.2   The **network** package 425
15.4   Diagrams 426
     15.4.1   The **diagram** and **shape** packages 428

16 3D Graphics   **431**
16.1   3D graphics concepts 431
16.2   The Canterbury earthquake 435
16.3   Traditional graphics 435
16.4   **lattice** graphics 439
16.5   The **scatterplot3d** package 440
16.6   The **rgl** package 443
16.7   The **vrmlgen** package 448

17 Dynamic and Interactive Graphics   **453**
17.1   Dynamic graphics 453
       17.1.1   The **animation** package 454
17.2   Interactive graphics 456
       17.2.1   Tools and techniques 459
       17.2.2   The **rggobi** package 461
       17.2.3   The **iplots** package 463
17.3   Graphics GUIs 467
       17.3.1   GUIs for R 467
       17.3.2   GUI toolkits 470
17.4   Interactive graphics for the web 475

18 Importing Graphics   **479**
18.1   The Moon and the tides 480
18.2   Importing raster graphics 481
       18.2.1   Manipulating raster images 483
18.3   Importing vector graphics 484
       18.3.1   The **grImport** package 484
       18.3.2   Manipulating vector images 486

19 Combining Graphics Systems   **489**
19.1   The **gridBase** package 489
       19.1.1   Annotating traditional graphics using **grid** 490
       19.1.2   Traditional graphics in **grid** viewports 491
       19.1.3   Problems and limitations 493

Bibliography   **497**

Index   **507**

List of Figures

1.1   A simple scatterplot   2
1.2   Some standard plots   7
1.3   A customized scatterplot   8
1.4   A dramatized barplot   9
1.5   A Trellis dotplot   10
1.6   A **ggplot2** facetted scatterplot with smoother   11
1.7   A map of New Zealand produced using R   12
1.8   A plot of financial data   13
1.9   A novel decision tree plot   14
1.10 A table-like plot   15
1.11 Didactic diagrams   16
1.12 A music score   17
1.13 An infographic   17
1.14 The structure of the R graphics system   19

2.1   Four variations on a scatterplot   28
2.2   Plotting an `"lm"` object   29
2.3   Plotting an `agnes` object   31
2.4   Plots for a single variable   33
2.5   Plots of two variables   36
2.6   Plots of many variables   39
2.7   Modifying default `barplot()` and `boxplot()` output   40
2.8   Standard arguments for high-level functions   43
2.9   Some specialized plots   45

3.1   The plot regions in traditional graphics   49
3.2   Multiple figure regions in traditional graphics   50
3.3   The user coordinate system in the plot region   51
3.4   Figure margin coordinate systems   52
3.5   Outer margin coordinate systems   53
3.6   Font families and font faces   61
3.7   Alignment of text in the plot region   63
3.8   Data symbols in traditional graphics   64
3.9   Basic plot types   66
3.10 Different axis styles   67
3.11 Graphics state settings controlling plot regions   70

3.12 Some basic layouts 75
3.13 Some complex layouts 77
3.14 Annotating the plot region 80
3.15 Drawing in the plot region 82
3.16 More examples of annotating the plot region 85
3.17 Drawing polygons 87
3.18 Annotating the margins 88
3.19 Some simple legends 91
3.20 Customizing axes 93
3.21 Custom coordinate systems 95
3.22 Overlaying plots 98
3.23 Overlaying output 101
3.24 Special-case annotations 103
3.25 A panel function example 105
3.26 Annotating a 3D surface 107
3.27 A back-to-back barplot 109
3.28 A graphics function template 114

4.1   A scatterplot using **lattice** 121
4.2   A modified scatterplot using **lattice** 122
4.3   Plot types available in **lattice** 125
4.4   A **lattice** scatterplot 127
4.5   A **lattice** multipanel conditioning plot 128
4.6   A **lattice** plot with multiple groups 130
4.7   Controlling the layout of **lattice** panels 131
4.8   Arranging multiple **lattice** plots 133
4.9   Modifying **lattice** axes 134
4.10 Annotating **lattice** plots 136
4.11 A **lattice** panel function 137
4.12 Some default **lattice** settings 141
4.13 Plot types available in **latticeExtra**. 143

5.1   A scatterplot using qplot() 147
5.2   A modified scatterplot using qplot() 147
5.3   Core components in **ggplot2** 149
5.4   Examples of geoms and aesthetics 153
5.5   Examples of scales 156
5.6   Scale components in **ggplot2** 156
5.7   Stat components in **ggplot2** 157
5.8   Examples of statistical transformations 158
5.9   Examples of grouping 160
5.10 Examples of position adjustment 162
5.11 Examples of coordinate transformations 165
5.12 Coord components in **ggplot2** 165
5.13 Examples of facetting 166

5.14 Examples of themes 168
5.15 Example annotations 170

6.1 A simple scatterplot using **grid**. 177
6.2 Primitive **grid** output 180
6.3 Drawing curves 181
6.4 Drawing arrows 182
6.5 Drawing polygons 183
6.6 Drawing paths 184
6.7 A demonstration of **grid** units 189
6.8 Graphical parameters for graphical primitives. 195
6.9 Recycling graphical parameters. 197
6.10 Recycling graphical parameters for polygons 198
6.11 A diagram of a simple viewport 200
6.12 Pushing a viewport 201
6.13 Pushing several viewports 202
6.14 Popping a viewport 203
6.15 Navigating between viewports 204
6.16 Clipping output in viewports 206
6.17 Viewport lists, stacks, and trees 209
6.18 The inheritance of viewport graphical parameters 211
6.19 Layouts and viewports 214
6.20 Layouts and units 215
6.21 Nested layouts 217
6.22 Non-finite values for line-tos, polygons, and arrows 218
6.23 Adding **grid** output to a **lattice** plot 221
6.24 Embedding a **lattice** plot within **grid** output 222
6.25 Adding **grid** output to **ggplot2** 223
6.26 Embedding a **ggplot2** plot within **grid** output 224

7.1 Modifying a `circle` grob 229
7.2 Editing grobs 231
7.3 The structure of a gTree 232
7.4 Editing a gTree 234
7.5 Using a gTree to group grobs 238
7.6 Packing grobs by hand 241
7.7 Calculating the size of a grob 244
7.8 Grob dimensions by reference 245
7.9 Calculating grob locations 246
7.10 Calculating null grob locations 248
7.11 Editing the graphical context 249
7.12 Editing the grobs in a **lattice** plot 252
7.13 Working with **ggplot2** grobs 253

8.1 A plot of oceanographic data 256

16.6   A 3D scatterplot from **lattice** 441
16.7   A 3D scatterplot from **scatterplot3d** 444
16.8   A 3D surface plot from **rgl** 445
16.9   A 3D contour plot of earthquakes 447
16.10 A 3D image constructed from basic shapes 448
16.11 A 3D image in VRML format 450

17.1   A simple animation 454
17.2   An animation in a web page 456
17.3   Two static plots 457
17.4   A static **lattice** plot 457
17.5   Two interactive plots 458
17.6   Two linked plots 459
17.7   The GGobi main window 462
17.8   Two **iplots** interactive plots 464
17.9   Two **iplots** custom plots 466
17.10 An **Rcmdr** dialog box 468
17.11 The Lattice Explorer 468
17.12 A **latticist** window 469
17.13 Two **playwith** plots 471
17.14 A clock function 472
17.15 An interactive clock 473
17.16 Linked plots in a web browser 477

18.1   Two images of the Moon 480
18.2   Adding an imported image to a plot 482
18.3   Two raster images 484
18.4   Manipulating raster images 485
18.5   Breaking a vector image into paths 487
18.6   Variations on the Moon 487

19.1   Annotating a traditional plot with **grid** 491
19.2   Embedding a traditional plot within **lattice** output 494

List of Tables

2.1 Plots for a single variable 32
2.2 Plots of two variables 35
2.3 Plots of many variables 37

3.1 High-level traditional graphics state settings 55
3.2 Low-level traditional graphics state settings 56
3.3 Read-only traditional graphics state settings 57
3.4 Traditional graphical primitives 79
3.5 Traditional coordinate systems 97

4.1 Plotting functions in **lattice** 124
4.2 **lattice** Panel functions 138
4.3 Plotting functions in **latticeExtra** 143
4.4 **latticeExtra** Panel functions 144

5.1 Geoms and Aesthetics in **ggplot2** 152
5.2 Scales in **ggplot2** 157
5.3 Stats in **ggplot2** 159
5.4 Plot elements in **ggplot2** 167

6.1 Graphical primitives in **grid** 179
6.2 Coordinate systems in **grid** 186
6.3 Graphical parameters in **grid** 193
6.4 **grid** Font faces 196

7.1 Functions for working with grobs 230

9.1 Graphical output formats 308
9.2 Graphics device packages 319

10.1 Functions to generate color sets 325
10.2 Font families 332
10.3 Font faces 332

18.1 Packages for reading raster images 481

Preface

R is a popular open source software tool for statistical analysis and graphics. This book focuses on the very powerful graphics facilities that R provides for the production of publication-quality diagrams and plots.

What this book is about

This book describes the graphics system in R. The first chapter provides an overview of the R graphics facilities. There are several figures that demonstrate the variety and complexity of plots and diagrams that can be produced using R and there is a description of the overall organization of the R graphics facilities, so that the user has some idea of where to find a function for a particular purpose.

The most important feature of the R graphics setup is the existence of two distinct *graphics systems* within R: the traditional graphics system and the **grid** graphics system. Section 1.2.2 offers some advice on which system to use.

Part I of this book is concerned with the traditional graphics system, which implements many of the "traditional" graphics facilities of the S language (originally developed at Bell Laboratories and available in a commercial implementation as S-PLUS). The majority of R graphics functions are based upon this system. The chapters in this part of the book describe how to work with the traditional graphics functions, with a particular emphasis on how to modify or add output to a plot to produce exactly the right final output. Chapter 2 describes the functions that are available to produce complete plots and Chapter 3 focuses on how to customize the details of plots, combine multiple plots, and add further output to plots.

Part II describes the **grid** graphics system, which is unique to R and is much more powerful than the traditional system. The graphics facilities that are based on the **grid** graphics system are further split into three major graphics packages.

Deepayan Sarkar's **lattice** package provides a complete and coherent set of graphics functions for producing plots, based on Bill Cleveland's Trellis graphics paradigm. This is described in Chapter 4.

Hadley Wickhams' **ggplot2** package provides another complete and coherent set of graphics functions for producing plots, this time based on Leland Wilkinson's Grammar of Graphics paradigm. This is described in Chapter 5.

Finally, there is the **grid** package itself, which provides a low-level, general-purpose graphics system for producing a wide variety of images, including plots. Both **lattice** and **ggplot2** use **grid** to draw plots, but both can be used without directly encountering **grid**. The **grid** package can be used on its own, or as a low-level way to work with plots produced by **lattice** or **ggplot2**. The remaining chapters in Part II describe how the **grid** system can be used to produce graphical scenes starting from a blank page. In particular, there is a discussion of how to use **grid** to develop new graphical functions that are easy for other people to use and build on.

Underlying both traditional and **grid** graphics systems is a *graphics engine*, which represents a common set of fundamental graphics facilities, such as color management and support for different graphical output formats. These facilities are described in Part III of this book.

Part IV provides a series of brief overviews of some of the major areas of application for R graphics and some of the major extensions of R graphics:

- Chapter 11 describes some packages that provide additional low-level graphics facilities, such as unusual shapes or fancy fill patterns.
- Chapter 12 describes some packages that provide special types of plots, such as Venn diagrams and Chernoff faces.
- Chapter 13 describes functions for drawing plots of categorical data.
- Chapter 14 describes packages for drawing maps.
- Chapter 15 describes packages for drawing node-and-edge graphs.
- Chapter 16 describes functions for drawing three-dimensional plots.
- Chapter 17 describes packages that provide dynamic and interactive plots.
- Chapter 18 describes packages for reading external graphics files into R and including them in plots.
- Chapter 19 describes functions that allow traditional graphics and **grid** graphics to be used together.

Changes in the second edition

Six years have elapsed since *R Graphics* was first published, which is a very long time in the world of computing.

Much of the graphics system in R that was described in the first edition of this book still exists and is still being heavily used, but there have been numerous changes in some of the details. One purpose of this second edition is to provide updated information on the core graphics engine, the traditional graphics system, the **grid** graphics system, and the **lattice** package. The material has been slightly rearranged so that the fine details of the graphics engine, which are common to all graphics systems, are collected together in the new Part III.

The largest changes have occurred in the number and variety of graphics packages that extend the graphics capabilities of R. One particularly important addition is Hadley Wickham's **ggplot2** package, which provides another complete graphics system, based on Leland Wilkinson's Grammar of Graphics ideas. This package is afforded an entire new chapter of its own, alongside the existing chapter on **lattice**.

The entirely new Part IV has been added to provide an overview of the plethora of graphics extension packages. These chapters cover such topics as geographic maps, dynamic and interactive graphics, and node-and-edge graphs.

The one major deletion from the first edition involves the removal of Appendix A, which used to provide a brief introduction to the R language itself. There are now many books that introduce R and it is now assumed that the reader of this book has already gained at least a basic familiarity with the R language and with R data structures.

What this book is *still* not about

This book does *not* contain discussions about which sort of plot is most appropriate for a particular sort of data, nor does it contain guidelines for correct graphical presentation. In fact, instructions are provided for producing some types of plots and graphical elements that are generally disapproved of, such as pie charts and cross-hatched fill patterns.

The information in this book is meant to be used to produce a plot once the format of the plot has been decided upon and to experiment with different ways of presenting a set of data. No plot types are deliberately excluded, partly because no plot type is all bad (e.g., a pie chart can be a very effective way to represent a simple proportion) and partly because some graphical elements, such as cross-hatching, might be required by a particular publisher.

The flexibility of R graphics encourages the user *not* to be constrained to thinking in terms of just the traditional types of plots. The aim of this book is to provide lots of useful tools and to describe how to use them. There are many other sources of information on graphical guidelines and recommended plot types, some of which are mentioned below.

Most introductory statistics text books will contain basic guidelines for selecting an appropriate type of plot. Examples of books that deal specifically with the construction of effective plots and are aimed at a general audience are *Creating More Effective Graphs* by Naomi Robbins and Edward Tufte's *Visual Display of Quantitative Information* and *Envisioning Information.* For more technical discussions of these issues, see *Graphics For Statistics and Data Analysis With R* by Kevin Keene, *Visualizing Data* and *Elements of Graphing Data* by Bill Cleveland, and *The Grammar of Graphics* by Leland Wilkinson.

For ideas on appropriate graphical displays for particular types of analysis or particular types of data, some starting points are *Data Analysis and Graphics Using R* by John Maindonald and John Braun, *An R and S-Plus Companion to Applied Regression* by John Fox, *Statistical Analysis and Data Display* by Richard Heiberger and Burt Holland, and *Visualizing Categorical Data* by Michael Friendly.

This book is also *not* a complete reference to the R system. There are many freely available documents that provide both introductory and in-depth explanations of the R system. The best place to start is the "Documentation" section on the home page of the R project web site (see "On the web" on page xxvi). Two examples of introductory texts are *Introductory Statistics with R* by Peter Dalgaard and *Using R for Introductory Statistics* by John Verzani.

While a significant amount of new material has been added in this edition, this volume does not come close to providing a comprehensive description of all of the graphics facilities in the R ecosystem. It can only be said that this edition is less uncomprehensive than the first.

In particular, this edition is still missing a description of plots for specific types of data analysis. It could be said that the plotting functions in this book describe graphics that are driven by data, but not graphics that are driven by statistical models.

This means that some quite large and well-known packages with significant graphical content have been omitted, including Frank Harrell's **Design** package, John Fox's **car** and **effects** packages, Heiberger and Holland's **HH** package, and a whole raft of packages from the Bioconductor project. These omissions are a reflection of the time, space, and intellectual constraints of the author of this book and bear no reflection on the packages themselves.

Who should read this book

This book should be of interest to a variety of R users. For people who are new to R, this book provides an overview of the graphics facilities, which is useful for understanding what to expect from R's graphics functions and how to modify or add to the output they produce. For this purpose, Chapter 1 is the place to start. In particular, the discussion of which graphics system to use in Section 1.2.2 will be of interest. Chapters 2, 4, and 5 provide relatively brief introductions to the major packages that produce standard plots, so it should be possible to get started fairly quickly using one of those chapters.

For intermediate-level R users, this book provides all of the information necessary to perform sophisticated customizations of plots produced in R. As with many software applications, it is possible to work with R for years and remain unaware of important and useful features. This book will be useful in making users aware of the full scope of R graphics, and in providing a description of the correct model for working with R graphics. Chapters 3, 6, and 7 contain a lot of this detailed information about how R graphics works.

For readers who are familiar with R, but wish to use R to produce specific sorts of graphics, there may be a chapter in Part IV that provides relevant information to get started.

For advanced R users, this book contains vital information for producing coherent, reusable, and extensible graphics functions. Advanced users should pay particular attention to Chapters 6, 7, and 8.

Conventions used in this book

This book describes a large number of R functions and there are many code examples. Samples of code that could be entered interactively at the R command line are formatted as follows:

```
> 1:10
```

where the > denotes the R command-line prompt and everything else is what the user should enter. When an expression is longer than a single line it will look like the following, with the additional lines indented appropriately:

```
> plot(1:10, 1:10, col="blue", lty="dashed",
       axes=FALSE, type="l")
```

Often, the functions described in this book are used for the side effect of producing graphical output, so the result of running a function is represented by a figure. In cases where the result of a function is a value that we might be interested in, the result will be shown below the code that produced it and will be formatted as follows:

```
[1]  1  2  3  4  5  6  7  8  9 10
```

In some places, an entirely new R function is defined. Such code would normally be entered into a script file and loaded into R in one step (rather than being entered at the command line), so the code for new R functions will be presented in a figure and formatted as follows:

```
1   myfun <- function(x, y) {
2     plot(x ,y)
3   }
```

with line numbers provided for easy reference to particular parts of the code from the main text.

When referring to a function within the main text, it will be formatted in a `typewriter` font and will have parentheses after the function name, e.g., `plot()`.

When referring to the arguments to a function or the values specified for the arguments, they will also be formatted in a `typewriter` font, but they will not have any parentheses at the end, e.g., `x`, `y`, or `col="red"`.

When referring to an S3 class, statements will be of the form: "the `"classname"` class," using a typewriter font with the class name in double quotes. However, when referring to an object that is an instance of a class, statements will be of the form: "the `classname` object," using a typewriter font, but without the double quotes around the class name.

All package names are in **bold** and names of software and computer languages and formats are in Sans Serif.

On the web

There is a web site with errata and links to pages of PNG versions of all figures from the book and the R code used to produce them:

`http://www.stat.auckland.ac.nz/~paul/RG2e/`

There is also an **RGraphics** package containing functions to produce the figures in this book and all functions, classes, and methods defined in the book.

The **RGraphics** package and most of the packages mentioned in this book are available from the Comprehensive R Archive Network (CRAN):

`http://cran.r-project.org/`

Packages that are not on CRAN will be found on the Bioconductor web site, or the Omegahat Project web site, or, in rare cases, on the R-Forge web site:

`http://www.bioconductor.org/`
`http://www.omegahat.org/`
`http://r-forge.r-project.org/`

Version information

Software development is an ongoing process and this book can only provide a snapshot of R's graphics facilities. The descriptions and code samples in this book are accurate for R version 2.13.0, but future changes are inevitable. Much of the content of Parts I, II, and III is also accurate for earlier versions of R, but specific areas of incompatibility are not indicated in the text.

A new "minor" version of R is released approximately every six months. The most up-to-date information on the most recent versions of R and **grid** are available in the on-line help pages and at the home page for the R Project:

http://www.R-project.org/

Acknowledgments

One advantage of writing a new edition of this book is that it provides one of the rare opportunities to express in print my thanks to colleagues in the R Core team of developers for the work that they do to make R the powerful, reliable, and fun system that it is.

An enormous debt of gratitude is also due to the wider group of people responsible for the smooth and sane expansion of the R universe, including places like CRAN and the R-Forge web site.

Wider still is the group of enthusiastic useRs who have made the transition to developeRs and produced a staggering number of graphics extension packages for R. Thanks for giving me so much interesting software to write about and apologies for not doing them full justice. I'm sorry I couldn't include more.

Special thanks to Roger Bivand for helpful comments and suggestions for the maps chapter of this book.

Last, and most, and always, thank you Ju.

Paul Murrell
The University of Auckland
New Zealand

This manuscript was generated on a CentOS 5 Linux system using LaTeX, numerous GNU tools, the GIMP, ghostscript, ImageMagick, **Sweave**, and about 100 R packages. Kudos to them all!

1

An Introduction to **R** Graphics

Chapter preview

This chapter provides the most basic information to get started producing plots in R. First of all, there is a three-line code example that demonstrates the fundamental steps involved in producing a plot. This is followed by a series of figures to demonstrate the range of images that R can produce. There is also a section on the organization of R graphics giving information on where to look for a particular function.

The following code provides a simple example of how to produce a plot using R (see Figure 1.1).

```
> plot(pressure)
> text(150, 600,
        "Pressure (mm Hg)\nversus\nTemperature (Celsius)")
```

The expression `plot(pressure)` produces a scatterplot of pressure versus temperature, including axes, labels, and a bounding rectangle. The call to the `text()` function adds the label at the data location (150, 600) within the plot.

This example is basic R graphics in a nutshell. In order to produce graphical output, the user calls a series of graphics functions, each of which produces either a complete plot or adds some output to an existing plot. R graphics follows a "painters model," which means that graphics output occurs in steps, with later output drawn on top of any previous output.

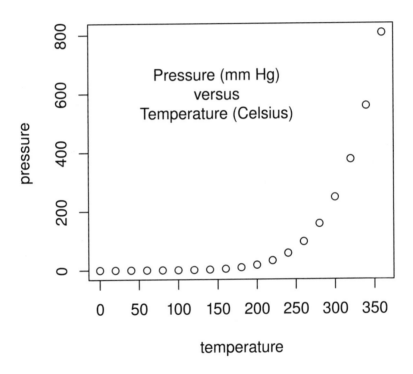

Figure 1.1
A simple scatterplot of vapor pressure of mercury as a function of temperature. The plot is produced from two simple R expressions: one expression to draw the basic plot, consisting of axes, data symbols, and bounding rectangle, and another expression to add the text label within the plot.

There are very many graphical functions provided by R and the extension packages for R so, before describing individual functions, Section 1.1 demonstrates the variety of results that can be achieved. This should provide some idea of what users can expect to be able to achieve with R graphics.

Section 1.2 gives an overview of how the graphics functions in R are organized. This should provide users with some basic ideas of where to look for a function to do a specific task. By the end of this chapter, the reader will be in a position to start understanding in more detail the core R functions that produce graphical output.

1.1 R graphics examples

This section provides an introduction to R graphics by way of a series of examples. None of the code used to produce these images is shown, but it is available from the web site for this book. The aim for now is simply to provide an overall impression of the range of graphical images that can be produced using R. The figures are described over the next few pages and the images themselves are all collected together on pages 7 to 17.

1.1.1 Standard plots

R provides the usual range of standard statistical plots, including scatterplots, boxplots, histograms, barplots, pie charts, and basic 3D plots. Figure 1.2 shows some examples.

In R, these basic plot types can be produced by a single function call (e.g., `pie(pie.sales)` will produce a pie chart), but plots can also be considered merely as starting points for producing more complex images. For example, in the top-left scatterplot in Figure 1.2, a text label has been added within the body of the plot (in this case to show a subject identification number) and a secondary y-axis has been added on the right-hand side of the plot. Similarly, in the histogram, lines have been added to show a theoretical normal distribution for comparison with the observed data. In the barplot, labels have been added to the elements of the bars to quantify the contribution of each element to the total bar and, in the boxplot, a legend has been added to distinguish between the two data sets that have been plotted.

This ability to add several graphical elements together to create the final result is a fundamental feature of R graphics. The flexibility that this allows is demonstrated in Figure 1.3, which illustrates the estimation of the original number of vessels based on broken fragments gathered at an archaeological site: a measure of "completeness" is obtained from the fragments at the site; a theoretical relationship is used to produce an estimated range of "sampling fraction" from the observed completeness; and another theoretical relationship dictates the original number of vessels from a sampling fraction. This plot is based on a simple scatterplot, but requires the addition of many extra lines, polygons, and pieces of text, and the use of multiple overlapping coordinate systems to produce the final result.

R graphics allows fine control of very low-level aspects of a plot and these features can be used to produce some dramatic effects (at the risk of detracting from the message in the data). Figure 1.4 demonstrates one such example, where a simple barplot of tiger population levels has been embellished with an image of the head of a tiger.

For more information on the R functions that produce these standard plots, see Chapter 2. Chapter 3 describes the various ways that further output can be added to a plot.

1.1.2 Trellis plots

In addition to the traditional statistical plots, R provides an implementation of Trellis plots via the package lattice by Deepayan Sarkar. Trellis plots embody a number of design principles proposed by Bill Cleveland that are aimed at ensuring accurate and faithful communication of information via statistical plots. These principles are evident in a number of new plot types in Trellis and in the default choice of colors, symbol shapes, and line styles provided by Trellis plots. Furthermore, Trellis plots provide a feature known as *multipanel conditioning*, which creates multiple plots by splitting the data being plotted according to the levels of other variables.

Figure 1.5 shows an example of a Trellis plot. The data are yields of several different varieties of barley at six sites, over two years. The plot consists of six *panels*, one for each site. Each panel consists of a dotplot showing yield for each variety with different symbols used to distinguish different years, and a *strip* showing the name of the site.

For more information on the Trellis system and how to produce Trellis plots using the **lattice** package, see Chapter 4.

1.1.3 The Grammar of Graphics

Leland Wilkinson's Grammar of Graphics provides another completely differ-ent paradigm for producing statistical plots and this approach to plotting has been implemented for R by Hadley Wickham's **ggplot2** package.

One advantage of this package is that it makes it possible to create a very wide variety of plots from a relatively small set of fundamental components. The **ggplot2** package also has a feature called *facetting*, which is similar to **lattice**'s multipanel plots.

Figure 1.6 shows an example of a plot that has been produced using **ggplot2**. For more information on the **ggplot2** package, see Chapter 5.

1.1.4 Specialized plots

As well as providing a wide variety of functions that produce complete plots, R provides a set of functions for producing graphical output primitives, such as lines, text, rectangles, and polygons. This makes it possible for users to write their own functions to create plots that occur in more specialized ar-eas. There are many examples of special-purpose plots in extension packages for R. For example, Figure 1.7 shows a map of New Zealand produced using R and the extension packages **maps**, **mapdata**, and **mapproj**. Figure 1.8 shows another example: a financial chart produced by the **quantmod** pack-age. The chapters in Part IV describe many different packages and functions that produce different sorts of plots. Chapter 14 provides more information on drawing maps with R.

In some cases, researchers are inspired to produce a totally new type of plot for their data. R is not only a good platform for experimenting with novel plots, but it is also a good way to deliver new plotting techniques to other researchers. Figure 1.9 shows a novel display for decision trees, visualizing the distribution of the dependent variable in each terminal node (produced using the **party** package).

For more information on how to generate a plot starting from an empty page with traditional graphics functions, see Chapter 3. The **grid** package provides even more power and flexibility for producing customized graphical output (see Chapters 6 and 7), especially for the purpose of producing functions for others to use (see Chapter 8).

1.1.5 General graphical scenes

The generality and flexibility of R graphics make it possible to produce graphical images that go beyond what is normally considered to be statistical graphics, although the information presented can usually be thought of as data of some kind. A good mainstream example is the ability to embed tabular arrangements of text as graphical elements within a plot as in Figure 1.10. This is a standard way of presenting the results of a meta-analysis.

R has also been used to produce figures that help to visualize important concepts or teaching points. Figure 1.11 shows two examples that provide a geometric representation of extensions to F-tests (provided by Arden Miller). A more unusual example of a general diagram is provided by the musical score in Figure 1.12 (provided by Steven Miller). R graphics can even be used to produce infographics like Figure 1.13. These examples tend to require more effort to achieve the final result as they cannot be produced from a single function call.

These examples present only a tiny taste of what R graphics (and clever and enthusiastic users) can do. They highlight the usefulness of R graphics not only for producing what are considered to be standard plot types, for little effort, but also for providing tools to produce final images that are well beyond the standard plot types, including going beyond the boundaries of what is normally considered statistical graphics.

The R Graph Gallery web site provides a much more comprehensive range of graphics examples and the interested reader might also like to browse the R Graphical Manual web site. The R Wiki also has a "Tips & Tricks" section on graphics, with lots of examples.[*]

---

[*]`http://addictedtor.free.fr/graphiques/`; `http://rgm2.lab.nig.ac.jp/RGM2/` `http://rwiki.sciviews.org/`.

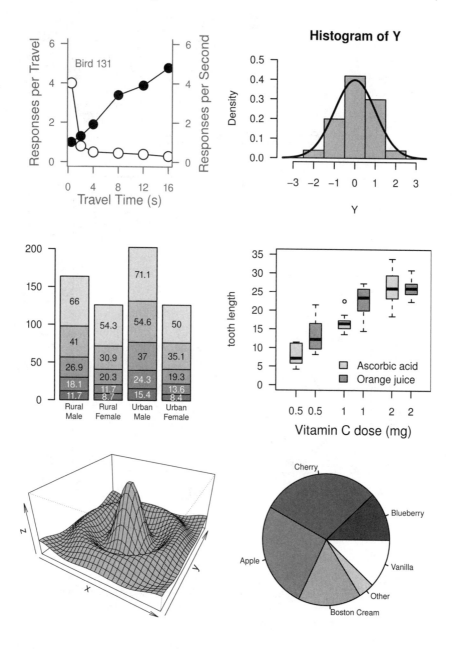

Figure 1.2

Some standard plots produced using R: (from left-to-right and top-to-bottom) a scatterplot, a histogram, a barplot, a boxplot, a 3D surface, and a pie chart. In the first four cases, the basic plot type has been augmented by adding additional labels, lines, and axes.

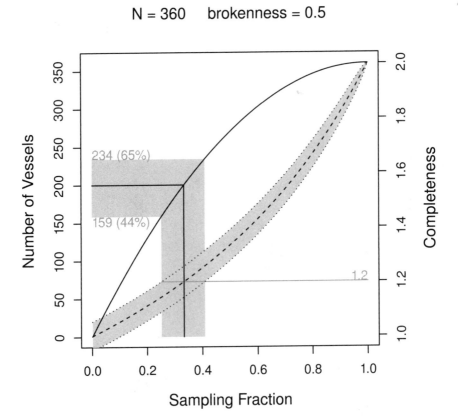

Figure 1.3
A customized scatterplot produced using R. This is created by starting with a simple scatterplot and augmenting it by adding an additional y-axis and several additional sets of lines, polygons, and text labels.

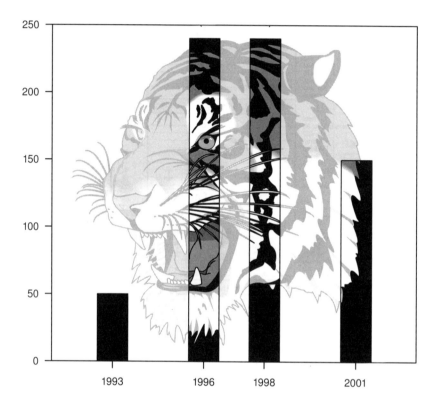

Figure 1.4
A dramatized barplot produced using R. This is created by starting with a simple barplot and augmenting it by adding a background image in light gray, with bolder sections of the image drawn in each bar.

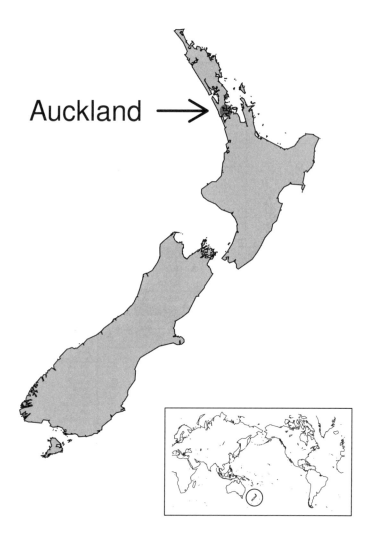

Figure 1.7
A map of New Zealand produced using the **maps** package, the **mapdata** package,
and the **mapproj** package. The map (of New Zealand) is drawn as a series of
polygons, and then text, an arrow, and a data point have been added to indicate the
location of Auckland, the birthplace of R. A separate world map has been drawn in
the bottom-right corner, with a circle to help people locate New Zealand.

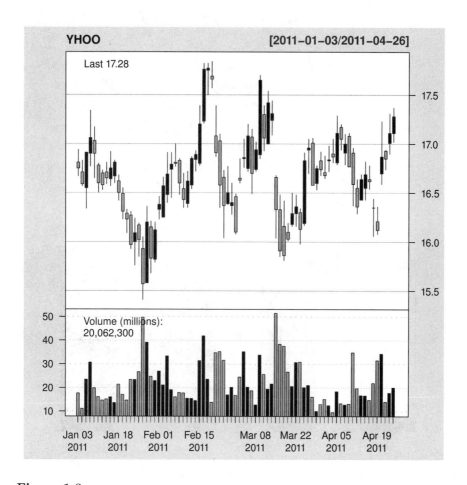

Figure 1.8
A financial chart produced with the `chartSeries()` function from the **quantmod** package.

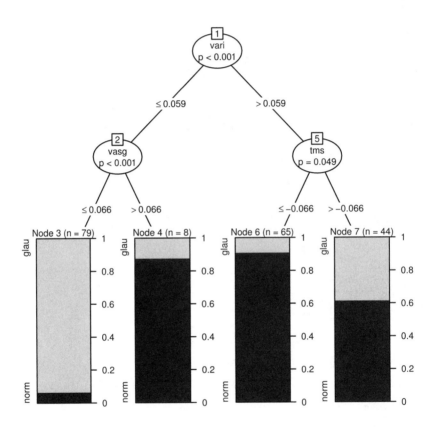

Figure 1.9
A novel decision tree plot, visualizing the distribution of the dependent variable in
each terminal node. Produced using the **party** package.

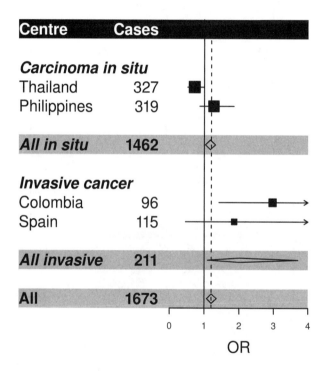

Figure 1.10
A table-like plot produced using R. This is a typical presentation of the results from
a meta-analysis.

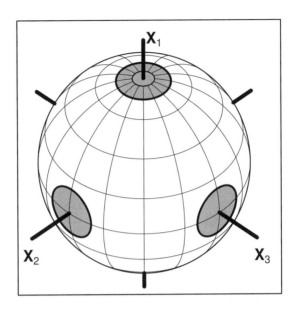

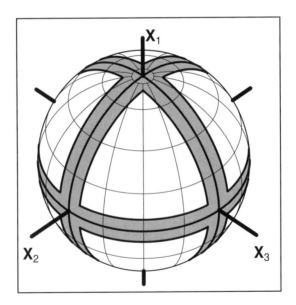

Figure 1.11
Didactic diagrams produced using R and functions provided by Arden Miller. The
figures show a geometric representation of extensions to F-tests.

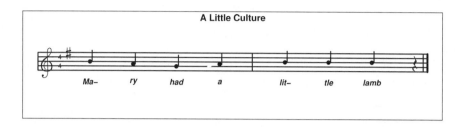

Figure 1.12

A music score produced using R (code by Steven Miller).

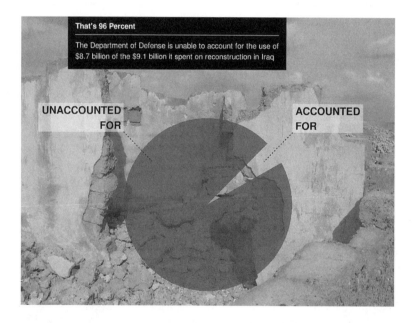

Figure 1.13

An infographic showing the proportion of aid money unaccounted for in the reconstruction of Iraq. This image is a remix of a blog post: http://www.good.is/post/infographic-where-did-the-money-to-rebuild-iraq-go/. The background image is from Adam Henning's flickr photostream: http://www.flickr.com/photos/adamhenning/66822173/.

1.2   The organization of R graphics

This section briefly describes how the functions in the R graphics universe are organized so that the user knows where to start looking for a particular function.

At the heart of the graphics facilities in R lies the package **grDevices**, which will be referred to as the *graphics engine*. This provides fundamental infrastucture for graphics in R, such as selecting colors and fonts and selecting a graphics output format. Although almost all graphics applications in R make use of this package, a lot can be achieved with just basic knowledge, so a detailed description of the functions in this package is delayed until Part III of this book.

Two packages build directly on top of the graphics engine: the **graphics** package and the **grid** package. These represent two largely incompatible *graphics systems* and they divide the bulk of graphics functionality in R into two separate worlds.

The **graphics** package, which will be referred to as the *traditional* graphics system, provides a complete set of functions for creating a wide variety of plots *plus* functions for customizing those plots in very fine detail. It is described in Part I of this book.

The **grid** package provides a separate set of basic graphics tools. It does not provide functions for drawing complete plots, so it is not often used directly to produce statistical plots. It is more common to use functions from one of the graphics packages that are built on top of **grid**, especially either the **lattice** package or the **ggplot2** package. These three packages, which make up the majority of the **grid** graphics world in R, are described in Part II of this book.

Many other *graphics packages* are built on top of either the **graphics** package or the **grid** package. For example, the **maps** package provides functions for drawing maps in the traditional graphics world, and the **pixmap** package provides functions for including external raster images, particularly within plots produced by the **graphics** package, while the **grImport** package provides functions for including external vector images, particularly within pictures that have been produced within the **grid** world. Only a small selection of these graphics packages is shown in Figure 1.14. Part IV of this book provides an introduction to a broader selection of extension packages.

Several graphics package exist largely independently of the main graphics facilities in R. These provide interfaces between R and third-party graphics

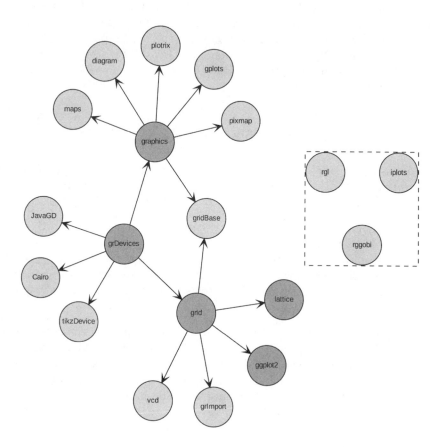

Figure 1.14

The structure of the R graphics system. The packages with darker gray backgrounds form the core of the graphics system. The **graphics** package is described in Part I, **grid**, **lattice**, and **ggplot2** are described in Part II, and **grDevices** is described in Part III. The packages with lighter gray backgrounds are examples of the packages that extend the core system and those are described in Part IV. Some packages provide a stand-alone graphics system and these are represented by the trio of packages in the diagram that have no connection to the core graphics packages.

systems, such as OpenGL for sophisticated 3D images (the **rgl** package), and dynamic and interactive graphics systems (the **rggobi** and **iplots** packages). Basic introductions to these packages are included in Part IV.

Finally, there are several packages that provide additional *graphics devices* for R (see Section 9.1). These allow R graphics output to be embedded nicely within other systems, such as the **tikzDevice** package for producing plots for inclusion within a LATEX document, and the **JavaGD** package for incorporating R graphics output within a Java application. These packages receive a mention in Part III of this book.

1.2.1   Types of graphics functions

Functions in the graphics systems and graphics packages can be broken down into two main types: *high-level* functions that produce complete plots and *low-level* functions that add further output to an existing plot.

The traditional system, or graphics packages built on top of it, provide the majority of the high-level functions currently available in R. The most significant exceptions are the **lattice** package (see Chapter 4) and the **ggplot2** package (see Chapter 5), which provide complete plots based on the **grid** system. Both the traditional and **grid** systems provide many low-level graphics functions.

Most functions in graphics packages produce complete plots and typically offer specialized plots for a specific sort of analysis or a specific field of study. For example: the **hexbin** package has functions for producing hexagonal binning plots for visualizing large amounts of data (see Section 12.5); the **maps** package provides functions for visualizing geographic data (see Section 14.1.1); and the package **scatterplot3d** produces a variety of three-dimensional plots (see Section 16.5). If there is a need for a particular sort of plot, there is a reasonable chance that someone has already written a function to do it. For example, a common request on the R-help mailing list is for a way to add error bars to scatterplots or barplots and this can be achieved in many different ways, for example, using the traditional `arrows()` function, using the `plotCI()` function from the **gplots** package, or the `errbar()` function from the **Hmisc** package. There are some search facilities linked off the main R home page web site to help to find a particular function for a particular purpose. The **sos** package provides a nice web interface for the search with the `findFn()` function.

1.2.2 Traditional graphics versus grid graphics

The existence of two distinct graphics systems in R, the traditional graphics world versus the **grid** graphics world, raises the issue of when to use each system.

For the purpose of producing complete plots from a single function call, which graphics system to use will largely depend on what type of plot is required. The choice of graphics system is largely irrelevant if no further output needs to be added to the plot.

If it is necessary to add further output to a plot, the most important thing to know is which graphics system was used to produce the original plot. In general, the same graphics system should be used to add further output (though the **gridBase** package, described in Chapter 19, provides one way to get around this restriction).

For a wide range of standard plots, it will be possible to produce the same sort of plot in three different styles, using functions from any one of the **lattice**, **ggplot2**, or **graphics** packages. As a general rule, the default style of the **lattice** and **ggplot2** packages will often be superior because they are both motivated by principles of human perception and designed to make it easier to extract information from a plot.

Both the **lattice** and **ggplot2** packages also provide more sophisticated support for visualizing multivariate data sets where, for example, a simple scatterplot between two continuous variables may be augmented by having separate lines or distinct plotting symbols for different subgroups within the data, or by having entire separate plots for different subgroups.

The price of the additional advanced features of both **lattice** and **ggplot2** is that there is a steeper learning curve required to master their respective conceptual frameworks. For **lattice**, there is a particular effort required to learn how to make significant customizations of the default style, while for **ggplot2**, the overall philosophy takes some getting used to, although once grasped it provides a more coherent and powerful paradigm.

In summary, given the choice, it may be quicker to get going with traditional graphics, but both **lattice** and **ggplot2** offer more efficient and sophisticated options in the long run.

For more specialized plots, such as geographic maps or node-and-edge graphs, the choice will be driven by which packages provide the relevant functionality (see the chapters in Part IV of this book).

A different problem is that of producing an image for which there is no existing function, which requires resorting to low-level graphics functions. For this situation, the **grid** system offers the benefit of a much wider range of possibilities than the low-level function in the traditional system, at the cost of having to learn a few additional concepts.

If the goal is to create a new graphical function for others to use, **grid** again provides better support, compared to the traditional system, for producing more general output that can be combined with other output more easily.

One final consideration is speed. None of the graphics systems could be described as blindingly fast, but the **grid**-based systems are noticeably slower than traditional graphics and that performance penalty may be important in some applications.

Chapter summary

The R graphics system consists of a core graphics engine and two low-level graphics systems: traditional and **grid**. The traditional system also includes high-level functions for producing complete plots. The **lattice** package and the **ggplot2** package provide high-level plotting systems on top of **grid**. Many extension packages provide further graphical facilities for both graphics systems, which means that it is possible to create a very wide range of plots and general graphical images with R.

Part I

TRADITIONAL GRAPHICS

```
> help(barplot)
```

Another useful way of learning about a graphics function is to use the `example()` function. This runs the code in the "Examples" section of the help page for a function. The following code runs the examples for `barplot()`.

```
> example(barplot)
```

2.1   The traditional graphics model

As described at the start of Chapter 1, a plot is created in traditional graphics by first calling a high-level function that creates a complete plot, then calling low-level functions to add more output if necessary.

If there is only one plot per page, then a high-level function starts a new plot on a new page. There may be multiple plots on a page, in which case a high-level function starts the next plot on the same page, only starting a new page when the number of plots per page is exceeded (see Section 3.3). All low-level functions add output to the current plot. It is not generally possible to go back to a previous plot in the traditional graphics system (see Section 3.3.3 for an exception).

2.2   The plot() function

The most important high-level function in traditional graphics is the `plot()` function. In many situations, this provides the simplest way to produce a complete plot in R.

The first argument to `plot()` provides the data to plot and there is a reasonable amount of flexibility in the way that the data can be specified. For example, each of the following calls to `plot()` can be used to produce the scatterplot in Figure 1.1 (with small variations in the axis labels). In the first case, all of the data to plot are specified in a single data frame. In the second case, separate x and y variables are specified as two separate arguments. In the third case, the data to plot are specified as a formula of the form `y ~ x`, plus a data frame that contains the variables mentioned in the formula.

```
> plot(pressure)
> plot(pressure$temperature, pressure$pressure)
> plot(pressure ~ temperature, data=pressure)
```

Traditional graphics does not make a major distinction between, for example, scatterplots that only plot data symbols at each (x, y) location and scatterplots that draw straight lines connecting the (x, y) locations (line plots). These are just variations on the basic scatterplot, controlled by a `type` argument. This is demonstrated by the following code, which produces four different plots by varying the value of the `type` argument (see Figure 2.1).

```
> y <- rnorm(20)
> plot(y, type="p")
> plot(y, type="l")
> plot(y, type="b")
> plot(y, type="h")
```

Traditional graphics also does not make a distinction between a plot of a single set of data and a plot containing multiple series of data. Additional data series can be added to a plot using low-level functions such as `points()` and `lines()` (see Section 3.4.1; also see the function `matplot()` in Section 2.5).

The `plot()` function is *generic*. One consequence of this has just been described; the `plot()` function can cope with the same data being specified in several different *formats* (and it will produce the same result). However, the fact that `plot()` is generic also means that if `plot()` is given different *types* of data, it will produce different types of plots. For example, the `plot()` function will produce boxplots, rather than a scatterplot, if the x variable is a factor, rather than a numeric vector. Another example is shown in the code below. Here an "lm" object is created from a call to the `lm()` function. When this object is passed to the `plot()` function, the special plot method for "lm" objects produces several regression diagnostic plots (see Figure 2.2).*

```
> lmfit <- lm(sr ~ pop15 + pop75 + dpi + ddpi,
              data = LifeCycleSavings)
> plot(lmfit)
```

In order to learn more about the "lm" method for the `plot()` function, type `help(plot.lm)`.

---

*The data used in this example are measures relating to the savings ratio (aggregate personal saving divided by disposable income) averaged over the period 1960-1970 for 50 countries, available as the data set `LifeCycleSavings` in the `datasets` package.

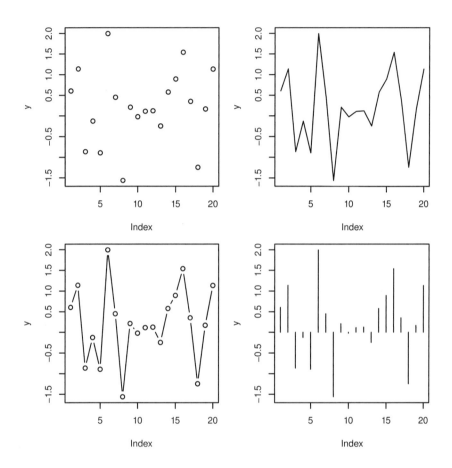

Figure 2.1

Four variations on a scatterplot. In each case, the plot is produced by a call to the
`plot()` function with the same data; all that changes is the value of the `type` argument. At top-left, `type="p"` to give points (data symbols), at top-right, `type="l"`
to give lines, at bottom-left, `type="b"` to give both, and at bottom-right, `type="h"`
to give histogram-like vertical lines.

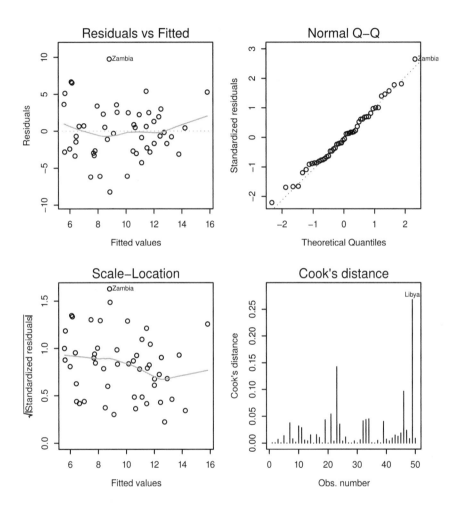

Figure 2.2
Plotting an `"lm"` object. There is a special `plot()` method for `"lm"` objects that produces a number of diagnostic plots from the results of a linear model analysis.

Table 2.1
High-level traditional graphics plotting functions for producing plots
of a single variable.

Function	Data	Description
plot()	Numeric	Scatterplot
plot()	Factor	Barplot
plot()	1-D table	Barplot
barplot()	Numeric (bar heights)	Barplot
pie()	Numeric	Pie chart
dotchart()	Numeric	Dotplot
boxplot()	Numeric	Boxplot
hist()	Numeric	Histogram
stripchart()	Numeric	1-D scatterplot
stem()	Numeric	Stem-and-leaf plot

will produce a scatterplot of the numeric values as a function of their indices,
while both a factor and a table produce a barplot of the counts for each level
of the factor. The plot() function will also accept a formula of the form ˜ x
and if the variable x is numeric, the result is a one-dimensional scatterplot
(stripchart). If x is a factor, the result is a barplot.

A barplot can also be produced explicitly with the barplot() function. The
difference is that this function requires a numeric vector, rather than a factor,
as input — the numeric values are treated as the heights of the bars to be
plotted.

One issue with producing a barplot is providing a meaningful label below each
bar. The plot() function uses the levels of the factor being plotted for bar
labels and barplot() will use the names attribute of the numeric vector if it
is available.

As alternatives to a barplot, the pie() function plots the values in a numeric
vector as a pie chart, and dotchart() produces a dotplot.

Several functions provide a variety of ways to view the distribution of values in
a single numeric vector. The boxplot() function produces a boxplot (or box-
and-whisker plot), the hist() function produces a histogram, stripchart()
produces a one-dimensional scatterplot (stripchart), and stem() produces a
stem-and-leaf plot (but as text, on the console, rather than graphical output).

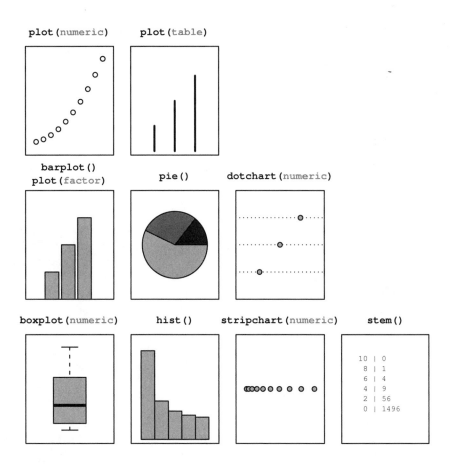

Figure 2.4
High-level traditional graphics plotting functions for producing plots of a single
variable. Where the function can be used to produce more than one type of plot,
the relevant data type is shown (in gray).

2.4   Plots of two variables

Table 2.2 and Figure 2.5 show the traditional graphics functions that produce plots of two variables.

The `plot()` function will accept two variables in a variety of formats: a pair of numeric vectors; one numeric vector and one factor; two factors; a list of two vectors or factors (named x and y); a two-dimensional table; a matrix or data frame with two columns (the first column is treated as x); or a formula of the form y ~ x.

If both variables are numeric, the result is a scatterplot. If x is a factor and y is numeric, the result is a boxplot for each level of x. If x is numeric and y is a factor, the result is a (grouped) stripchart, and if both variables are factors, the result is a spineplot. If `plot()` is given a table of counts, the result is a mosaic plot.

Two functions provide alternatives to the scatterplot, both motivated by the problem of overplotting, which occurs when values repeat or when there are very many points to plot. The `sunflowerplot()` function draws a special symbol at each location to indicate how many points are overplotted and the `smoothScatter()` function draws a representation of the density of points in the scatterplot (rather than drawing individual points). Another way to produce multiple stripcharts is to provide `stripchart()` with a list of numeric vectors.

When x is a factor and y is numeric, another way to produce multiple boxplots is with the `boxplot()` function, with the data provided either as a list of numeric vectors or as a formula of the form y ~ x, where x is a factor.

If the data consist of a numeric matrix, where each column or row represents a different group, the `barplot()` function will produce a stacked or side-by-side barplot from the numeric values and `dotchart()` will produce a dotplot.

When x is numeric and y is a factor, the `spineplot()` function will produce a spinogram, and `cdplot()` will produce a conditional density plot. Both functions will also accept the data as a formula of the form y ~ x.

For plotting two factors, there are also several options. Given the raw factors, the `spineplot()` function will produce a spineplot, just like `plot()` produces from two factors. An alternative is to work with a table of counts of the two factors. Given a table, the `mosaicplot()` function produces a mosaic plot, just like `plot()` does. The `mosaicplot()` function will also accept a formula of the form y ~ x where both y and x are factors.

Table 2.2

High-level traditional graphics plotting functions for producing plots of two variables.

Function	Data	Description
plot()	Numeric, numeric	Scatterplot
plot()	Numeric, factor	Stripcharts
plot()	Factor, numeric	Boxplots
plot()	Factor, factor	Spineplot
plot()	2-D table	Mosaic plot
sunflowerplot()	Numeric, numeric	Sunflower scatterplot
smoothScatter()	Numeric, numeric	Smooth scatterplot
boxplot()	List of numeric	Boxplots
barplot()	Matrix	Stacked/side-by-side barplot
dotchart()	Matrix	Dotplot
stripchart()	List of numeric	Stripcharts
spineplot()	Numeric, factor	Spinogram
cdplot()	Numeric, factor	Conditional density plot
fourfoldplot()	2x2 table	Fourfold display
assocplot()	2-D table	Association plot
mosaicplot()	2-D table	Mosaic plot

In the special case where both factors have only two levels, assocplot() produces a Cohen-Friendly association plot and fourfoldplot() produces a fourfold display. See Chapter 13 for more plots that are designed specifically for displaying categorical variables.

In addition to the numeric vector and factor data types, another important basic data type is *dates* (or *date-times*). If plot() is given either x or y as a "Date" or "POSIXt" object then the corresponding axis will be labeled with date descriptions (e.g., using month names).

2.5 Plots of many variables

Table 2.3 and Figure 2.6 show the traditional graphics functions that produce plots of many variables.

Given a data frame, with all columns numeric, the plot() function will produce a scatterplot matrix, plotting all pairs of variables against each other.

range of symbols is provided, some of which allow multiple variables to be represented within the symbol, for example, a rectangle symbol can encode separate variables as the width and height of the rectangle.

When the data consist of two numeric variables and one or two grouping factors, the `coplot()` function can be used to produce a conditioning plot, which draws a separate plot for each level of the grouping factors. The data must be given to this function as a formula of the form y ~ x | g or y ~ x | g*h, where g and h are factors. This idea is implemented on a much grander scale in the **lattice** package (see Chapter 4) and in the **ggplot2** package (see Chapter 5).

For data consisting of multiple factors, the `mosaicplot()` function will produce a multidimensional mosaic plot, given a multidimensional table of counts (see Chapter 13 for other options for plotting multiple factors).

2.6 Arguments to graphics functions

It is often the case, especially when producing graphics for publication, that the output produced by a single call to a high-level graphics function is not exactly right in all its details. There are many ways in which the output of graphics functions may be modified and Chapter 3 addresses this topic in full detail. This section will only consider the possibility of specifying arguments to high-level graphics functions in order to modify their output.

Many of these arguments are specific to a particular function. For example, the `boxplot()` function has `width` and `boxwex` arguments (among others) for controlling the width of the boxes in the plot, and the `barplot()` function has a `horiz` argument for controlling whether bars are drawn horizontally rather than vertically. The following code shows examples of the use of the `boxwex` argument for `boxplot()` and the `horiz` argument for `barplot()` (see Figure 2.7).

In the first example, there are two calls to `boxplot()`, which are identical except that the second specifies that the individual boxplots should be half as wide as they would be by default (`boxwex=0.5`).

```
> boxplot(decrease ~ treatment, data = OrchardSprays,
          log = "y", col="light gray")
> boxplot(decrease ~ treatment, data = OrchardSprays,
          log = "y", col="light gray",
          boxwex=0.5)
```

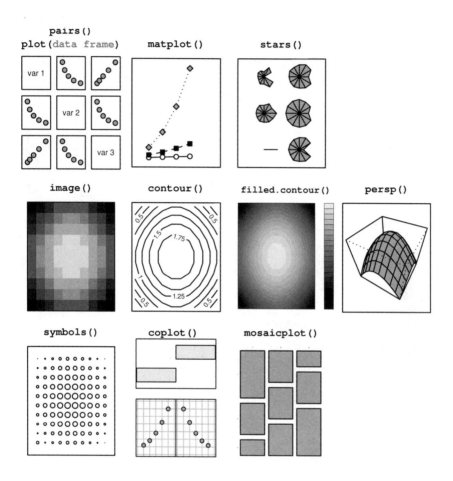

Figure 2.6

High-level traditional graphics plotting functions for producing plots of many variables. Where the function can be used to produce more than one type of plot, the relevant data type is shown (in gray).

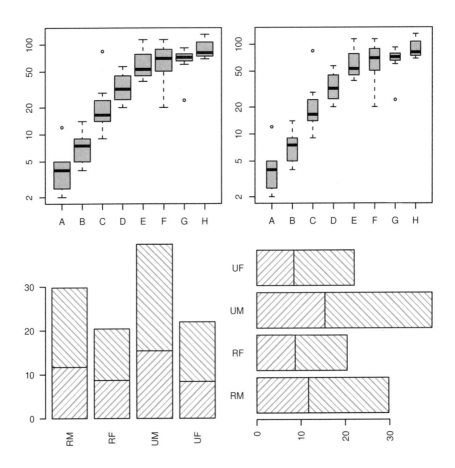

Figure 2.7

Modifying default `barplot()` and `boxplot()` output. The top two plots are produced by calls to the `boxplot()` function with the same data, but with different values of the `boxwex` argument. The bottom two plots are both produced by calls to the `barplot()` function with the same data, but with different values of the `horiz` argument.

In the second example, there are two calls to `barplot()`, which are identical except that the second specifies that the bars should be drawn horizontally rather than vertically (`horiz=TRUE`).

```
> barplot(VADeaths[1:2,], angle = c(45, 135),
          density = 20, col = "gray",
          names=c("RM", "RF", "UM", "UF"))
> barplot(VADeaths[1:2,], angle = c(45, 135),
          density = 20, col = "gray",
          names=c("RM", "RF", "UM", "UF"),
          horiz=TRUE)
```

In general, the user should consult the documentation for a specific function to determine which arguments are available and what effect they have.

2.6.1 Standard arguments to graphics functions

Despite the existence of many arguments that are specific only to a single graphics function, there are several arguments that are "standard" in the sense that many high-level traditional graphics functions will accept them.

Most high-level functions will accept *graphical parameters* that control such things as color (`col`), line type (`lty`), and text font (`font` and `family`). Section 3.2 provides a full list of these arguments and describes their effects.

Unfortunately, because the interpretation of these standard arguments may vary in some cases, some care is necessary. For example, if the `col` argument is specified for a standard scatterplot, this only affects the color of the data symbols in the plot (it does not affect the color of the axes or the axis labels), but for the `barplot()` function, `col` specifies the color for the fill or pattern used within the bars.

In addition to the standard graphical parameters, there are standard arguments to control the appearance of axes and labels on plots. It is usually possible to modify the range of the axis scales on a plot by specifying `xlim` or `ylim` arguments in the call to the high-level function, and often there is a set of arguments for specifying the labels on a plot: `main` for a title, `sub` for a subtitle, `xlab` for an x-axis label and `ylab` for a y-axis label.

Although there is no guarantee that these standard arguments will be accepted by high-level functions in graphics extension packages, in many cases they will be accepted, and they will have the expected effect.

The following code shows examples of setting some of these standard arguments for the `plot()` function (see Figure 2.8). All of the calls to `plot()`

One interesting case is the display of a parametric curve where, rather than specifying explicit data points, a *relationship* between x and y is provided. This can be achieved in two ways: via the plot() method for function objects and via the curve() function. The following code shows both approaches to draw a sine wave (see Figure 2.9).

```
> plot(function(x) {
          sin(x)/x
       },
       from=-10*pi, to=10*pi,
       xlab="", ylab="", n=500)

> curve(sin(x)/x, -10*pi, 10*pi)
```

There are also some functions that produce quite different sorts of plots. The plot() method for dendrogram objects is provided for drawing hierarchical or tree-like structures, such as the results from clustering or a recursive partitioning regression tree. The bottom two plots in Figure 2.9 show examples of output from the plot() method for dendrogram objects.* Part IV of this book contains several chapters that describe how to produce specialized plots of various kinds. For example, Chapter 15 describes other functions that draw this sort of node-and-edge graph.

2.8 Interactive graphics

The strength of the traditional graphics system lies in the production of static graphics and there are only limited facilities for interacting with graphical output.

The locator() function allows the user to click within a plot and returns the coordinates where the mouse click occurred. It will also optionally draw data symbols at the clicked locations or draw lines between the clicked locations.

The identify() function can be used to add labels to data symbols on a plot. The data point closest to the mouse click gets labeled.

There is also a more general-purpose getGraphicsEvent() function that allows capture of mouse and keyboard events (mouse button down, mouse up,

---

*The data used in these examples are measures of crime rates in various US states in 1973, available as the data set USArrests in the datasets package.

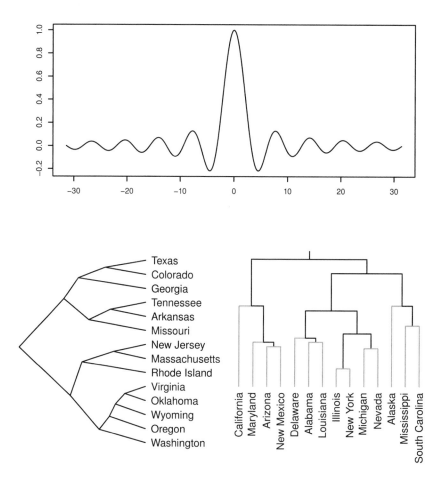

Figure 2.9
Some specialized plots. At the top is a plot of an R function and along the bottom are two variations on a dendrogram.

The real power of the traditional graphics system lies in the ability to control many aspects of the appearance of a plot, to add extra output to a plot, and even to build a plot from scratch in order to produce precisely the right final output.

Section 3.1 introduces important concepts of *drawing regions, coordinate systems,* and *graphics state* that are required for properly working with traditional graphics at a lower level. Section 3.2 describes how to control aspects of output such as colors, fonts, line styles, and plotting symbols, and Section 3.3 addresses the problem of placing several plots on the same page. Section 3.4 describes how to customize a plot by adding extra output and Section 3.5 looks at ways to develop entirely new types of plots.

3.1   The traditional graphics model in more detail

In order to explain some of the facilities for customizing plots, it is necessary to describe more about the model underlying traditional graphics plots.

3.1.1   Plotting regions

In the traditional graphics system, every page is split up into three main regions: the *outer margins,* the current *figure region,* and the current *plot region.* Figure 3.1 shows these regions when there is only one figure on the page and Figure 3.2 shows the regions when there are multiple figures on the page.

The region obtained by removing the outer margins from the device is called the *inner region.* When there is only one figure, this usually corresponds to the figure region, but when there are multiple figures the inner region corresponds to the union of all figure regions.

The area outside the plot region, but inside the figure region is referred to as the *figure margins.* A typical high-level function draws data symbols and lines within the plot region and axes and labels in the figure margins or outer margins (see Section 3.4 for information on the functions used to draw output in the different regions).

The size and location of the different regions are controlled either via the `par()` function, or using special functions for arranging plots (see Section 3.3). Specifying an arrangement of the regions does not usually affect the current

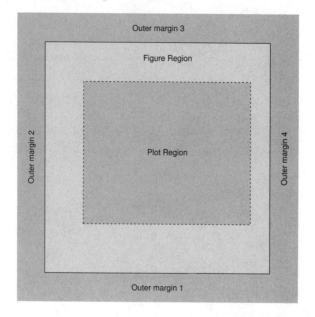

Figure 3.1

The plot regions in traditional graphics — the outer margins, figure region, and plot region — when there is a single plot on the page.

plot as the settings only come into effect when the next plot is started.

Coordinate systems

Each plotting region has one or more coordinate systems associated with it. Drawing in a region occurs relative to the relevant coordinate system. The coordinate system in the plot region, referred to as *user coordinates*, is probably the easiest to understand as it simply corresponds to the range of values on the axes of the plot (see Figure 3.3). The drawing of data symbols, lines, and text in the plot region occurs relative to this user coordinate system.

The scales on the axes of a plot are often set up automatically by R, but Sections 2.6 and 3.4.4 describe ways to set the scales manually.

The figure margins contain the next most commonly used coordinate systems. The coordinate systems in these margins are a combination of x- or y-ranges (like user coordinates) and lines of text away from the boundary of the plot region. Figure 3.4 shows two of the four figure margin coordinate systems. Axes are drawn in the figure margins using these coordinate systems.

There is a further set of "normalized" coordinate systems available for the

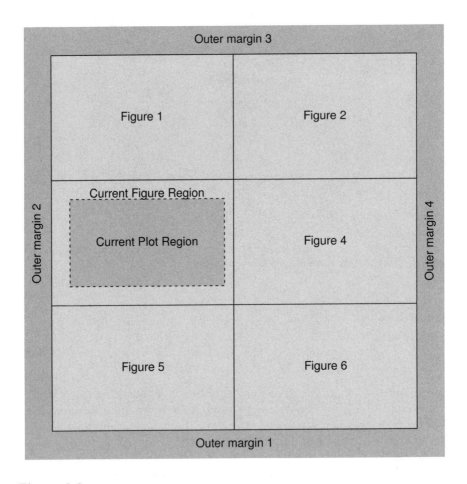

Figure 3.2

Multiple figure regions in traditional graphics — the outer margins, *current* figure region, and *current* plot region — when there are multiple plots on the page.

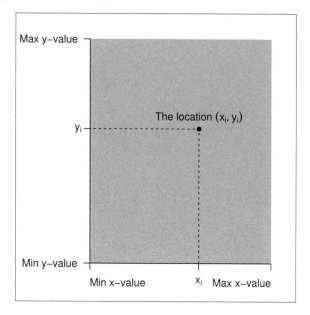

Figure 3.3
The user coordinate system in the plot region. Locations within this coordinate system are relative to the scales on the plot axes.

figure margins in which the x- and y-ranges are replaced with a range from 0 to 1. In other words, it is possible to specify locations along the axes as a proportion of the total axis length. Axis labels and plot titles are drawn relative to this coordinate system. All of these figure margin coordinate systems are created implicitly from the arrangement of the figure margins and the setting of the user coordinate system.

The outer margins have similar sets of coordinate systems, but locations along the boundary of the inner region can only be specified in normalized coordinates (always relative to the extent of the complete outer margin). Figure 3.5 shows two of the four outer margin coordinate systems.

Sections 3.4.2 and 3.4.4 describe functions that draw output relative to the figure margin and outer margin coordinate systems.

3.1.2 The traditional graphics state

The traditional graphics system maintains a graphics "state" for each graphics device and, when drawing occurs, this state is consulted to determine where output should be drawn, what color to use, what fonts to use, and so on.

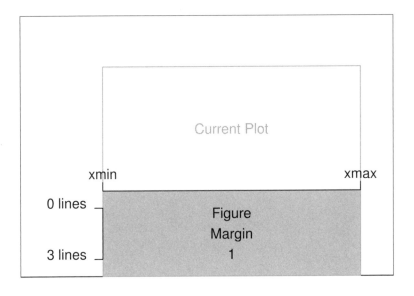

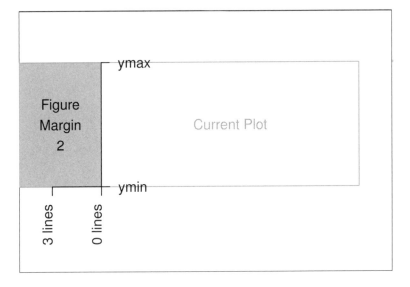

Figure 3.4

Figure margin coordinate systems. The typical coordinate systems for figure margin 1 (top plot) and figure margin 2 (bottom plot). Locations within these coordinate systems are a combination of position along the axis scale and distance away from the axis in multiples of lines of text.

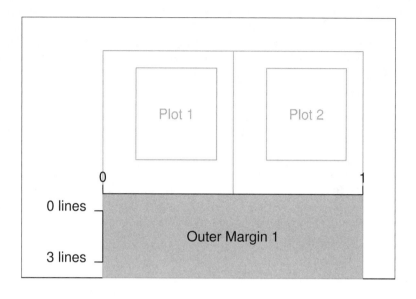

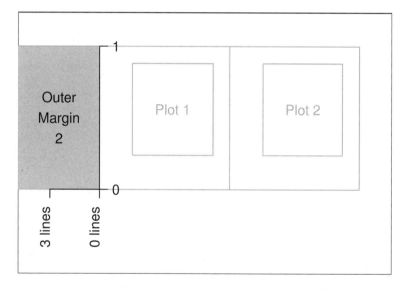

Figure 3.5

Outer margin coordinate systems. The typical coordinate systems for outer margin 1 (top plot) and outer margin 2 (bottom plot). Locations within these coordinate systems are a combination of a proportion along the inner region and distance away from the inner region in multiples of lines of text.

The graphics state consists of a large number of settings. Some of these settings describe the size and placement of the plot regions and coordinate systems that were described in the previous section. Some settings describe the general appearance of graphical output (e.g., the colors and line types that are used to draw lines and the fonts that are used to draw text) and some settings describe aspects of the output device (e.g., the physical size of the device and the current clipping region).

Tables 3.1 to 3.3 together provide a list of all of the graphics state settings and a very brief indication of their meaning. Most of the settings are described in detail in Sections 3.2 and 3.3.

The main function used to access the graphics state is the **par()** function. Simply typing **par()** will result in a complete listing of the current graphics state. A specific state setting can be queried by supplying specific setting names as arguments to **par()**. The following code queries the current state of the **col** and **lty** settings.

```
> par(c("col", "lty"))
```

```
$col
[1] "black"

$lty
[1] "solid"
```

The **par()** function can be used to modify traditional graphics state settings by specifying a value via an argument with the appropriate setting name. The following code sets new values for the **col** and **lty** settings.

```
> par(col="red", lty="dashed")
```

Modifying traditional graphics state settings via **par()** has a persistent effect. Settings specified in this way will hold until a different setting is specified. Settings may also be *temporarily* modified by specifying a new value in a call to a graphics function such as **plot()** or **lines()**. The following code demonstrates this idea. First of all, the line type is permanently set using **par()**, then a plot is drawn and the lines drawn between data points in this plot are dashed. Next, a plot is drawn with a temporary line type setting of **lty="solid"** and the lines in this plot are solid. When the third plot is drawn, the permanent line type setting of **lty="dashed"** is back in effect so the lines are again dashed.

Table 3.1

High-level traditional graphics state settings. This set of graphics state settings can be queried and set via the `par()` function *and* can be used as arguments to other graphics functions (e.g., `plot()` or `lines()`). Each setting is described in more detail in the relevant **Section**.

Setting	Description	Section
adj	Justification of text	3.2.3
ann	Draw plot labels and titles?	3.2.3
bg	Background color	3.2.1
bty	Type of box drawn by `box()`	3.2.5
cex	Size of text (multiplier)	3.2.3
also `cex.axis, cex.lab, cex.main, cex.sub`		
col	Color of lines and data symbols	3.2.1
also `col.axis, col.lab, col.main, col.sub`		
family	Font family for text	3.2.3
fg	Foreground color	3.2.1
font	Font face (bold, italic) for text	3.2.3
also `font.axis, font.lab, font.main, font.sub`		
lab	Number of ticks on axes	3.2.5
las	Rotation of text in margins	3.2.3
lend	Line end/join style	3.2.2
also `ljoin, lmitre`		
lty	Line type (solid, dashed)	3.2.2
lwd	Line width	3.2.2
mgp	Placement of axis ticks and tick labels	3.2.5
pch	Data symbol type	3.2.4
srt	Rotation of text in plot region	3.2.3
tck	Length of axis ticks (relative to plot size)	3.2.5
tcl	Length of axis ticks (relative to text size)	3.2.5
xaxp	Number of ticks on x-axis	3.2.5
xaxs	Calculation of scale range on x-axis	3.2.5
xaxt	X-axis style (standard, none)	3.2.5
xpd	Clipping region	3.2.7
yaxp	Number of ticks on y-axis	3.2.5
yaxs	Calculation of scale range on y-axis	3.2.5
yaxt	Y-axis style (standard, none)	3.2.5

Table 3.2
Low-level traditional graphics state settings. This set of graphics
state settings can only be queried and set via the `par()` function.
Each setting is described in more detail in the relevant **Section**.

Setting	Description	Section
fig	Location of figure region (normalized)	3.2.6
fin	Size of figure region (inches)	3.2.6
lheight	Line spacing (multiplier)	3.2.3
mai	Size of figure margins (inches)	3.2.6
mar	Size of figure margins (lines of text)	3.2.6
mex	Line spacing in margins	3.2.6
mfcol	Number of figures on a page	3.3.1
mfg	Which figure is used next	3.3.1
mfrow	Number of figures on a page	3.3.1
new	Has a new plot been started?	3.2.8
oma	Size of outer margins (lines of text)	3.2.6
omd	Location of inner region (normalized)	3.2.6
omi	Size of outer margins (inches)	3.2.6
pin	Size of plot region (inches)	3.2.6
plt	Location of plot region (normalized)	3.2.6
ps	Size of text (points)	3.2.3
pty	Aspect ratio of plot region	3.2.6
usr	Range of scales on axes	3.4.5
xlog	Logarithmic scale on x-axis?	3.2.5
ylog	Logarithmic scale on y-axis?	3.2.5

```
> y <- rnorm(20)
> par(lty="dashed")
> plot(y, type="l") # line is dashed
> plot(y, type="l", lty="solid") # line is solid
> plot(y, type="l") # line is dashed
```

Only some of the graphics state settings can be set temporarily in calls to
graphics functions. For example, the `mfrow` setting may not be set in this way
and can only be set using `par()`. The "low-level" settings are listed in Table
3.2.

A small set of graphics state settings cannot be modified at all and can only
be queried using `par()`. For example, there is no function to allow the user
to modify the size of the current device (after the device has been created),
but its size (in inches) may be obtained using `par("din")`. These "read-only"
settings are listed in Table 3.3.

Table 3.3

Read-only traditional graphics state settings. This set of graphics state settings can only be queried (via the `par()` function). Each setting is described in more detail in the relevant **Section**.

Setting	Description	Section
cin	Size of a character (inches)	3.4.5
cra	Size of a character ("pixels")	3.4.5
cxy	Size of a character (user coordinates)	3.4.5
din	Size of graphics device (inches)	3.4.5

Every graphics device has its own graphics state and calls to `par()` only affect the traditional graphics state of the currently active graphics device (see Chapter 9.1).

3.2 Controlling the appearance of plots

This section is concerned with the appearance of plots, which means the colors, line types, fonts and so on that are used to draw a plot. As described in Section 3.1.2, these features are controlled via traditional graphics state settings and values are specified for the settings either with a call to the `par()` function or as arguments to a specific graphics function such as `plot()`. For example, there is a setting called `col` to control the color of output (see Section 3.2.1). This can be set permanently using `par()` with an expression of the form:

```
par(col="red")
```

This will affect all subsequent graphical output. Alternatively, the setting can be specified as an argument to a high-level function using an expression of the form:

```
plot(..., col="red")
```

This will affect the output just for that plot. Finally, the setting can be used as an argument to a low-level function, as in the expression below.

```
lines(..., col="red")
```

This demonstrates that the setting can be used to control the appearance of just a single piece of graphical output.

There are many individual settings that affect the appearance of a plot, but they can be grouped in terms of what aspects of a plot the settings affect. Each of the following sections details a particular group of settings, including a description of the role of individual settings. There are sections on specifying colors; how to control the appearance of lines, text, data symbols, and axes; how to control the size and location of the various plotting regions; clipping (only drawing output on certain parts of the page); and specifying what should happen when a high-level function is called to start a new plot.

The appearance of plots is also affected by the location and size of the plotting regions, but this is dealt with separately in Section 3.3.

The following sections provide some simple examples of how to specify the settings for the traditional graphical parameters, but much more detail is provided in Chapter 10.

3.2.1   Colors

There are three main color settings in the traditional graphics state: `col`, `fg`, and `bg`.

The `col` setting is the most commonly used. The primary use is to specify the color of data symbols, lines, text, and so on that are drawn in the plot region. Unfortunately, when specified via a graphics function, the effect can vary. For example, a standard scatterplot produced by the `plot()` function will use `col` for coloring data symbols and lines, but the `barplot()` function will use `col` for filling the contents of its bars. In the `rect()` function (see Section 3.4), the `col` argument provides the color to fill the rectangle and there is a `border` argument specific to `rect()` that gives the color to draw the border of the rectangle. The effect of `col` on graphical output drawn in the margins also varies. It does not affect the color of axes and axis labels, but it does affect the output from the `mtext()` function. There are specific settings for affecting axes, labels, titles, and subtitles called `col.axis`, `col.lab`, `col.main`, and `col.sub`.

The `fg` setting is primarily intended for specifying the color of axes and borders on plots. There is some overlap between this and the specific `col.axis`, `col.main`, etc. settings described above.

The `bg` setting is primarily intended to specify the color of the background for traditional graphics output. This color is used to fill the entire page. As with the `col` setting, when `bg` is specified in a graphics function it can have a quite different meaning. For example, the `plot()` and `points()` functions use `bg` to specify the color for the interior of the data symbols, which can have different colors on the border (`pch` values 21 to 25; see Section 3.2.4).

Colors may be specified in a number of different ways. The most simple is to use a color name, such as `"red"` and `"blue"`, but there are many alternatives, including generating sets of colors by calling a function. Section 10.1 describes the specification of colors in R in complete detail (also see Section 11.3).

Fill patterns

In some cases (e.g., when printing in black and white), it is difficult to make use of different colors to distinguish between different elements of a plot. Using different levels of gray can be effective, but another option is to make use of some sort of fill pattern, such as cross-hatching. These should be used with caution because it is very easy to create visual effects that are distracting. Nevertheless, some journals actively encourage their use, so the facility has some purpose.

In traditional graphics, there is only limited support for fill patterns and they can only be applied to rectangles and polygons. It is possible to fill a rectangle or polygon with a set of lines drawn at a certain angle, with a specific separation between the lines. A `density` argument controls the separation between the lines (in terms of lines per inch) and an `angle` argument controls the angle of the lines (in terms of degrees anti-clockwise from 3 o'clock). Examples of the use of fill patterns are given in Figures 2.7, 3.19, and their associated code.

These settings can only be controlled via arguments to the functions `rect()`, `polygon()`, `hist()`, `barplot()`, `pie()`, and `legend()` (and *not* via `par()`).

Section 11.2 describes some functions in add-on packages that provide other ways to produce fill-pattern effects.

3.2.2 Lines

There are five graphics state settings for controlling the appearance of lines. The `lty` setting describes the type of line to draw (e.g., solid, dashed, or dotted), the `lwd` setting describes the width of lines, and the `ljoin`, `lend`, and `lmitre` settings control how the ends and corners in lines are drawn (rounded or pointy).

The line type can be specified as a character value, for example, `"solid"`, `"dashed"`, or `"dotted"`, and the line width is given as a number, where 1 corresponds to 1/96 inch (which is roughly 1 pixel on many computer screens).

The scope of these settings again differs depending on the graphics function being called. For example, for standard scatterplots, the setting only applies to lines drawn within the plot region. In order to affect the lines drawn as part of the axes, the `lty` setting must be passed directly to the `axis()` function (see Section 3.4.4).

Section 10.2 describes the specification of lines in R in complete detail.

3.2.3   Text

There are a large number of traditional graphics state settings for controlling the appearance of text. The size of text is controlled via `ps` and `cex`; the font is controlled via `font` and `family`; the justification of text is controlled via `adj`; and the angle of rotation is controlled via `srt`.

There is also an `ann` setting, which indicates whether titles and axis labels should be drawn on a plot. This is intended to apply to high-level functions, but is not guaranteed to work with all such functions (especially functions from extension packages). There are examples of the use of `ann` as an argument to high-level plotting functions in Section 3.4.1.

Text size

The size of text is ultimately a numerical value specifying the size of the font in "points." The font size is controlled by two settings: `ps` specifies an absolute font size setting (e.g., `ps=9`), and `cex` specifies a multiplicative modifier (e.g., `cex=1.5`). The final font size specification is simply `fontsize * cex`.

As with specifying color, the scope of a `cex` setting can vary depending on where it is given. When `cex` is specified via `par()`, it affects most text. However, when `cex` is specified via `plot()`, it only affects the size of data symbols. There are special settings for controlling the size of text that is drawn as axis tick labels (`cex.axis`), text that is drawn as axis labels (`cex.lab`), text in the title (`cex.main`), and text in the subtitle (`cex.sub`).

Specifying fonts

The font used for drawing text is controlled by the settings `family` and `font`.

The `family` setting is a character value giving the name of a specific font family, such as `"Times Roman"`, or a generic family style, such as `"serif"`, `"sans"` (sans-serif), or `"mono"` (monospaced). Specific font families will only be available if they are installed on the operating system that R is run on, but

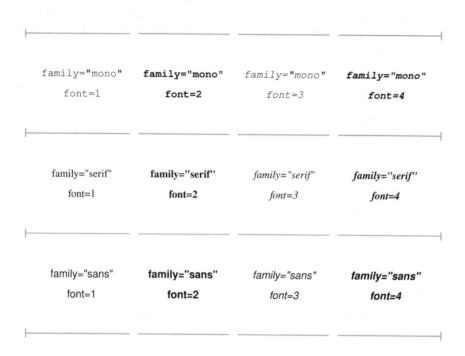

Figure 3.6
Font families and font faces. The appearance of the twelve font family and font face
combinations that are available in the traditional graphics system.

the generic family styles are always available.

The `font` is a numeric value that selects between normal text (1), **bold** (2),
italic (3), and ***bold-italic*** (4). Similar to color and text size, the `font` setting
applies mostly to text drawn in the plot region. There are additional settings
specifically for labels (`font.lab`), and titles (`font.main` and `font.sub`). Figure 3.6 demonstrates the 12 basic font family and face combinations.

Section 10.4 contains more details about how to specify text fonts in R graphics.

Justification of text

The `adj` setting is a value from 0 to 1 indicating the horizontal justification of text strings (0 means left-justified, 1 means right-justified and a value of 0.5 centers text).

The meaning of the `adj` setting depends on whether text is being drawn in the plot region, in the figure margins, or in the outer margins. In the plot region, the justification is relative to the (x, y) location at which the text is being drawn. In this context, it is also possible to specify two values for the setting and the second value is taken as a vertical justification for the text. Furthermore, non-finite values (`NA`, `NaN`, or `Inf`) may be specified for the justification and this is taken to mean "exact" centering (see below).

There is only a difference between a justification value of 0.5 and a non-finite justification value for vertical justification. In this case, a setting of 0.5 means text is vertically centered based on the height of the text above the text baseline (i.e., ignoring "descenders" like the tail on a "y"). A non-finite value means that text is vertically centered based on the full height of the text (including descenders). Figure 3.7 shows how various `adj` settings affect the alignment of text in the plot region.

In the figure margins and outer margins, the meaning of the `adj` setting depends on the `las` setting (see below). When margin text is parallel to the axis, `adj` specifies *both* the location and the justification of the text. For example, a value of 0 means that the text is left-justified *and* that the text is located at the left end of the margin. When text is perpendicular to the axis, the `adj` setting only affects justification. Furthermore, the `adj` setting only affects "horizontal" justification (justification in the reading direction) for text in the margins. Section 3.4.2 contains more information about the justification of text in the plot margins.

Rotating text

The `srt` setting specifies a rotation angle anti-clockwise from the positive x-axis, in degrees. This will only affect text drawn in the plot region (text drawn by the `text()` function; see Section 3.4.1). Text can be drawn at any angle within the plot region.

In the figure and outer margins, text may only be drawn at angles that are multiples of 90°, and this angle is controlled by the `las` setting. A value of 0 means text is always drawn parallel to the relevant axis (i.e., horizontal in margins 1 and 3, and vertical in margins 2 and 4). A value of 1 means text is always horizontal, 2 means text is always perpendicular to the relevant axis, and 3 means text is always vertical.

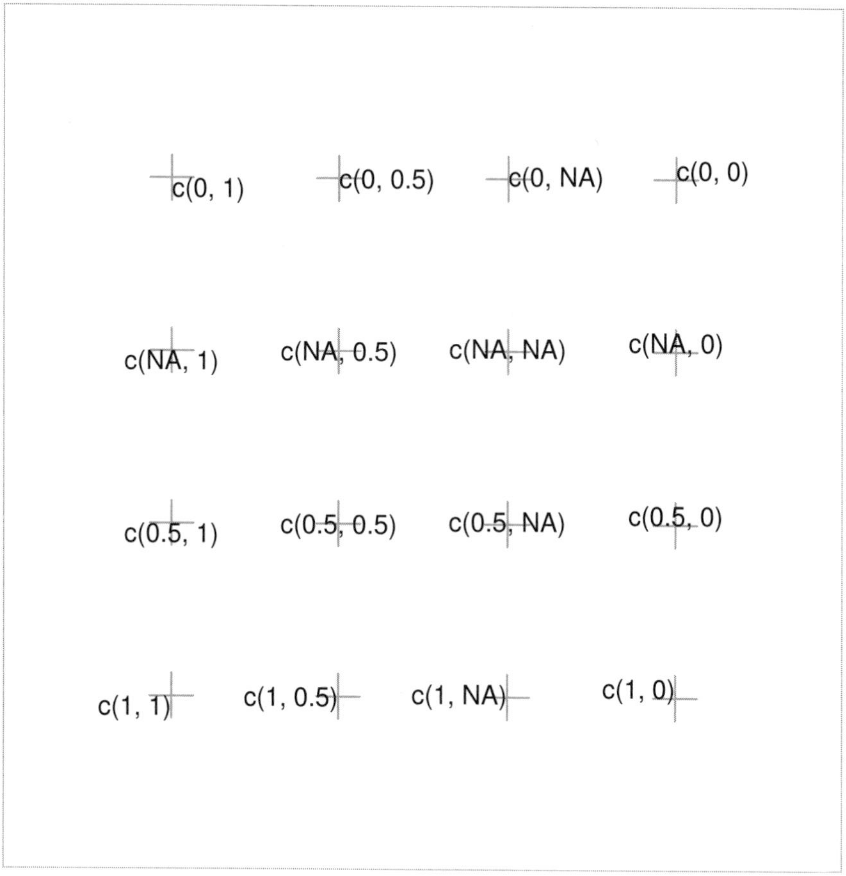

Figure 3.7

Alignment of text in the plot region. The `adj` graphical setting may be given two values, c(*hjust, vjust*), where *hjust* specifies horizontal justification and *vjust* specifies vertical justification. Each piece of text in the diagram is justified relative to a gray cross to represent the effect of the relevant `adj` setting. The vertical adjustment for `NA` is subtly different from the vertical adjustment for 0.5.

Figure 3.8
The first six data symbols that are available in traditional graphics. In the diagram, the relevant integer value for the pch setting is shown in gray to the left of the corresponding symbol.

Multi-line text

The spacing between multiple lines of text is controlled by the lheight setting, which is a multiplier applied to the natural height of a line of text. For example, lheight=2 specifies double-spaced text. This setting can only be specified via par().

3.2.4   Data symbols

The data symbol used for plotting points is controlled by the pch setting. This can be an integer value to select one of a fixed set of data symbols, or a single character. For example, specifying pch=0 produces an open square, pch=1 produces an open circle, and pch=2 produces an open triangle (see Figure 3.8). Specifying pch="#" means that a hash character will be plotted at each data location.

Some of the predefined data symbols (pch between 21 and 25) allow a fill color separate from the border color, with the bg setting controlling the fill color in these cases.

Section 10.3 describes the possible set of data symbols in more detail.

The size of the data symbols is linked to the size of text and is affected by the cex setting. If the data symbol is a character, the size will also be affected by the ps setting.

The type setting controls how data are represented in a plot. A value of "p" means that data symbols are drawn at each (x, y) location. The value "l" means that the (x, y) locations are connected by lines. A value of "b" means that both data symbols and lines are drawn. The type setting may also have the value "o", which means that data symbols are "over-plotted" on lines (with the value "b", the lines stop short of each data symbol). It is also possible to specify the value "h", which means that vertical lines are drawn from the x-axis to the (x, y) locations (the appearance is like a barplot with very thin bars). Two further values, "s" and "S" mean that (x, y) locations are joined in a city-block fashion with lines going horizontally then vertically (or vertically then horizontally) between each data location. Finally, the value "n" means that nothing is drawn at all.

Figure 3.9 shows simple examples of the different plot types. This setting is most often specified within a call to a high-level function (e.g., plot()) rather than via par().

3.2.5 Axes

By default, the traditional graphics system produces axes with sensible labels and tick marks at sensible locations. If the axis does not look right, there are a number of graphical state settings specifically for controlling aspects such as the number of tick marks and the positioning of labels. These are described below. If none of these gives the desired result, the user may have to resort to drawing the axis explicitly using the axis() function (see Section 3.4.4).

The lab setting in the traditional graphics state is used to control the number of tick marks on the axes. The setting is only used as a starting point for the algorithm R uses to determine sensible tick locations so the final number of tick marks that are drawn could easily differ from this specification. The setting takes two values: the first specifies the number of tick marks on the x-axis and the second specifies the number of tick marks on the y-axis.

The xaxp and yaxp settings also relate to the number and location of the tick marks on the axes of a plot. This setting is almost always calculated by R for each new plot so user settings are usually overridden (see Section 3.4.4 for an exception to this rule). In other words, it only makes sense to query this setting for its current value. The settings consist of three values: the first two specify the location of the left-most and right-most tick marks (bottom and top tick marks for the y-axis), and the third value specifies how many intervals there are between tick marks. When a log transformation is in effect for an axis, the three values have a different meaning altogether (see the on-line help page for par()).

The mgp setting controls the distance that the components of the axes are drawn away from the edge of the plot region. There are three values representing the positioning of the overall axis label, the tick mark labels, and the lines for the ticks. The values are in terms of lines of text away from the edges of the plot region. The default value is c(3, 1, 0). Figure 3.10 gives an example of different mgp settings.

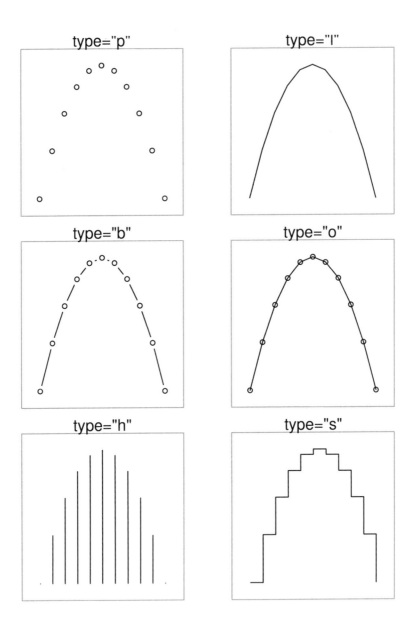

Figure 3.9

Basic plot types. Plotting the same data with different plot type settings. In each case, the output is produced by an expression of the form plot(x, y, type=*something*), where the relevant value of type is shown above each plot.

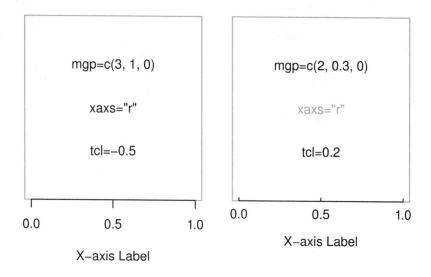

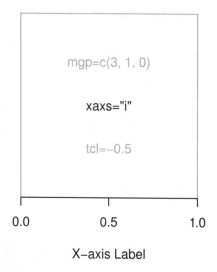

Figure 3.10
Different axis styles. The top-left plot demonstrates the default axis settings for an x-axis. The top-right plot shows the effects of specifying different positions for the axis labels (the tick labels and axis labels are closer to the plot region) and different lengths for the tick marks and the bottom-left plot shows the effect of specifying an "internal" axis range calculation.

The `tck` and `tcl` settings control the length of tick marks. The `tcl` setting specifies the length of tick marks as a fraction of the height of a line of text. The sign dictates the direction of the tick marks — a negative value draws tick marks outside the plot region and a positive value draws tick marks inside the plot region. The `tck` setting specifies tick mark lengths as a fraction of the smaller of the physical width or height of the plotting region, but it is only used if its value is not `NA` (and it is `NA` by default). Figure 3.10 gives an example of different `tcl` settings.

The `xaxs` and `yaxs` settings control the "style" of the axes of a plot. By default, the setting is `"r"`, which means that R calculates the range of values on the axis to be wider than the range of the data being plotted (so that data symbols do not collide with the boundaries of the plot region). It is possible to make the range of values on the axis exactly match the range of values in the data, by specifying the value `"i"`. This can be useful if the range of values on the axes are being explicitly controlled via `xlim` or `ylim` arguments to a function. Figure 3.10 gives an example of different `xaxs` settings.

The `xaxt` and `yaxt` settings control the "type" of axes. The default value, `"s"`, means that the axis is drawn. Specifying a value of `"n"` means that the axis is not drawn.

The `xlog` and `ylog` settings control the transformation of values on the axes. The default value is `FALSE`, which means that the axes are linear and values are not transformed. If this value is `TRUE` then a logarithmic transformation is applied to any values on the relevant dimension in the plot region. This also affects the calculation of tick mark locations on the axes.

When data of a special nature are being plotted (e.g., time series data), some of these settings may not apply (and may not have any sensible interpretation).

The `bty` setting is not strictly to do with axes, but it controls the output of the `box()` function, which is most commonly used in conjunction with drawing axes. This function draws a bounding box around the edges of the plot region (by default). The `bty` setting controls the type of box that the `box()` function draws. The value can be `"n"`, which means that no box is drawn, or it can be one of `"o"`, `"l"`, `"7"`, `"c"`, `"u"`, or `"]"`, which means that the box drawn resembles the corresponding uppercase character. For example, `bty="c"` means that the bottom, left, and top borders will be drawn, but the right border will not be drawn.

In addition to these graphics state settings, many high-level plotting functions, e.g., `plot()`, provide arguments `xlim` and `ylim` to control the range of the scale on the axes. Section 2.6.1 has an example.

3.2.6 Plotting regions

As described in Section 3.1.1, the traditional graphics system defines several different regions on the graphics device. This section describes how to control the size and layout of these regions using graphics state settings. Figure 3.11 shows a diagram of some of the settings that affect the widths and horizontal placement of the regions.

The size of each margin can be controlled independently, but R will check whether an overall specification is consistent. For example, if the margins are made too big, so that there is not room left on the page for the plot region, then R will give an error message like the following:

```
Error in plot.new() : figure margins too large
```

Outer margins

By default, there are no outer margins on a page. Outer margins can be specified using the `oma` graphics state setting. This consists of four values for the four margins in the order (bottom, left, top, right) and values are interpreted as lines of text (a value of 1 provides space for one line of text in the margin). The margins can also be specified in inches using `omi` or in normalized device coordinates (i.e., as a proportion of the device region) using `omd`. If `omd` is used, the margins are specified in the order (left, right, bottom, top).

Figure regions

By default, the figure region is calculated from the settings for the outer margins and the number of figures on the page. The figure region can be specified explicitly instead, using either the `fig` setting or the `fin` state setting. The `fig` setting specifies the location, (left, right, bottom, top), of the figure region where each value is a proportion of the "inner" region (the page less the outer margins). The `fin` setting specifies the size, (width, height), of the figure region in inches and the resulting figure region is centered within the inner region.

Figure margins

The figure margins can be controlled using the `mar` state setting. This consists of four values for the four margins in the order (bottom, left, top, right) where each value represents a number of lines of text. The default values are

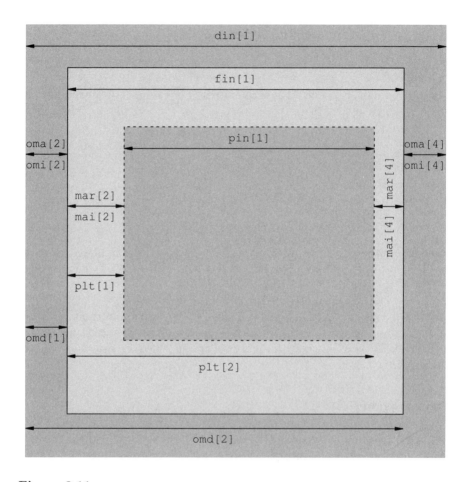

Figure 3.11
Graphics state settings controlling plot regions. These are some of the settings that
control the widths and horizontal locations of the plot regions. For ease of comparison, this diagram has the same layout as Figure 3.1: the central gray rectangle
represents the plot region, the lighter gray rectangle around that is the figure region,
and the darker gray rectangle around that is the outer margins. A similar diagram
could be produced for settings controlling heights and vertical locations.

c(5, 4, 4, 2) + 0.1. The margins may also be specified in terms of inches using `mai`.

The `mex` setting controls the size of a "line" in the margins. This does not affect the size of text drawn in the margins, but is used to multiply the size of text to determine the height of one line of text in the margins.

Plot regions

By default, the plot region is calculated from the figure region less the figure margins. The location and size of the plot region may be controlled explicitly instead, using the `plt`, `pin`, or `pty` settings. The `plt` setting allows the user to specify the location of the plot region, (`left, right, bottom, top`), where each value is a proportion of current figure region. The `pin` setting specifies the size of the plot region, (`width, height`), in terms of inches.

The `pty` setting controls how much of the available space (figure region less figure margins) the plot region occupies. The default value is "m", which means that the plot region occupies all of the available space. A value of "s" means that the plot region will take up as much of the available space as possible, but it must be "square" (i.e., its physical width will be the same as its physical height).

3.2.7 Clipping

Traditional graphics output is usually clipped to the plot region. This means that any output that would appear outside the plot region is not drawn. For example, in the default behavior, data symbols for (`x, y`) locations which lie outside the ranges of the axes are not drawn. Traditional graphics functions that draw in the margins clip output to the current figure region or to the device. Section 3.4 has information about which functions draw in which regions.

It can be useful to override the default clipping region. For example, this is necessary to draw a legend outside the plot region using the `legend()` function.

The traditional clipping region is controlled via the `xpd` setting. Clipping can occur either to the whole device (an `xpd` value of `NA`), to the current figure region (a value of `TRUE`), or to the current plot region (a value of `FALSE`, which is the default).

There is also a `clip()` function for setting the clipping region to be *smaller* than the plot region.

The function `layout.show()` may be helpful for visualizing the figure regions that are created. The following code creates a figure visualizing the layout created in the previous example (see Figure 3.12a).

```
> layout.show(6)
```

The contents of the layout matrix determine the order in which the resulting figure regions will be used. The following code creates a layout with exactly the same rows and columns as the previous one, but the figure regions will be used in the reverse order (see Figure 3.12b).

```
> layout(rbind(c(6, 5),
               c(4, 3),
               c(2, 1)))
```

By default, all row heights are the same and all column widths are the same size and the available inner region is divided up equally. The `heights` arguments can be used to specify that certain rows are given a greater portion of the available height (for all of what follows, the `widths` argument works analogously for column widths). When the available height is divided up, the proportion of the available height given to each row is determined by dividing the row heights by the sum of the row heights. For example, in the following layout there are two rows and one column. The top row is given two thirds of the available height, $2/(2+1)$, and the bottom row is given one third, $1/(2+1)$. Figure 3.12c shows the resulting layout.

```
> layout(matrix(c(1, 2)), heights=c(2, 1))
```

In the examples so far, the division of row heights has been completely independent of the division of column widths. The widths and heights can be forced to correspond as well so that, for example, a height of 1 corresponds to the same physical distance as a width of 1. This allows control over the aspect ratio of the resulting figure. The `respect` argument is used to force this correspondence. The following code is the same as the previous example except that the `respect` argument is set to TRUE (see Figure 3.12d).

```
> layout(matrix(c(1, 2)), heights=c(2, 1),
         respect=TRUE)
```

It is also possible to specify heights of rows and widths of columns in absolute terms. The `lcm()` function can be used to specify heights and widths for a layout in terms of centimeters. The following code is the same as the previous

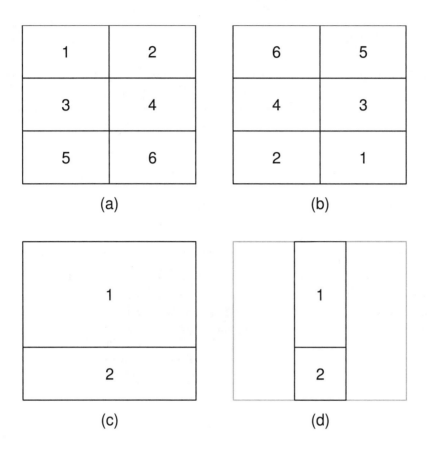

Figure 3.12
Some basic layouts: (a) A layout that is identical to `par(mfrow=c(3, 2))`; (b) Same as (a) except the figures are used in the reverse order; (c) A layout with unequal row heights; (d) Same as (c) except the layout widths and heights "respect" each other.

example, except that a third, empty region is created to provide a vertical gap of 0.5 cm between the two figures (see Figure 3.13a). The 0 in the first matrix argument means that no figure occupies that region.

```
> layout(matrix(c(1, 0, 2)),
         heights=c(2, lcm(0.5), 1),
         respect=TRUE)
```

This next piece of code demonstrates that a figure may occupy more than one row or column in the layout. This extends the previous example by adding a second column and creating a figure region that occupies both columns of the bottom row. In the matrix argument, the value 2 appears in both columns of row 3 (see Figure 3.13b).

```
> layout(rbind(c(1, 3),
               c(0, 0),
               c(2, 2)),
         heights=c(2, lcm(0.5), 1),
         respect=TRUE)
```

Finally, it is possible to specify that only certain rows and columns should respect each other's heights/widths. This is done by specifying a matrix for the respect argument. In the following code, the previous example is modified by specifying that only the first column and the last row should respect each other's widths/heights. In this case, the effect is to ensure that the width of figure region 1 is the same as the height of figure region 2, but the width of figure region 3 is free to expand to the available width (see Figure 3.13c).

```
> layout(rbind(c(1, 3),
               c(0, 0),
               c(2, 2)),
         heights=c(2, lcm(0.5), 1),
         respect=rbind(c(0, 0),
                       c(0, 0),
                       c(1, 0)))
```

3.3.3   The split-screen approach

The split.screen() function provides yet another way to divide the page into a number of figure regions. The first argument, figs, is either two values specifying a number of rows and columns of figures (i.e., like the

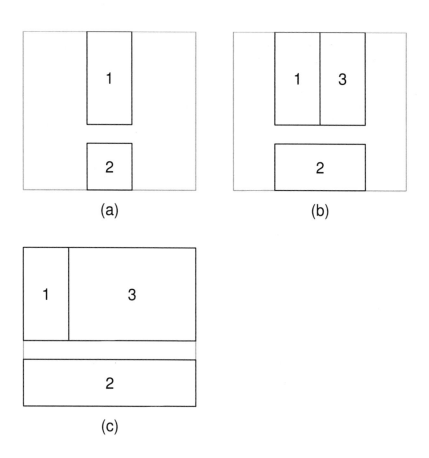

Figure 3.13
Some more complex layouts: (a) A layout with a row height specified in centimeters;
(b) A layout with a figure occupying more than one column; (c) Same as (b), but
with only column 1 and row 3 respected.

mfrow setting), or a matrix containing a figure region location, (left, right, bottom, top), on each row (i.e., like a par(fig) setting on each row).

Having established figure regions in this manner, a figure region is used by calling the screen() function to select a region. This means that the order in which figures are used is completely under the user's control, and it is possible to reuse a figure region, though there are dangers in doing so (the on-line help for split.screen() provides further discussion). The function erase.screen() can be used to clear a defined screen and close.screen() can be used to remove one or more screen definitions.

An even more useful feature of this approach is that each figure region can itself be divided up by a further call to split.screen(). This allows complex arrangements of plots to be created.

The downside to this approach is that it does not fit very nicely with the underlying traditional graphics system model (see Section 3.1). The recommended way to achieve complex arrangements of plots is via the layout() function from the previous section or by using the **grid** graphics system (see Part II), possibly in combination with traditional high-level functions (see Chapter 19). Section 11.4.2 describes yet more alternatives that are available in extension packages.

3.4 Annotating plots

Sometimes it is not enough to be able to modify the default output from high-level functions and further graphical output must be added, using low-level functions, to achieve the desired result (see, for example, Figure 1.3). R graphics in general is fundamentally oriented to supporting the annotation of plots — the ability to add graphical output to an existing plot. In particular, the regions and coordinate systems used in the construction of a plot remain available for adding further output to the plot. For example, it is possible to position a text label relative to the scales on the axes of a plot.

3.4.1 Annotating the plot region

Most low-level graphics functions that add output to an existing plot, add the output to the plot region. In other words, locations are specified relative to the user coordinate system (see Section 3.1.1).

Table 3.4

The low-level traditional graphics functions for drawing basic graphical primitives.

Function	Description
points()	Draw data symbols at locations (x, y)
lines()	Draw lines between locations (x, y)
segments()	Draw line segments between (x0, y0) and (x1, y1)
arrows()	Draw line segments with arrowheads at the end(s)
xspline()	Draw a smooth curve relative to control points (x, y)
rect()	Draw rectangles with bottom-left corner at (xl, yb) and top-right corner at (xr, yt)
polygon()	Draw one or more polygons with vertices (x, y)
polypath()	Draw a single polygon made up of one or more paths with vertices (x, y)
rasterImage()	Draw a bitmap image
text()	Draw text at locations (x, y)

Graphical primitives

This section describes the graphics functions that provide the most basic graphics output (lines, rectangles, text, etc). Table 3.4 provides a complete list.

The most common use of this facility is to add extra sets of data to a plot. The lines() function draws lines between (x, y) locations, and the points() function draws data symbols at (x, y) locations. The following code demonstrates a common situation where three different sets of y-values, recorded at the same set of x-values, are plotted together on the same plot (see the left-hand plot in Figure 3.14).

First some data are generated, consisting of one set of x-values and three sets of y-values, and the first set of y-values are plotted as a gray line (type="l" and col="gray"). The scale on the y-axis is set, using ylim, to ensure that there will be room on the plot for all of the data series.

```
> x <- 1:10
> y <- matrix(sort(rnorm(30)), ncol=3)
> plot(x, y[,1], ylim=range(y), ann=FALSE, axes=FALSE,
        type="l", col="gray")
> box(col="gray")
```

Now a set of points are added for the first set of y-values, then lines and points are added for the other two sets of y-values.

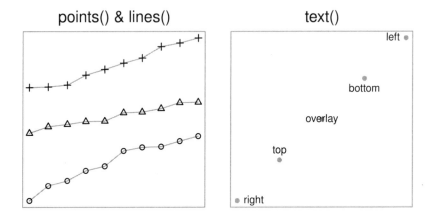

Figure 3.14
Annotating the plot region of a traditional graphics plot. The left-hand plot shows points and extra lines being added to an initial line plot. The right-hand plot shows text being added to an initial scatterplot.

```
> points(x, y[,1])
> lines(x, y[,2], col="gray")
> points(x, y[,2], pch=2)
> lines(x, y[,3], col="gray")
> points(x, y[,3], pch=3)
```

It is also possible to draw text at (x, y) locations with the text() function. This is useful for labeling data locations, particularly using the pos argument to offset the text so that it does not overlay the corresponding data symbols. The following code creates a diagram demonstrating the use of text() (see the right-hand plot in Figure 3.14). Again, some data are created and (gray) data symbols are plotted at the (x, y) locations.

```
> x <- 1:5
> y <- x
> plot(x, y, ann=FALSE, axes=FALSE, col="gray", pch=16)
> box(col="gray")
```

Now some text labels are added, with each one offset in a different way from the (x, y) location. Notice that the arguments to text() may be vectors so that several pieces of text are drawn by the one function call.

```
> text(x[-3], y[-3], c("right", "top", "bottom", "left"),
      pos=c(4, 3, 1, 2))
> text(3, 3, "overlay")
```

Like the `plot()` function, the `text()`, `lines()`, and `points()` functions are generic. This means that they have flexible interfaces for specifying the data for the (`x`, `y`) locations, or they produce different output when given objects of a particular class in the `x` argument. For example, both `lines()`, and `points()` will accept formulae for specifying the (`x`, `y`) locations and the `lines()` function will behave sensibly when given a `ts` (time series) object to draw.

The `text()` function normally takes a character value to draw, but it will also accept an R expression (as produced by the `expression()` function), which can be used to produce a mathematical formula with special symbols (e.g., Greek letters) and formatting (e.g., superscripts). Section 10.5 describes this facility in more detail.

As a parallel to the `matplot()` function (see Section 2.5), there are functions `matpoints()` and `matlines()` specifically for adding lines and data symbols to a plot, given `x` or `y` as matrices.

Having access to graphical primitives not only makes it easy to add new data series to a plot and to add labels, but it also makes it possible to add arbitrary drawing to a plot. In addition to lines, points, and text, there are graphical primitives for drawing more complex shapes.

In order to demonstrate these other graphical primitives, the following code produces a simple set of x- and y-values. These points will be plotted and used to draw a variety of shapes (see Figure 3.15).

```
> t <- seq(60, 360, 30)
> x <- cos(t/180*pi)*t/360
> y <- sin(t/180*pi)*t/360
```

The `lines()` function draws a single line through several points. Missing values in the (`x`, `y`) locations will create breaks in the line.

```
> lines(x, y)
```

An alternative is provided by the `segments()` function, which will draw several different straight lines between pairs of end points. In the following code, a straight line is drawn from (`0`, `0`) to each of the (`x`, `y`) locations. Notice that R's normal *recycling rule* behavior is applied to most arguments of graphics functions.

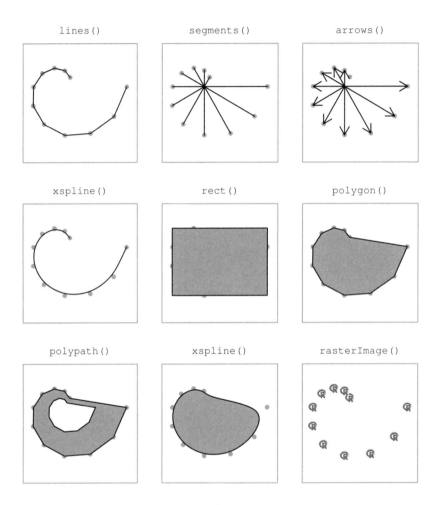

Figure 3.15
Drawing in the plot region of a traditional graphics plot. These pictures show some
of the functions that draw more complex graphical shapes. The shapes are based
on a set of (x, y) points which are drawn as light gray dots.

```
> segments(0, 0, x, y)
```

The `arrows()` function produces the same output as `segments()`, but also adds simple arrowheads at either end of the line segments. The `length` argument is used here to control the size of the arrowheads.

```
> arrows(0, 0, x[-1], y[-1], length=.1)
```

The `xspline()` function also produces a line, but the line is an X-spline, which treats the (x, y) locations as *control points* from which to produce a smooth curve. The smoothness of the curve is controlled by a `shape` parameter.

```
> xspline(x, y, shape=1)
```

There are also several functions for producing closed shapes. The simplest is `rect()`, which only requires a left, bottom, right, and top value to draw a rectangle (though all values can be vectors, which will result in several rectangles being drawn).

```
> rect(min(x), min(y), max(x), max(y), col="gray")
```

The `polygon()` function produces more complex shapes, using the (x, y) locations as vertices. Multiple polygons may be drawn using `polygon()` by inserting an `NA` value between each set of polygon vertexes. For both `rect()` and `polygon()`, the `col` argument specifies the color to *fill* the interior of the shape and the argument `border` controls the color of the line around the boundary of the shape.

```
> polygon(x, y, col="gray")
```

The `polygon()` function can draw self-intersecting polygons, but cannot represent polygons with holes. For the latter case, there is `polypath()`, which only draws a single polygon, but the polygon can be composed of more than one subpath. This allows for polygons consisting of distinct paths as well as polygons with holes.

```
> polypath(c(x, NA, .5*x), c(y, NA, .5*y),
           col="gray", rule="evenodd")
```

The `xspline()` function can also be used to create closed shapes, by specifying `open=FALSE`.

```
> xspline(x, y, shape=1, open=FALSE, col="gray")
```

Finally, there is a function, `rasterImage()`, for drawing bitmap images on a plot. The bitmap can be an external file, or it can just be a vector, matrix, or array. The following code draws the R logo at each of the (x, y) locations (code to read in the R logo is not shown; see Chapter 18 for more information).

```
> rasterImage(rlogo,
              x - .07, y - .07,
              x + .07, y + .07,
              interpolate=FALSE)
```

These examples only provide a tiny glimpse of what is possible with these graphical primitives (see Figure 3.15). The possibilities are endless and a number of the examples in the remainder of this chapter provide some further demonstrations of what can be achieved by adding basic graphical shapes to a plot (see, for example, Figure 3.23).

Graphical utilities

In addition to the low-level graphical primitives of the previous section, there are a number of utility functions that provide a set of slightly more complex shapes.

The `grid()` function adds a series of grid lines to a plot. This is simply a series of line segments, but the default appearance (light gray and dotted) is suited to the purpose of providing visual cues to the viewer without interfering with the primary data symbols.

The `abline()` function provides a number of convenient ways to add a line (or lines) to a plot. The line(s) can be specified either by a slope and y-axis intercept, or as a series of x-locations for vertical lines or as a series of y-locations for horizontal lines. The function will also accept the coefficients from a linear regression analysis (even as an `"lm"` object), thereby providing a simple way to add a line of best fit to a scatterplot.

The following code annotates a basic scatterplot with a line and arrows (see the left-hand plot of Figure 3.16).

First, some data are generated and plotted.

```
> x <- runif(20, 1, 10)
> y <- x + rnorm(20)
> plot(x, y, ann=FALSE, axes=FALSE, col="gray", pch=16)
> box(col="gray")
```

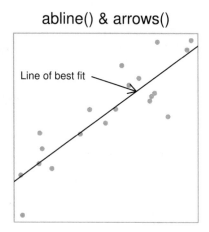
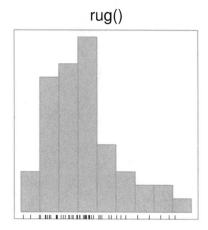

Figure 3.16
More examples of annotating the plot region of a traditional graphics plot. The left-hand plot shows a line of best fit (plus a text label and arrow) being added to an initial scatterplot. The right-hand plot shows a series of ticks being added as a rug plot on an initial histogram.

Now a line of best fit is drawn through the data using `abline()` and a text label and arrow are added using `text()` and `arrows()`.

```
> lmfit <- lm(y ~ x)
> abline(lmfit)
> arrows(5, 8, 7, predict(lmfit, data.frame(x=7)),
        length=0.1)
> text(5, 8, "Line of best fit", pos=2)
```

The `box()` function draws a rectangle around the boundary of the plot region. The `which` argument makes it possible to draw the rectangle around the current figure region, inner region, or outer region instead. The `box()` function has been used in many of the examples in this section.

The `rug()` function produces a "rug" plot along one of the axes, which consists of a series of tick marks representing data locations. This can be useful to represent an additional one-dimensional plot of data (e.g., in combination with a density curve). The following code uses this function to annotate a histogram (see the right-hand plot of Figure 3.16).

```
> y <- rnorm(50)
> hist(y, main="", xlab="", ylab="", axes=FALSE,
        border="gray", col="light gray")
> box(col="gray")
> rug(y, ticksize=0.02)
```

Missing values and non-finite values

R has special values representing missing observations (NA) and non-finite
values (NaN and Inf). Most traditional graphics functions allow such values
within (x, y) locations and handle them by not drawing the relevant location.
For drawing data symbols or text, this means the relevant data symbol or piece
of text will not be drawn. For drawing lines, this means that lines to or from
the relevant location are not drawn; a gap is created in the line. For drawing
rectangles, an entire rectangle will not be drawn if any of the four boundary
locations is missing or non-finite.

Polygons are a slightly more complex case. For drawing polygons, a missing
or non-finite value in x or y is interpreted as the end of one polygon and the
start of another. Figure 3.17 shows an example. On the left, a polygon is
drawn through 12 locations evenly spaced around a circle. On the right, the
first, fifth, and ninth locations have been set to NA so the output is split into
three separate polygons.

Missing or non-finite values can also be specified for some traditional graphics
state settings. For example, if a color setting is missing or non-finite then
nothing is drawn (this is a brute-force way to specify a completely transparent
color). Similarly, specifying a missing value or non-finite value for cex means
that the relevant data symbol or piece of text is not drawn.

3.4.2   Annotating the margins

There are only two functions that produce output in the figure or outer margins, relative to the margin coordinate systems (Section 3.1.1).

The mtext() function draws text at any location in any of the margins. The
outer argument controls whether output goes in the figure or outer margins.
The side argument determines which margin to draw in: 1 means the bottom
margin, 2 means the left margin, 3 means the top margin, and 4 means the
right margin.

Text is drawn a number of lines of text away from the edges of the plot region
for figure margins or a number of lines away from the edges of the inner region
for outer margins. In the figure margins, the location of the text along the

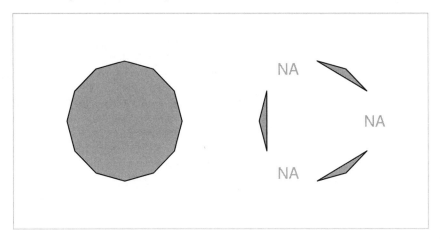

Figure 3.17
Drawing polygons using the `polygon()` function. On the left, a single polygon (dodecagon) is produced from multiple (`x, y`) locations. On the right, the first, fifth, and ninth values have been set to `NA`, which splits the output into three separate polygons. The `polygon()` function does not draw the gray `NA` values; those have been drawn using the `text()` function purely for the purposes of illustration.

margin can be specified relative to the user coordinates on the relevant axis using the `at` argument. In some cases it is possible to specify the location as a proportion of the length of the margin using the `adj` argument, but this is dependent on the value of the `las` state setting (see page 62). For certain `las` settings, the `adj` argument instead controls the justification of the text relative to a position chosen by the `las` argument. There is also a `padj` argument for controlling the "vertical" justification of text in the margins (the justification of the text **p**erpendicular to the reading direction of the text).

The `title()` function is essentially a specialized version of `mtext()`. It is more convenient for producing a few specific types of output, but much less flexible than `mtext()`. This function can be used to produce a main title for a plot (in the top figure margin), axis labels (in the left and bottom figure margins), and a subtitle for a plot (in the bottom margin below the x-axis label). The output from this function is heavily influenced by various graphics state settings, such as `cex.main` and `col.main`, which control the size and color of the title.

Just like the `text()` function, which draws text in the plot region, the functions that draw text in the margins all accept not only a character value, but also an R expression, so that axis labels and plot titles can include special symbols and formatting (see Section 10.5).

With a little extra effort, it is also possible to produce graphical output in the figure or outer margins using the functions that normally draw in the

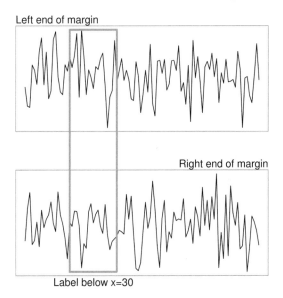

Figure 3.18
Annotating the margins of a traditional graphics plot. Text has been added in margin 3 of the top plot and in margins 1 and 3 in the bottom plot. Thick gray lines have been added to both plots (and overlapped so that it appears to be a single rectangle across the plots).

plot region (e.g., `points()` and `lines()`). In order to do this, the clipping region of the plot must first be set using the `xpd` state setting (see Section 3.2.7). This approach is not very convenient because the functions are drawing relative to user coordinates rather than locations relative to the margin co-ordinate systems. Nevertheless, it can sometimes be useful and the functions `grconvertX()` and `grconvertY()` can help with converting locations between coordinate systems.

The following code demonstrates the use of `mtext()` and a simple application of using `lines()` outside the plot region for drawing what appears to be a rectangle extending across two plots (see Figure 3.18).*

First of all, the `mfrow` setting is used to set up an arrangement of two figure regions, one above the other. The clipping region is set to the entire device using `xpd=NA`.

---

*This example was motivated by a question to R-help on December 14, 2004 with subject: "drawing a rectangle through multiple plots".

```
> y1 <- rnorm(100)
> y2 <- rnorm(100)
```

```
> par(mfrow=c(2, 1), xpd=NA)
```

The first data set is plotted as a line on the top plot and a label is added at the left end of figure margin 3. In addition, thick gray lines are drawn to represent the top of the rectangle, with the lines deliberately extending well below the bottom of the plot.

```
> plot(y1, type="l", axes=FALSE,
        xlab="", ylab="", main="")
> box(col="gray")
> mtext("Left end of margin", adj=0, side=3)
> lines(x=c(20, 20, 40, 40), y=c(-7, max(y1), max(y1), -7),
        lwd=3, col="gray")
```

The second data set is plotted as a line in the bottom plot, a label is added to this plot at the right end of figure margin 3, and another label is drawn beneath the x-location 30 in figure margin 1. Finally, thick gray lines are drawn to represent the bottom of the rectangle, again deliberately extending these above the plot. The thick gray lines overlap the lines drawn with respect to the top plot to create the impression of a single rectangle traversing both plots.

```
> plot(y2, type="l", axes=FALSE,
        xlab="", ylab="", main="")
> box(col="gray")
> mtext("Right end of margin", adj=1, side=3)
> mtext("Label below x=30", at=30, side=1)
> lines(x=c(20, 20, 40, 40), y=c(7, min(y2), min(y2), 7),
        lwd=3, col="gray")
```

3.4.3 Legends

The traditional graphics system provides the legend() function for adding a legend or key to a plot. The legend is usually drawn within the plot region, and is located relative to user coordinates. The function has many arguments, which allow for a great deal of flexibility in the specification of the contents and layout of the legend. The following code demonstrates a couple of typical uses.

The first example shows a scatterplot with a legend to relate group names to different symbols (see the top plot in Figure 3.19). The first two arguments give the position of the top-left corner of the legend, relative to the user coordinate system. The third argument provides labels for the legend and, because the pch argument is also specified, data symbols are drawn beside each label.

```
> with(iris,
        plot(Sepal.Length, Sepal.Width,
             pch=as.numeric(Species), cex=1.2))
> legend(6.1, 4.4, c("setosa", "versicolor", "virginica"),
         cex=1.5, pch=1:3)
```

The next example shows a barplot with a legend to relate group names to different fill patterns (see the bottom plot in Figure 3.19). In this example, the angle, density, and fill arguments are specified, so small rectangles with fill patterns are drawn beside each label in the legend.

```
> barplot(VADeaths[1:2,], angle=c(45, 135), density=20,
          col="gray", names=c("RM", "RF", "UM", "UF"))
> legend(0.4, 38, c("55-59", "50-54"), cex=1.5,
         angle=c(135, 45), density=20, fill="gray")
```

It should be noted that it is entirely the responsibility of the user to ensure that the legend corresponds to the plot. There is no automatic checking that data symbols in the legend match those in the plot, or that the labels in the legend have any correspondence with the data. This is one area where the **lattice** and **ggplot2** graphics systems provide a significant convenience (see Part II).

Some high-level functions draw their own legend specific to their purpose (e.g., filled.contour()).

3.4.4   Axes

In most cases, the axes that are automatically generated by the traditional graphics system will be sufficient for a plot. This is true even when the data being plotted on an axis are not numeric. For example, the axes of a boxplot or barplot are labeled appropriately using group names.

Section 3.2.5 describes ways in which the default appearance of automatically-generated axes can be modified, but it is more often the case that the user needs to inhibit the production of the automatic axis and draw a customized axis using the axis() function.

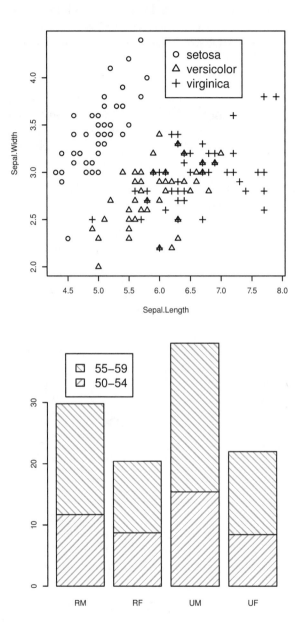

Figure 3.19
Some simple legends. Legends can be added to any kind of plot and can relate text labels to different symbols or different fill colors or patterns.

The first step is to inhibit the default axes. Most high-level functions should provide an `axes` argument which, when set to `FALSE`, indicates that the high-level function should not draw axes. Specifying the traditional graphics setting `xaxt="n"` (or `yaxt="n"`) may also do the trick.

The `axis()` function can draw axes on any side of a plot (chosen by the `side` argument), and the user can specify the location along the axis of tick marks and the text to use for tick labels (using the `at` and `labels` arguments, respectively). The following code demonstrates a simple example of a plot where the automatic axes are inhibited and custom axes are drawn, including a "secondary" y-axis on the right side of the plot (see Figure 3.20).

First of all, some temperature data are generated and an empty plot is created with no data symbols and no axes.

```
> x <- 1:2
> y <- runif(2, 0, 100)
> par(mar=c(4, 4, 2, 4))
> plot(x, y, type="n", xlim=c(0.5, 2.5), ylim=c(-10, 110),
        axes=FALSE, ann=FALSE)
```

Next, the main y-axis is drawn with specific tick locations to represent the Centigrade scale. The number 2 means that the axis should be drawn in margin 2 (the left margin) and the `at` argument specifies the locations of the tick marks for the axis.

```
> axis(2, at=seq(0, 100, 20))
> mtext("Temperature (Centigrade)", side=2, line=3)
```

Now the bottom axis is drawn with special labels and a secondary y-axis is drawn to represent the Fahrenheit scale. In the first expression, the `labels` argument is used to draw special tick mark labels on the x-axis. The second expression draws the secondary y-axis to the right of the plot by specifying 4 as the axis margin number.

```
> axis(1, at=1:2, labels=c("Treatment 1", "Treatment 2"))
> axis(4, at=seq(0, 100, 20), labels=seq(0, 100, 20)*9/5 + 32)
> mtext("Temperature (Fahrenheit)", side=4, line=3)
> box()
```

Finally, some thermometer-like symbols are drawn to represent the actual temperatures.

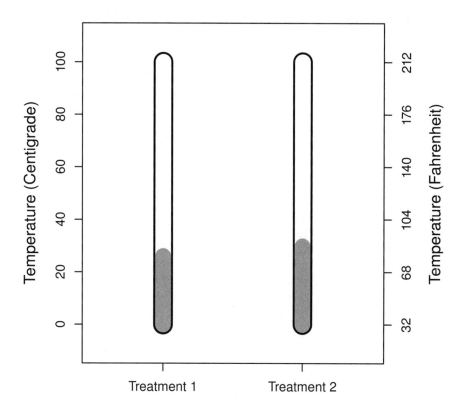

Figure 3.20
Customizing axes. An initial plot is drawn with a y-scale in degrees Centigrade, then a secondary y-axis is drawn with a scale in degrees Fahrenheit. The x-axis is drawn using special text labels, rather than the default numeric locations of the tick marks.

```
> segments(x, 0, x, 100, lwd=20)
> segments(x, 0, x, 100, lwd=16, col="white")
> segments(x, 0, x, y, lwd=16, col="gray")
```

The axis() function is not generic, but there are special alternative functions for plotting time-related data. The functions axis.Date() and axis.POSIXct() take an object containing dates and produce an axis with appropriate labels representing times, days, months, and years (e.g., 10:15, Jan 12 or 1995).

In some cases, it may be useful to draw tick marks at the locations that the default axis would use, but with different labels. The axTicks() function can be used to calculate these default locations. This function is also useful for enforcing an xaxp (or yaxp) graphics state setting, which control the number and placement of tick marks. If these settings are specified via par(), they usually have no effect because the traditional graphics system almost always calculates the settings itself. The user can choose these settings by passing them as arguments to axTicks(), then passing the resulting locations via the at argument to axis().

3.4.5   Coordinate systems

The traditional graphics system provides a number of coordinate systems for conveniently locating graphical output (see Section 3.1.1). Graphical output in the plot region is automatically positioned relative to the scales on the axes and text in the figure margins is placed in terms of a number of lines away from the edge of the plot (i.e., a scale that naturally corresponds to the size of the text).

It is also possible to locate output according to other coordinate systems that are not automatically supplied, but a little more work is required from the user. The basic principle is that the traditional graphics state can be queried to determine features of existing coordinate systems, then new coordinate systems can be calculated from this information.

The par() function

As well as being used to enforce new graphics state settings, the function par() can also be used to query current graphics state settings. The most useful settings are: din, fin, and pin, which reflect the current size, (width, height), of the graphics device, figure region, and plot region, in inches; and usr, which reflects the current user coordinate system (i.e., the ranges on the axes). The values of usr are in the order (xmin, xmax, ymin, ymax). When

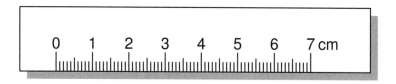

Figure 3.21
Custom coordinate systems. The lines and text are drawn relative to real physical centimeters (rather than the default coordinate system defined by the scales on plot axes).

a scale has a logarithmic transformation, the values are (10^xmin, 10^xmax, 10^ymin, 10^ymax).

There are also settings that reflect the size, (width, height), of a "standard" character. The setting cin gives the size in inches, cra in "rasters" or pixels, and cxy in "user coordinates." However, these values are not very useful because they only refer to a cex value of 1 (i.e., they ignore the current cex setting) *and* they only refer to the ps value when the current graphics device was first opened. Of more use are the strheight() function and the strwidth() function. These calculate the height and width of a given piece of text in inches, or in terms of user coordinates, or as a proportion of the current figure region (taking into account the current cex and ps settings).

The following code demonstrates a simple example of making use of customized coordinates where a ruler is drawn showing centimeter units (see Figure 3.21).

A blank plot region is set up first and calculations are performed to establish the relationship between user coordinates in the plot and physical centimeters.*

```
> plot(0:1, 0:1, type="n", axes=FALSE, ann=FALSE)
> usr <- par("usr")
> pin <- par("pin")
> xcm <- diff(usr[1:2])/(pin[1]*2.54)
> ycm <- diff(usr[3:4])/(pin[2]*2.54)
```

---

*R graphics relies on having accurate information on the physical size of the natural units on the page or screen (e.g., the physical size of pixels on a computer screen). The physical size of output for PostScript and PDF files should always be correct, but small inaccuracies may occur when specifying output with an physical size (such as inches) on screen devices such as Windows and X Window windows.

Now drawing can occur with positions expressed in terms of centimeters. First of all a "drop shadow" is drawn to give a three-dimensional effect by drawing a gray rectangle offset by 2 mm from the main ruler. The call to par() makes sure that the gray rectangle is not clipped to the plotting region (see Section 3.2.7).

```
> par(xpd=NA)
> rect(0 + 0.2*xcm, 0 - 0.2*ycm,
       1 + 0.2*xcm, 1 - 0.2*ycm,
       col="gray", border=NA)
```

The ruler itself is drawn with a call to rect() to draw the edges of the ruler, a call to segments() to draw the scale, and calls to text() to label the scale.

```
> rect(0, 0, 1, 1, col="white")
> segments(seq(1, 8, 0.1)*xcm, 0,
           seq(1, 8, 0.1)*xcm,
           c(rep(c(0.5, rep(0.25, 4),
                   0.35, rep(0.25, 4)),
                 7), 0.5)*ycm)
> text(1:8*xcm, 0.6*ycm, 0:7, adj=c(0.5, 0))
> text(8.2*xcm, 0.6*ycm, "cm", adj=c(0, 0))
```

There are utility functions, xinch() and yinch(), for performing the inches-to-user coordinates transformation (plus xyinch() for converting a location in one step and cm() for converting inches to centimeters). More powerful still are the grconvertX() and grconvertY() functions, which can be used to convert locations between any of the coordinate systems that the traditional graphics engine recognizes (see Table 3.5).

One problem with performing coordinate transformations like these is that the locations and sizes being drawn have no memory of how they were calculated. They are specified as locations and dimensions in user coordinates. This means that if the graphics window is resized (so that the relationship between physical dimensions and user coordinates changes), the locations and sizes will no longer have their intended meaning. If, in the above example, the graphics window is resized, the ruler will no longer accurately represent centimeter units. This problem will also occur if output is copied from one device to another device that has different physical dimensions. The legend() function performs calculations like these when arranging the components of a legend and its output is affected by device resizes and copying between devices.[*]

---

[*]It is possible to work around these problems in by using the recordGraphics() function, although this function should be used with extreme care.

Table 3.5

The coordinate systems recognized by the traditional graphics system.

Name	Description
"user"	The scales on the plot axes
"inches"	Inches, with $(0, 0)$ at bottom-left
"device"	Pixels for screen or bitmap output, otherwise 1/72"
"ndc"	Normalized coordinates, with $(0, 0)$ at bottom-left and $(1, 1)$ at top-right, within the entire device
"nic"	Normalized coordinates within the inner region
"nfc"	Normalized coordinates within the figure region
"npc"	Normalized coordinates within the plot region

Overlaying output

It is sometimes useful to plot two data sets on the same plot where the data sets share a common x-variable, but have very different y-scales. This can be achieved in at least two ways. One approach is simply to use par(new=TRUE) to overlay two distinct plots on top of each other, though care must be taken to avoid conflicting axes overwriting each other. Another approach is to explicitly reset the usr state setting before plotting a second set of data. The following code demonstrates both approaches to produce exactly the same result (see the top plot of Figure 3.22).

The data are yearly numbers of drunkenness-related arrests* and mean annual temperature in New Haven, Connecticut from 1912 to 1971. The temperature data are available as the data set nhtemp in the datasets package. There are only arrests data for the first 9 years.

```
> drunkenness <- ts(c(3875, 4846, 5128, 5773, 7327,
                      6688, 5582, 3473, 3186,
                      rep(NA, 51)),
                    start=1912, end=1971)
```

The first approach is to draw a plot of the drunkenness data, call par(new=TRUE), then draw a complete second plot of the temperature data on top of the first plot. The second plot does not draw default axes (axes=FALSE), but uses the axis() function to draw a secondary y-axis to represent the temperature scale.

---

*These data were obtained from "Crime Statistics and Department Demographics" on the New Haven Police Department Web Site:
http://www.cityofnewhaven.com/police/html/stats/crime/yearly/1863-1920.htm.

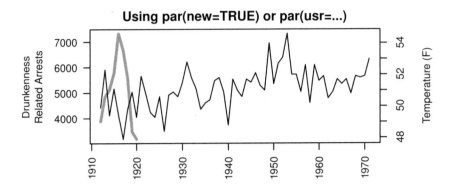

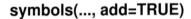

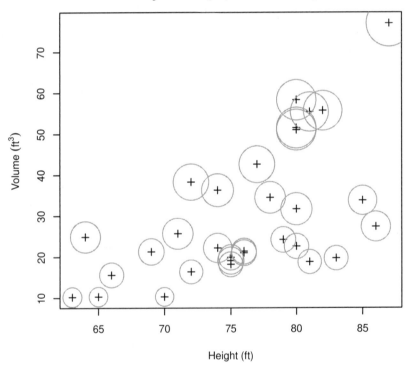

Figure 3.22
Overlaying plots. In the top plot, two line plots are drawn one on top of each other
to produce aligned plots of two data sets with very different scales. In the bottom
plot, the plotting function symbols() is used in "annotating mode" so that it adds
circles to an existing scatterplot rather than producing a complete plot itself.

```
> par(mar=c(5, 6, 2, 4))
> plot(drunkenness, lwd=3, col="gray", ann=FALSE, las=2)
> mtext("Drunkenness\nRelated Arrests", side=2, line=3.5)
> par(new=TRUE)
> plot(nhtemp, ann=FALSE, axes=FALSE)
> mtext("Temperature (F)", side=4, line=3)
> title("Using par(new=TRUE)")
> axis(4)
```

The second approach draws only one plot (for the drunkenness data). The user coordinate system is then redefined by specifying a new usr setting and the second "plot" is produced simply using lines(). Again, a secondary axis is drawn using the axis() function.

```
> par(mar=c(5, 6, 2, 4))
> plot(drunkenness, lwd=3, col="gray", ann=FALSE, las=2)
> mtext("Drunkenness\nRelated Arrests", side=2, line=3.5)
> usr <- par("usr")
> par(usr=c(usr[1:2], 47.6, 54.9))
> lines(nhtemp)
> mtext("Temperature (F)", side=4, line=3)
> title("Using par(usr=...)")
> axis(4)
```

Some high-level functions (e.g., symbols() and contour()) provide an argument called add which, if set to TRUE, will add the function output to the current plot, rather than starting a new plot. The following code shows the symbols() function being used to annotate a basic scatterplot (see the bottom plot of Figure 3.22). The data used in this example are physical measurements of black cherry trees available as the **trees** data frame from the **datasets** package.

```
> with(trees,
      {
        plot(Height, Volume, pch=3,
             xlab="Height (ft)",
             ylab=expression(paste("Volume ", (ft^3))))
        symbols(Height, Volume, circles=Girth/12,
                fg="gray", inches=FALSE, add=TRUE)
})
```

Another function of this type is the bxp() function. This function is called by boxplot() to draw the individual boxplots and is specifically set up to add boxplots to an existing plot (although it can also produce a complete plot).

It is also worth remembering that R follows a painters model, with later output obscuring earlier output. The following example makes use of this feature to fill a complex region within a plot (see Figure 3.23).

The first step is to generate some data and calculate some important features of the data.

```
> xx <- c(1:50)
> yy <- rnorm(50)
> n <- 50
> hline <- 0
```

The first thing to draw is a plot with a filled polygon beneath the y-values (see the top-left plot of Figure 3.23).

```
> plot (yy ~ xx, type="n", axes=FALSE, ann=FALSE)
> polygon(c(xx[1], xx, xx[n]), c(min(yy), yy, min(yy)),
          col="gray", border=NA)
```

The next step is to draw a rectangle over the top of the polygon up to a fixed y-value. The expression par("usr") is used to obtain the current x-scale and y-scale ranges (see the top-right plot of Figure 3.23).

```
> usr <- par("usr")
> rect(usr[1], usr[3], usr[2], hline, col="white", border=NA)
```

Now a line through the y-values is drawn over the top of the rectangle (see the bottom-left plot of Figure 3.23).

```
> lines(xx, yy)
```

Finally, a horizontal line is drawn to indicate the y-value cut-off, and axes are added to the plot (see the bottom-right plot of Figure 3.23).

```
> abline (h=hline,col="gray")
> box()
> axis(1)
> axis(2)
```

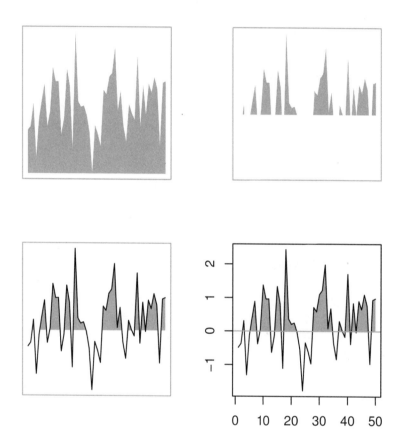

Figure 3.23

Overlaying output (making use of the painters model). The final complex plot, shown at bottom-right, is the result of overlaying several basic pieces of output: a gray polygon at top-left, with a white rectangle over the top (top-right), a black line on top of that (bottom-left), and a gray line on top of it all (plus axes and a bounding box).

3.4.6   Special cases

Some high-level functions are a little more difficult to annotate than others
because the plotting regions that they set up either are not immediately obvious or are not available after the function has run. This section describes
a number of high-level functions where additional knowledge is required to
perform annotations.

Obscure scales on axes

It is not immediately obvious how to add extra annotation to a barplot or a
boxplot in traditional R graphics because the scale on the categorical axis is
not obvious.

The difficulty with the `barplot()` function is that because the scale on the
x-axis is not labeled at all by default. the numeric scale is not obvious (and
calling `par("usr")` is not much help because the scale that the function sets
up is not intuitive either). In order to add annotations sensibly to a barplot
it is necessary to capture the value returned by the function. This return
value gives the x-locations of the mid-points of each bar that the function has
drawn. These midpoints can then be used to locate annotations relative to
the bars in the plot.

The code below shows an example of adding extra horizontal reference lines
to the bars of a barplot. The mid-points of the bars are saved to a variable
called `midpts`, then locations are calculated from those mid-points (and the
original counts) to draw horizontal white line segments within each bar using
the `segments()` function (see the left plot of Figure 3.24).

```
> y <- sample(1:10)
> midpts <- barplot(y, col=" light gray")
> width <- diff(midpts[1:2])/4
> left <- rep(midpts, y - 1) - width
> right <- rep(midpts, y - 1) + width
> heights <- unlist(apply(matrix(y, ncol=10),
                          2, seq))[-cumsum(y)]
> segments(left, heights, right, heights,
           col="white")
```

The `boxplot()` function is similar to the `barplot()` function in that the x-
scale is typically labeled with category names so the numeric scale is not obvious from looking at the plot. Fortunately, the scale set up by the `boxplot()`
function is much more intuitive. The individual boxplots are drawn at x-
locations `1:n`, where n is the number of boxplots being drawn.

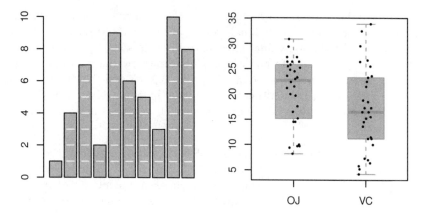

Figure 3.24
Special-case annotations. Some examples of functions where annotation requires special care. In the barplot at left, the value returned by the `barplot()` function is used to add horizontal white lines within the bars. Jittered points are added to the boxplot (right) using the knowledge that the *i*th box is located at position *i* on the x-axis.

The following code provides a simple example of annotating boxplots to add a jittered dotplot of individual data points on top of the boxplots. This provides a detailed view of the data as well as showing the main features via the boxplot. It is also a useful way to show how interesting features of the data, such as small clusters of points, can be hidden by a boxplot. In this example, the jittered data are centered upon the x-locations 1:2 to correspond to the centers of the relevant boxplots (see the right plot of Figure 3.24).

```
> with(ToothGrowth,
       {
         boxplot(len ~ supp, border="gray",
                 col="light gray", boxwex=0.5)
         points(jitter(rep(1:2, each=30), 0.5),
                unlist(split(len, supp)),
                cex=0.5, pch=16)
       })
```

Functions that draw several plots

The `pairs()` function is an example of a high-level function that draws more than one plot. This function draws a matrix of scatterplots. Such functions

tend to save the traditional graphics state before drawing, call `par(mfrow)` or
`layout()` to arrange the individual plots, and restore the traditional graphics
state once all of the individual plots have been drawn. This means that it
is not possible to annotate any of the plots drawn by the `pairs()` function
once the function has completed drawing. The regions and coordinate systems
that the function set up to draw the individual plots have been thrown away.
The only way to annotate the output from such functions is by way of *panel
functions*.

The `pairs()` function has a number of arguments that allow the user to
specify a function: `panel`, `diag.panel`, `upper.panel`, `lower.panel`, and
`text.panel`. The functions specified via these arguments are run as each
individual plot is drawn. In this way, the panel function has access to the plot
regions that are set up for each individual plot.

The following code shows a `pairs()` plot of the first two variables in the `iris`
data set. The `diag.panel` argument is used to draw boxplots in the diagonal
panels, instead of the default variable names. Notice that the panel function
must only add extra output, *not* start its own plot and this is achieved in this
case by called `boxplot()` with `add=TRUE`. Because `axes=FALSE`, the normal
boxplot axes are not drawn, and the `at` argument is used to make sure the
boxplots are centered horizontally within the panels. Because the normal
diagonal panels have variable names drawn in them, a `text.panel` function
is also specified. This panel function calls `mtext()` so that the normal text is
drawn in the top margin of the panel instead. The resulting plot is shown in
Figure 3.25.

```
> pairs(iris[1:2],
        diag.panel=function(x, ...) {
            boxplot(x, add=TRUE, axes=FALSE,
                    at=mean(par("usr")[1:2]))
        },
        text.panel=function(x, y, labels, ...) {
            mtext(labels, side=3, line=0)
        })
```

The `filled.contour()` function and the `coplot()` function have the same
problem as `pairs()` because the legends that they draw are actually separate
plots. Again, those functions allow annotation via panel function arguments.

The `panel.smooth()` function provides a predefined panel function to add a
smoothed trend line to a scatterplot of points.

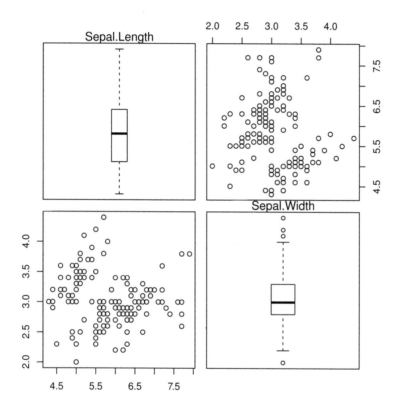

Figure 3.25
A panel function example. An example of using a panel function to add customized output to each the diagonal panels of a `pairs()` plot.

3D plots

It is possible to annotate a plot that was produced using the `persp()` function, but it is more difficult than for most other high-level functions. The important step is to acquire the transformation matrix that the `persp()` function returns. This can be used to transform 3D locations into 2D locations, using the `trans3d()` function. The result can then be given to the standard annotation functions such as `lines()` and `text()`. The `persp()` function also has an `add` argument, which allows multiple `persp()` plots to be over-plotted.

The following code demonstrates annotation of `persp()` output to add a contour plot beneath a 3D plot of the Maunga Whau volcano in Auckland New Zealand (see Figure 3.26). The data are from the `volcano` matrix in the **datasets** package.

The first step is to draw the 3D surface. The important features of this code are that the `zlim` is specified to leave room for the contour plot and the result of the call to `persp()` is assigned to a variable called `trans`.

```
> z <- 2 * volcano
> x <- 10 * (1:nrow(z))
> y <- 10 * (1:ncol(z))
> trans <- persp(x, y, z, zlim=c(0, max(z)),
                 theta = 150, phi = 12, lwd=.5,
                 scale = FALSE, axes=FALSE)
```

The next code calculates contour lines from the 3D data and then adds them to the plot. The result of `contourLines()` is a list, so `lapply()` is used to draw each contour line separately. The locations of the contour lines in the 3D plot are calculated using `trans3d()`, which is given the x and y vertices for a contour line, plus the z-position of zero (below the 3D surface). The `trans3d()` function converts the 3D locations into 2D locations which are drawn with the `lines()` function.

```
> clines <- contourLines(x, y, z)
> lapply(clines,
         function(contour) {
             lines(trans3d(contour$x, contour$y, 0, trans))
         })
```

A major limitation with annotating `persp()` output is that there is no support for automatically hiding output that should not be seen. In the above example, the view point was carefully chosen so that the entire contour plot was visible beneath the 3D surface. If the viewing angle is changed so that the surface

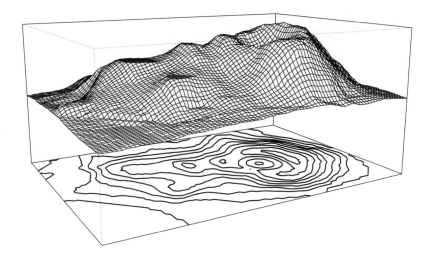

Figure 3.26
Annotating a 3D surface created by `persp()`. The contour lines are added to the
3D plot using the transformation matrix returned by the `persp()` function.

and the contour lines overlap, the contour lines will be drawn *on top of* the 3D
surface. In simple cases, this sort of problem can be worked around through
careful ordering of drawing operations, but in the general case something more
sophisticated is required (see Chapter 16).

3.5 Creating new plots

There are cases where no existing plot provides a sensible starting point for
creating the final plot that the user requires, situations where simply drawing more shapes on the plot is not sufficient. This section describes how to
construct a new plot entirely from scratch for such cases.

The `plot.new()` function is the most basic starting point for producing a
traditional graphics plot (the `frame()` function is equivalent). This function
starts a new plot and sets up the various plotting regions described in Section

3.1.1, with both the x-scale and y-scale set to $(0, 1)$.* The size and position of the regions that are set up depend on the current graphics state settings.

The `plot.window()` function resets the scales in the user coordinate system, given x- and y-ranges via the arguments `xlim` and `ylim`, and the `plot.xy()` function draws data symbols and lines between locations within the plot region.

3.5.1   A simple plot from scratch

In order to demonstrate the use of these functions, the following code produces the simple scatterplot in Figure 1.1 from scratch.

```
> plot.new()
> plot.window(range(pressure$temperature),
              range(pressure$pressure))
> plot.xy(pressure, type="p")
> box()
> axis(1)
> axis(2)
```

The call to `plot.new()` starts a new, completely blank, plot and the call to `plot.window()` sets the scales on the axes to fit the range of the data to be plotted. At this point, there is still nothing drawn. The `plot.xy()` function draws data symbols (`type="p"`) at the data locations, then `box()` draws a rectangle around the plot region, and `axis()` is used to draw the axes.

The output could be produced by the simple expression `plot(pressure)`, but this code shows that the steps in building a plot are available as separate functions as well, which allows the user to have fine control over the construction of a plot.

3.5.2   A more complex plot from scratch

This section describes a slightly more complex example of creating a plot from scratch. The final goal is represented in Figure 3.27 and the steps involved are described below.

The first chunk of code generates some data to plot. These are the counts of (adult) male and female survivors of the sinking of the *Titanic*.

---

*The actual scale setup depends on the current settings for `xaxs` and `yaxs`. With the default settings, the scales are $(-0.04, 1.04)$.

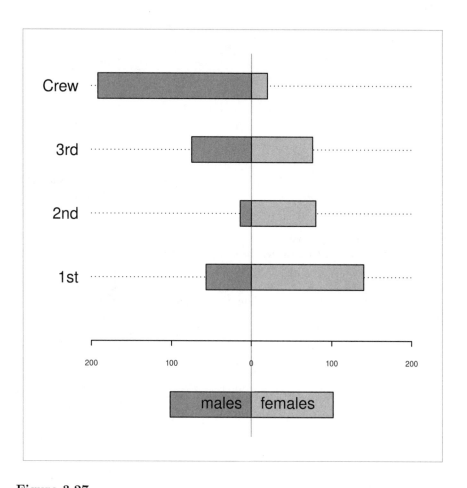

Figure 3.27

A back-to-back barplot from scratch. This demonstrates the use of lower-level plotting functions to produce a novel plot that cannot be produced by an existing high-level function.

```
> groups <- dimnames(Titanic)[[1]]
> males <- Titanic[, 1, 2, 2]
> females <- Titanic[, 2, 2, 2]

> males

 1st  2nd  3rd Crew
  57   14   75  192

> females

 1st  2nd  3rd Crew
 140   80   76   20
```

There are several ways that the plot could be created, the main idea being that it fundamentally consists of just a collection of graphical primitives that have been arranged in a meaningful way.

For this example, the approach will be to create a single plot. The labels to the left of the plot will be drawn in the margins of the plot, but everything else will be drawn inside the plot region. This next bit of code sets up the figure margins so that there is enough room for the labels in the left margin, but all other margins are nice and small (to avoid lots of empty space around the plot).

```
> par(mar=c(0.5, 3, 0.5, 1))
```

Inside the plot region there are six different rows of output to draw: the four main pairs of bars, the x-axis, and the legend at the bottom. The axis will be drawn at a y-location of 0, the main bars at the y-locations 1:4, and the legend at -1. The following code starts the plot and sets up the appropriate y-scale and x-scale.

```
> plot.new()
> plot.window(xlim=c(-200, 200), ylim=c(-1.5, 4.5))
```

This next bit of code assigns some useful values to variables, including the x-locations of tick marks on the x-axis, the y-locations of the main bars, and a value representing half the height of the bars.

```
> ticks <- seq(-200, 200, 100)
> y <- 1:4
> h <- 0.2
```

Now some drawing can occur. This next code draws the main part of the plot. Everything is drawn using calls to the low-level functions such as `lines()`, `segments()`, `mtext()`, and `axis()`. In particular, the main bars are just rectangles produced using `rect()`. Notice that the x-axis is drawn within the plot region (`pos=0`).

```
> lines(rep(0, 2), c(-1.5, 4.5), col="gray")
> segments(-200, y, 200, y, lty="dotted")
> rect(-males, y-h, 0, y+h, col="dark gray")
> rect(0, y-h, females, y+h, col="light gray")
> mtext(groups, at=y, adj=1, side=2, las=2)
> par(cex.axis=0.5, mex=0.5)
> axis(1, at=ticks, labels=abs(ticks), pos=0)
```

The final step is to produce the legend at the bottom of the plot. Again, this is just a series of calls to low-level functions, although the bars are sized using `strwidth()` to ensure that they contain the labels.

```
> tw <- 1.5*strwidth("females")
> rect(-tw, -1-h, 0, -1+h, col="dark gray")
> rect(0, -1-h, tw, -1+h, col="light gray")
> text(0, -1, "males", pos=2)
> text(0, -1, "females", pos=4)
```

This example is particularly customized to the data set involved. It could be made much more general by replacing some constants with variable values (e.g., instead of using 4 because there are four groups in the data set, the code could have a variable `numGroups`). If more than one such plot needs to be made, it makes good sense to also wrap the code within a function. That task is discussed in the next section.

3.5.3   Writing traditional graphics functions

Having made the effort to construct a plot from scratch, it is usually worthwhile encapsulating the calls within a new function and possibly even making it available for others to use. This section briefly describes some of the things to consider when creating a new graphics function built on the traditional graphics system.

There are many advantages to developing new graphics functions in the **grid** graphics system (see Part II) rather than using traditional graphics. Consequently, Chapter 8 contains a more complete discussion of the issues involved in developing new graphics functions.

Helper functions

There are some helper functions that do no drawing, but are used by the predefined high-level plots to do some of the work in setting up a plot.

The `xy.coords()` function is useful for allowing `x` and `y` arguments to your new function to be flexibly specified (just like the `plot()` function where `y` can be left unspecified and `x` can be a `data.frame`, and so on). This function takes `x` and `y` arguments and creates a standard object containing x-values, y-values, and sensible labels for the axes. There is also an `xyz.coords()` function.

If your plotting function generates multiple subplots, the `n2mfrow()` function may be helpful to generate a sensible number of rows and columns of plots, based on the total number of plots to fit on a page.

Another set of useful helper functions are those that calculate values to plot from the raw data (but do no actual drawing). Examples of these sorts of functions are: `boxplot.stats()` used by `boxplot()` to generate five-number summaries; `contourLines()` used by `contour()` to generate contour lines; `nclass.Sturges()`, `nclass.scott()`, and `nclass.FD()` used by `hist()` to generate the number of intervals for a histogram; and `co.intervals()` used by `coplot()` to generate ranges of values for conditioning a data set into panels.

Some high-level functions invisibly return this sort of information too. For example, `boxplot()` returns the combined results from `boxplot.stats()` for all of the boxplots that it produces and `hist()` returns information on the intervals that it creates including the number of data values in each interval. The `hist()` function is also useful (with `plot=FALSE`) simply to perform binning of continuous data.

Argument lists

A common technique when writing a traditional graphics function is to provide an ellipsis argument (. . .) instead of individual graphics state arguments (such as `col` and `lty`). This allows users to specify any state settings (e.g., `col="red"` and `lty="dashed"`) and the new function can pass them straight on to the traditional graphics functions that the new function calls. This avoids having to specify all individual state settings as arguments to the new function. Some care must be taken with this technique because sometimes different graphics functions interpret the same graphics state setting in different ways (the `col` setting is a good example; see Section 3.2). In such cases, it becomes necessary to name the individual graphics state setting as an argument and explicitly pass it on only to other graphics calls that will accept it and respond to it in the desired manner.

Sometimes it is useful for a graphics function to deliberately override the current graphics state settings. For example, a new plot may want to force the xpd setting to be NA in order to draw lines and text outside of the plot region. In such cases, it is polite for the graphics function to revert the graphics state settings at the end of the function so that users do not get a nasty surprise! A standard technique is to put the following expressions at the start of the new function to restore the graphics state to the settings that existed before the function was called.

```
opar <- par(no.readonly=TRUE)
on.exit(par(opar))
```

Because some of the traditional graphics state settings interact with each other, such a wholesale save-and-replace approach is actually unlikely to return the graphics state to exactly what it was before, so an even better solution is to save and restore only those parameters that the function modifies.

Care should be taken to ensure that a new graphics function takes notice of appropriate graphics state settings (e.g., ann). This can be a little complicated to implement because it is necessary to be aware of the possibility that the user might specify a setting in the call to the function and that such a setting should override the main graphics state setting. The standard approach is to name the state setting explicitly as an argument to the graphics function and provide the permanent state setting as a default value. See the new graphics function template below for an example of this technique using the ann argument. An additional complication is that now there is a state setting that will not be part of the . . . argument, so the state setting must be explicitly passed on to any other functions that might make use of it.

Another good technique is to provide arguments that users are used to seeing in other graphics functions — the main, sub, xlim, and ylim arguments are good examples of this sort of thing — and a new graphics function should be able to handle missing and non-finite values. The functions is.na(), is.finite(), and na.omit() may be useful for this purpose.

Plot methods

If a new function is for use with a particular type of data, then it is convenient for users if the function is provided as a method for the generic plot() function. This allows users to simply call the new function by calling plot(x), where x is an object of the relevant class.

```
 1 plot.newclass <-
 2   function(x, y=NULL,
 3             main="", sub="",
 4             xlim=NULL, ylim=NULL,
 5             axes=TRUE, ann=par("ann"),
 6             col=par("col"),
 7             ...) {
 8   xy <- xy.coords(x, y)
 9   if (is.null(xlim))
10     xlim <- range(xy$x[is.finite(xy$x)])
11   if (is.null(ylim))
12     ylim <- range(xy$y[is.finite(xy$y)])
13   opar <- par(no.readonly=TRUE)
14   on.exit(par(opar))
15   plot.new()
16   plot.window(xlim, ylim, ...)
17   points(xy$x, xy$y, col=col, ...)
18   if (axes) {
19     axis(1)
20     axis(2)
21     box()
22   }
23   if (ann)
24     title(main=main, sub=sub,
25           xlab=xy$xlab, ylab=xy$ylab, ...)
26 }
```

Figure 3.28
A graphics function template. This code provides a starting point for producing a
new graphics function for others to use.

A graphics function template

The code in Figure 3.28 is a simple shell that combines some of the basic
guidelines from this section. This is just a simplified version of the default
`plot()` method. It is far from complete and will not gracefully accept all
possible inputs (especially via the ... argument), but it could be used as the
starting template for writing a new traditional graphics function.

Chapter summary

High-level traditional graphics functions produce complete plots and low-level traditional graphics functions add output to existing plots. There are low-level functions for producing simple output such as lines, rectangles, text, and polygons and also functions for producing more complex output such as axes and legends.

The traditional graphics system creates several regions for drawing the various components of a plot: a plot region for drawing data symbols and lines, figure margins for axes and labels, and so on. Each low-level graphics function produces output in a particular drawing region and most work in the plot region.

There is a traditional graphics system state that consists of settings to control the appearance of output and the arrangement of the drawing regions. There are settings for controlling color, fonts, line styles, data symbol style, and the style of axes. There are several mechanisms for arranging multiple plots on a single page.

It is straightforward to create a complete plot using only low-level graphics functions. This makes it possible to produce a completely new type of plot. It is also possible for the user to define an entirely new graphics function.

Part II

GRID GRAPHICS

4

Trellis Graphics: The lattice Package

Chapter preview

This chapter describes how to produce Trellis plots using R. There is a description of what Trellis plots are as well as a description of the functions used to produce them. Trellis plots are designed to be easy to interpret and at the same time provide some modern and sophisticated plotting styles, such as multipanel conditioning. The **grid** graphics system provides no high-level plotting functions itself, so this chapter also describes one way to produce a complete plot using the **grid** system.

This part of the book concerns the major graphics packages that are related to the **grid** graphics system. This graphics system exists in parallel with the traditional graphics system and the two worlds do not interact at all well (see Section 1.2, but also Chapter 19).

The **grid** package only provides low-level graphics functions; it does not provide any functions for drawing complete plots. Such high-level functions are provided instead by other packages. This chapter and the next describe two major packages of this type: Deepayan Sarkar's **lattice** and Hadley Wickham's **ggplot2**.

The **lattice** package implements the Trellis Graphics system with some novel extensions. This represents a complete and coherent graphics system, which can in most cases be used without encountering any concepts of the underlying **grid** system.

This chapter deals with **lattice** as a self-contained system consisting of functions for producing complete plots and functions for controlling the appear-

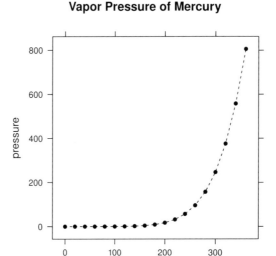

Figure 4.2
A modified scatterplot using **lattice**. Many of the standard high-level traditional graphics arguments also work with **lattice**.

Adding further lines and text to a plot is a little more complex in **lattice** compared to traditional graphics, so that topic is discussed later in Section 4.7.

One important difference compared to traditional graphics functions is that **lattice** graphics functions do not produce graphical output directly. Instead they produce an object of class `"trellis"`, which contains a description of the plot. The `print()` method for objects of this class does the actual drawing of the plot. This can be demonstrated quite easily. For example, the following code creates a `trellis` object, but does not draw anything.

```
> tplot <- xyplot(pressure ~ temperature, pressure)
```

The result of the call to `xyplot()` is assigned to the variable `tplot` so it is not printed. The plot can be drawn by calling print on the `trellis` object (the result is exactly the same as Figure 4.1).

```
> print(tplot)
```

This explicit printing is necessary when calling **lattice** functions within a loop or from another function.

4.1.1   Why another graphics system?

A number of functions in **lattice** produce output that is very similar to the output of functions in the traditional graphics system, but there are several reasons for using **lattice** functions instead of the traditional counterparts:

- The default appearance of the **lattice** plots is superior in some areas. For example, the default colors and the default data symbols have been deliberately chosen to make it easy to distinguish between groups when more than one data series is plotted, based on visual perception experiments. There are also some subtle things such as the fact that tick labels on the y-axes are written horizontally by default, which makes them easier to read.

- The arrangement of plot components is more automated in **lattice**. For example, the right amount of space is automatically created for axis labels and the plot title (it is usually not necessary to set figure margins manually).

- Legends can be automatically generated by the **lattice** system, so it is not the user's responsibility to ensure that the content of the legend corresponds correctly to the colors and data symbols used in the plot.

- The **lattice** plot functions can be extended in several very powerful ways. For example, several data series can be plotted at once in a convenient manner and multiple panels of plots can be produced easily (see Section 4.3).

- The output from **lattice** functions is **grid** output, so many powerful **grid** features are available for annotating, editing, and saving the graphics output. See Sections 6.8 and 7.7 for examples of these features.

4.2   lattice plot types

The **lattice** package provides functions to produce a number of standard plot types, plus some more modern and specialized plots. Table 4.1 describes the functions that are available and Figure 4.3 provides a basic idea of the sort of output that they produce.

Most of the **lattice** plotting functions provide a very long list of arguments and produce a wide range of different types of output. However, because **lattice** provides a single coherent system, many of the arguments are the

Table 4.1
The plotting functions available in **lattice**.

lattice Function	Description	Traditional Analog
barchart()	Barcharts	barplot()
bwplot()	Boxplots Box-and-whisker plots	boxplot()
densityplot()	Conditional kernel density plots Smoothed density estimate	plot.density
dotplot()	Dotplots Continuous versus categorical	dotchart()
histogram()	Histograms	hist()
qqmath()	Quantile–quantile plots Data set versus theoretical distribution	qqnorm()
stripplot()	Stripplots One-dimensional scatterplot	stripchart()
qq()	Quantile–quantile plots Data set versus data set	qqplot()
xyplot()	Scatterplots	plot()
levelplot()	Level plots	image()
contourplot()	Contour plots	contour()
cloud()	3D scatterplot	-
wireframe()	3D surfaces	persp()
splom()	Scatterplot matrices	pairs()
parallel()	Parallel coordinate plots	-

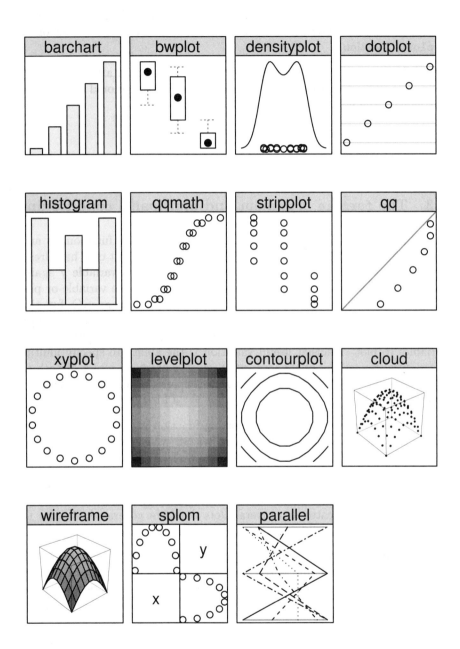

Figure 4.3
Plot types available in **lattice**. The name of the function used to produce the different plot types is shown in the strip above each plot.

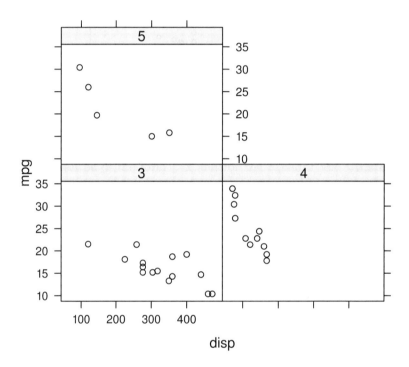

Figure 4.5
A **lattice** multipanel conditioning plot. A single function call produces several
scatterplots of the relationship between engine size and fuel efficiency for cars with
different numbers of forward gears.

the concept of a *shingle*. This is a continuous variable with a number of ranges associated with it. The ranges are used to split the continuous values into (possibly overlapping) groups. The `shingle()` function can be used to explicitly control the ranges, or the `equal.count()` function can be used to generate ranges automatically given a number of groups.

4.4   The `group` argument and legends

Another important argument in high-level **lattice** functions is the `group` argument, which allows multiple data series to be drawn on the same plot (or in each panel). The following code shows an example and the result is shown in Figure 4.6.

```
> xyplot(mpg ~ disp, data=mtcars,
         group=gear,
         auto.key=list(space="right"))
```

By specifying a variable via the `group` argument, a different plotting symbol will be used for cars with different numbers of gears. The `auto.key` argument is set so that **lattice** automatically generates an appropriate legend to show the mapping between data symbols and number of gears. This argument can either be just `TRUE` or a list of values specifying the appearance of the legend. In this case, the legend is positioned to the right of the plot. Notice that the page is automatically arranged to provide space for the plot legend.

In addition to the `auto.key`, there are arguments `key` and `legend` which provide progressively greater flexibility at the cost of increased complexity.

4.5   The `layout` argument and arranging plots

There are two types of arrangements to consider when dealing with **lattice** plots: the arrangement of panels and strips within a single **lattice** plot; and the arrangement of several complete **lattice** plots together on a single page.

In the first case (the arrangement of panels and strips within a single plot) there are two useful arguments that can be specified in a call to a **lattice**

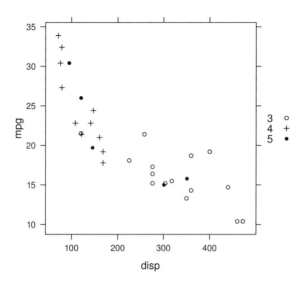

Figure 4.6
A **lattice** plot with multiple groups and an automatically generated legend. Different data symbols are used for cars with different numbers of gears.

plotting function: the `layout` argument and the `aspect` argument.

The `layout` argument consists of up to three values. The first two indicate
the number of columns and rows of panels on each page and the third value
indicates the number of pages. It is not necessary to specify all three values,
as **lattice** provides sensible default values for any unspecified values. The
following code produces a variation on Figure 4.5 by explicitly specifying that
there should be a single column of three panels, via the `layout` argument, and
that each panel must be "square," via the `aspect` argument. The final result
is shown in Figure 4.7.

```
> xyplot(mpg ~ disp | factor(gear), data=mtcars,
          layout=c(1, 3), aspect=1)
```

The `aspect` argument specifies the aspect ratio (height divided by width) for
the panels. The default value is `"fill"`, which means that panels expand to
occupy as much space as possible. In the example above, the panels were all
forced to be square by specifying `aspect=1`. This argument will also accept
the special value `"xy"`, which means that the aspect ratio is calculated to
satisfy the "banking to 45 degrees" rule proposed by Bill Cleveland.

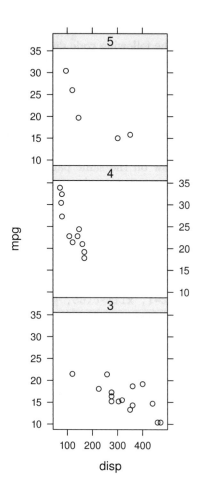

Figure 4.7

Controlling the layout of **lattice** panels. The **lattice** package arranges panels in a sensible way by default, but there are several ways to force the panels to be arranged in a particular layout. This figure shows a custom arrangement of the panels in the plot from Figure 4.5.

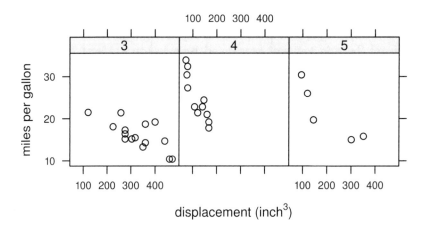

Figure 4.9
Modifying **lattice** axes. The placement of tick marks on the y-axis and the axis
labels have been customized in this plot.

pearance of axes. The list can have sublists, named x and y, if the settings
are intended to affect only the x-axes or only the y-axes.

In the following code, the `scales` argument is used to specify exactly where
tick marks should appear on y-axes. This code also demonstrates that the
`xlab` and `ylab` arguments can be expressions to allow the use of special formatting and special symbols. The plot produced by this code is shown in
Figure 4.9.

```
> xyplot(mpg ~ disp | factor(gear), data=mtcars,
         layout=c(3, 1), aspect=1,
         scales=list(y=list(at=seq(10, 30, 10))),
         ylab="miles per gallon",
         xlab=expression(paste("displacement (", inch^3, ")")))
```

Besides specifying the location and labels for tick marks, the `scales` argument
can also be used to control the font used for tick labels (`font`), the rotation
of the labels (`rot`), the range of values on the axes (`limits`), and whether
these ranges should be the same for all panels (`relation="same"`) or allowed
to vary between panels (`relation="free"`).

4.7 The `panel` argument and annotating plots

One advantage of the **lattice** graphics system is that it can produce extremely sophisticated plots from relatively simple expressions, especially with its multipanel conditioning feature. However, the cost of this is that the task of adding simple annotations of a **lattice** plot, such as adding extra lines or text, is more complex compared to the same task in traditional graphics.

Extra drawing can be added to the panels of a **lattice** plot via the `panel` argument. The value of this argument is a function, which gets called to draw the contents of each panel.

The following code shows an example panel function. The main plot is once again of the automobile fuel efficiency data, with three panels corresponding to different numbers of gear. The panel function consists of calls to various predefined functions that are designed to add graphics to **lattice** panels. The first function is very important. The `panel.xyplot()` function does the drawing that `xyplot()` would normally have done if the `panel` argument had not been specified. In this case, it draws a data symbol for each car. The other functions called in this panel function are `panel.abline()` and `panel.text()`, which add a dashed horizontal line and a label to indicate an efficiency criterion of 29 miles per gallon. The final result is shown in Figure 4.10.

```
> xyplot(mpg ~ disp | factor(gear), data=mtcars,
          layout=c(3, 1), aspect=1,
          panel=function(...) {
              panel.xyplot(...)
              panel.abline(h=29, lty="dashed")
              panel.text(470, 29.5, "efficiency criterion",
                          adj=c(1, 0), cex=.7)
      })
```

That panel function is a very simple one because it does exactly the same thing in each panel. Things get more complicated if the panel function has to produce different output for each panel. In that case, more attention has to be paid to the arguments of the panel function.

In the simple example above, the panel function is defined with just an ellipsis (`...`) argument. This means that any information that **lattice** sends to this panel function is captured by the ellipsis argument and the panel function simply passes the information on to `panel.xyplot()`.

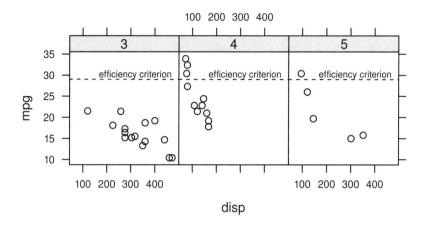

Figure 4.10
Adding annotations to **lattice** plots. The dashed horizontal lines and the labels
have been added to a standard `xyplot()` using a panel function.

Another common situation is that the extra graphics in a panel need to depend
on the x- and y-values that are plotted in that panel. The code below shows
an example, where the `panel.lmline()` function is called as part of the panel
function to draw a line of best fit to the data in each panel (see Figure 4.11).
The panel function now has explicit x- and y-arguments, which capture the
data values that **lattice** passes to each panel. These x- and y-values are passed
to `panel.lmline()` and to `panel.xyplot()` to produce the relevant output in
each panel. There is a lot of other information that **lattice** passes to the panel
function (see the argument list on the help page for `panel.xyplot()`), but
that is all simply passed through to `panel.xyplot()` via an ellipsis argument.

```
> xyplot(mpg ~ disp | factor(gear), data=mtcars,
         layout=c(3, 1), aspect=1,
         panel=function(x, y, ...) {
             panel.lmline(x, y)
             panel.xyplot(x, y, ...)
         })
```

As these examples have demonstrated, there are a number of predefined panel
functions available for adding output to a **lattice** panel, including both low-
level graphical primitives like points, and text and more high-level graphics

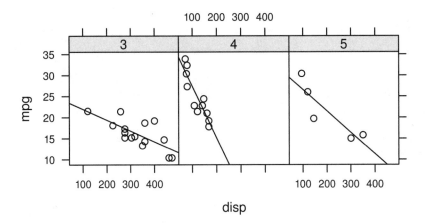

Figure 4.11
An example of a **lattice** panel function. A line of best fit has been added to each
panel in a standard `xyplot()` using a panel function.

like grids and lines of best fit. For every high-level **lattice** plotting function
(see Table 4.1) there is also a corresponding default panel function, for example, `panel.xyplot()`, `panel.bwplot()`, and `panel.histogram()`. Table 4.2
provides a list of some other predefined panel functions.

One other important panel function is `panel.superpose()`, which is the default panel function whenever multiple groups are drawn within a panel (e.g.,
when the **group** argument is used). When writing a custom panel function
for a **lattice** plot that has multiple groups in each panel, this function must
be called to reproduce the default plotting behavior.

In addition to the **panel** argument for adding further drawing to **lattice**
panels, there is a **strip** argument, which allows customization of the strips
above each panel.

4.7.1 Adding output to a lattice plot

Unlike in the original Trellis implementation, it is also possible to add output
to a complete **lattice** plot *after* the plot has been drawn (i.e., without using
a panel function).

Table 4.2

A selection of predefined panel functions for adding graphical output to the panels of **lattice** plots.

Function	Description
panel.points()	Draw data symbols at locations (x, y)
panel.lines()	Draw lines between locations (x, y)
panel.segments()	Draw line segments between (x0, y0) and (x1, y1)
panel.arrows()	Draw line segments and arrowheads to the end(s)
panel.rect()	Draw rectangles with bottom-left corner at (xl, yl) and top-right corner at (xr, yt)
panel.polygon()	Draw one or more polygons with vertices (x, y)
panel.text()	Draw text at locations (x, y)
panel.abline()	Draw a line with intercept a and slope b
panel.curve()	Draw a function given by expr
panel.rug()	Draw axis ticks at x- or y-locations
panel.grid()	Draw a (gray) reference grid
panel.loess()	Draw a loess smooth through (x, y)
panel.violin()	Draw one or more violin plots
panel.smoothScatter()	Draw a smoothed 2D density of (x, y)

The function `trellis.focus()` can be used to return to a particular panel or strip of the current **lattice** plot in order to add further output using, for example, `panel.lines()` or `panel.points()`. The `trellis.unfocus()` function should be called after the extra drawing is complete. The function `trellis.panelArgs()` may be useful for retrieving the arguments (including the data) that were used to originally draw the panel.

Sections 6.8 and 7.7 show how **grid** provides more flexibility for navigating to different parts of a **lattice** plot and for adding further output.

4.8 `par.settings` and graphical parameters

An important feature of Trellis Graphics is the careful selection of default settings that are provided for many of the features of **lattice** plots. For example, the default data symbols and colors used to distinguish between different data series have been chosen so that it is easy to visually discriminate between them. Nevertheless, it is still sometimes desirable to be able to make alterations to the default settings for aspects like color and text size.

The examples at the start of this chapter demonstrated that many of the familiar standard arguments from traditional graphics, such as `col`, `lty`, and `lwd`, do the same job in **lattice** plots. These graphical parameters can also be set via a `par.settings` argument. For example, the following code is an alternative way to produce Figure 4.2.

```
> xyplot(pressure ~ temperature, pressure,
         type="o",
         par.settings=list(plot.symbol=list(pch=16),
                           plot.line=list(lty="dashed")),
         main="Vapor Pressure of Mercury")
```

This approach works because **lattice** maintains a graphics state similar to the traditional graphics state: a large set of graphical parameter defaults.

The **lattice** graphical parameter settings consist of a large list of parameter groups and each parameter group is itself a list of parameter settings. For example, there is a `plot.line` parameter group consisting of `alpha`, `col`, `lty`, and `lwd` settings to control the color, line type, and line width for lines drawn between data locations. There is a separate `plot.symbol` group consisting of `alpha`, `cex`, `col`, `font`, `pch`, and `fill` settings to control the size, shape, and color of data symbols.

The settings in each parameter group affect some aspect of a **lattice** plot: some have a "global" effect, for example, the `fontsize` settings affect all text in a plot; some are more specific, for example, the `strip.background` setting affects the background color of strips; and some only affect a certain aspect of a certain sort of plot, for example, the `box.dot` settings affect only the dot that is plotted at the median value in boxplots.

The function `show.settings()` produces a picture representing some of the current graphical parameter settings. Figure 4.12 shows the settings for a black-and-white PostScript device.

The `par.settings` argument to high-level **lattice** plots allows specific graphical parameters to be set for a single plot, but, similar to `par()` in traditional graphics, the global default values can also be changed.

The current value of graphical parameter settings can be obtained using the `trellis.par.get()` function. For a list of all of the names of the parameter groups, type `names(trellis.par.get())`. If one of these group names is specified as the argument to `trellis.par.get()`, then only the relevant settings are returned. The following code shows how to obtain only the `add.text` group of settings.

```
> trellis.par.get("add.text")

$alpha
[1] 1

$cex
[1] 1

$col
[1] "#000000"

$font
[1] 1

$lineheight
[1] 1.2
```

The `trellis.par.set()` function can be used to specify new default values for graphical parameters. The value given to this function should be a list of lists. Only the components and groups that are to be changed need to be specified.

The following code demonstrates how to use `trellis.par.set()` to specify a new value for the `"col"` component of the `add.text` settings.

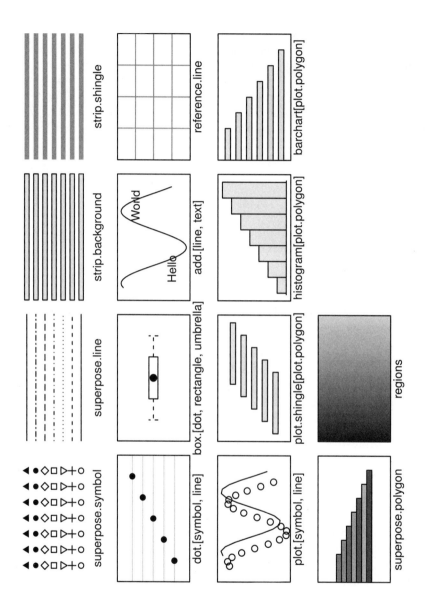

Figure 4.12
Some default **lattice** settings for a black-and-white PostScript device. This figure
was produced by the **lattice** function show.settings().

```
> trellis.par.set(list(add.text=list(col="red")))
```

A full set of **lattice** graphical parameter settings is called a *theme*. It is possible to specify such a theme and enforce a new "look and feel" for a plot, although choosing a complete set of defaults that all work together nicely is a difficult task. The **lattice** package currently provides one custom theme via the `col.whitebg()` function. It is also possible to obtain the default theme for a particular device using the `canonical.theme()` function. The **lattice** package maintains a separate set of these graphical parameter settings for each graphics device (see Section 9.1).

4.9 Extending lattice plots

This section briefly looks at the task of developing new **lattice** functions. This is not as simple as developing a new traditional graphics function because **lattice** is a more sophisticated and consistent graphics system; any new function has a lot to live up to.

The simplest case involves just writing a special panel function that others can call. Slightly better, in the case where a new function is designed for use with a specific class of data, is to define a method for an existing high-level function, like `xyplot()`. All high-level **lattice** plotting functions have been made generic so that this sort of customization can occur.

It is also possible to create an entirely new function, but the form of that function has to be relatively complex to retain the standard behavior of **lattice** plots. For a template example of this approach, see the code for the `dotplot()` function, which is really only a call to the `bwplot()` function with a different panel function supplied (type `lattice:::dotplot.formula` to see the code).

Users wanting to develop a new **lattice** plotting function along these lines are advised to read Chapter 6 to gain an understanding of the **grid** system that is used in the production of **lattice** output. Deepayan Sarkar's book on **lattice** has further discussion of this topic.

4.9.1 The latticeExtra package

The **latticeExtra** package provides a number of new **lattice** plots and new panel functions. Table 4.3 and Figure 4.13 show four of the new plots and Table 4.4 describes some of the new panel functions.

Table 4.3
Some plotting functions available in **latticeExtra**.

Function	Description
ecdfplot()	Empirical cumulative distribution
rootogram()	Tukey's hanging rootogram
segplot()	Segment plot
tileplot()	Voronoi mosaic or Dirichlet tesselation
marginal.plot()	Graphical summary of data frame

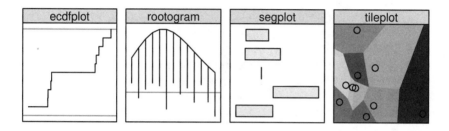

Figure 4.13
Plot types available in **latticeExtra**. The name of the function used to produce the different plot types is shown in the strip above each plot.

```
> library(latticeExtra)
```

In addition, the **latticeExtra** package provides a new **lattice** theme, via theEconomist.theme(), and it provides functions for manipulating trellis plot objects. For example, it provides a method for the c() function so that separate **lattice** plots can be combined to form a single plot (in certain circumstances). There is also a layer() function that allows a **lattice** plot to be annotated in separate steps, rather than using a panel function (in a style similar to that used by **ggplot2**; see Chapter 5). For example, the following code is an alternative way, using the **latticeExtra** package, to produce Figure 4.10.

Table 4.4

A range of panel functions from the **latticeExtra** package for adding
graphical output to the panels of **lattice** plots.

Function	Description
panel.key()	Draw legend within a panel
panel.ellipse()	Draw confidence ellipsoid
panel.xyarea()	Draw area below curve
panel.2dsmoother()	Draw level plot on irregular (x, y)
panel.3dbars()	Draw 3D bars (in wireframe() plot)
panel.3dpolygon()	Draw 3D planes
panel.3dtext()	Draw text at 3D location

```
> xyplot(mpg ~ disp | factor(gear), data=mtcars,
       layout=c(3, 1), aspect=1) +
     layer(panel.abline(h=29, lty="dashed")) +
     layer(panel.text(470, 29.5, "efficiency criterion",
                     adj=c(1, 0), cex=.7))
```

Chapter summary

The **lattice** package implements and extends the Trellis Graphics system for producing complete statistical plots. This system provides
most standard plot types and a number of modern plot types with
several important extensions. For a start, the layout and appearance
of the plots is designed to maximize readability and comprehension of
the information represented in the plot. Also, the system provides a
feature called multipanel conditioning, which produces multiple panels
of plots from a single data set, where each panel contains a different
subset of the data. The **lattice** functions provide an extensive set of
arguments for customizing the detailed appearance of a plot and there
are functions that allow the user to add further output to a plot.

5

The Grammar of Graphics:
The ggplot2 Package

Chapter preview

This chapter describes how to produce plots using the **ggplot2** package. There is a brief introduction to the concepts underlying the Grammar of Graphics paradigm as well as a description of the functions used to produce plots within this paradigm. The distinguishing feature of the **ggplot2** package is its ability to produce a very wide range of different plots from a relatively small set of fundamental components. Because **ggplot2** uses **grid** to draw plots, this chapter describes another way to produce a complete plot using the **grid** system.

The **ggplot2** package provides an interpretation and extension of the ideas in Leland Wilkinson's book *The Grammar of Graphics*. The **ggplot2** package represents a complete and coherent graphics system, completely separate from both traditional and **lattice** graphics.

The **ggplot2** package is built on **grid**, so it provides another way to generate complete plots within the **grid** world, but as with **lattice**, the package has so many features that it is unnecessary to encounter **grid** concepts for most applications.

The graphics functions that make up the graphics system are provided in an extension package called **ggplot2**. This package is not part of a standard R installation, so it must first be installed, then it can be loaded into R as follows.

```
> library(ggplot2)
```

This chapter presents a very brief introduction to **ggplot2**. Hadley Wickham's book, *ggplot2: Elegant Graphics for Data Analysis*, provides much more detail about the package.

5.1   Quick plots

For very simple plots, the qplot() function in **ggplot2** serves a similar purpose to the plot() function in traditional graphics. All that is required is to specify the relevant data values and the qplot() function produces a complete plot.

For example, the following code produces a scatterplot of pressure versus temperature using the **pressure** data set (see Figure 5.1).

```
> qplot(temperature, pressure, data=pressure)
```

This plot should be compared with Figures 1.1 and 4.1. The main differences between this scatterplot and what is produced by the traditional plot() function, or **lattice**'s xyplot(), are just the default settings used for things like the background grid, the plotting symbols, and the axis labeling.

There are also similarities in how the appearance of the plot can be modified. For example, the following code adds a title to the plot using the argument main.

```
> qplot(temperature, pressure, data=pressure,
        main="Vapor Pressure of Mercury")
```

However, **ggplot2** diverges quite rapidly from the other graphics systems if further customizations are desired. For example, in order to plot both points and lines on the plot, the following code is required (see Figure 5.2). Notice that, like **lattice**, the **ggplot2** result has automatically resized the plot region to provide room for the title.

```
> qplot(temperature, pressure, data=pressure,
        main="Vapor Pressure of Mercury",
        geom=c("point", "line"), lty=I("dashed"))
```

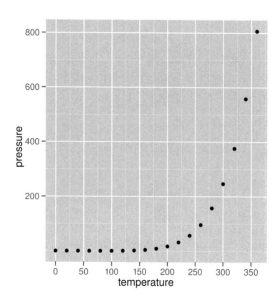

Figure 5.1

A scatterplot produced by the qplot() function from the **ggplot2** package. This plot is comparable to the traditional graphics plot in Figure 1.1.

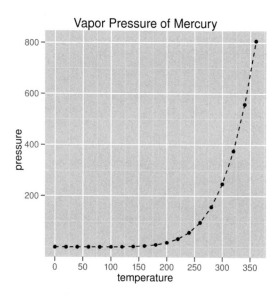

Figure 5.2

A scatterplot produced by the qplot() function from the **ggplot2** package, with a title and lines added. This plot is a modified version of 5.1.

The lty argument in this code is familiar, but the value "dashed" is wrapped inside a call to the I() function. The geom argument is also unique to **ggplot2**.

In order to understand how this code works, rather than spending a lot of time on the qplot() function, it is useful to move on instead to the conceptual structure, the grammar of graphics, that underlies the **ggplot2** package.

5.2   The ggplot2 graphics model

The **ggplot2** package implements the Grammar of Graphics paradigm. This means that, rather than having lots of different functions, each of which produces a different sort of plot, there is a small set of functions, each of which produces a different sort of plot *component*, and those components can be *combined* in many different ways to produce a huge variety of plots.

The steps in creating a plot with **ggplot2** often come down to the following essentials:

- Define the data that you want to plot and create an empty plot object with ggplot().

- Specify what graphics shapes, or *geoms*, that you are going to use to view the data (e.g., data symbols or lines) and add those to the plot with, for example, geom_point() or geom_line().

- Specify which features, or *aesthetics*, of the shapes will be used to represent the data values (e.g., the x- and y-locations of data symbols) with the aes() function.

In summary, a plot is created by mapping data values via aesthetics to the features of geometric shapes (see Figure 5.3).

For example, to produce the simple plot in Figure 5.1, the data set is the pressure data frame, and the variables temperature and pressure are used as the x and y locations of data symbols. This is expressed by the following code.

```
> ggplot(pressure) +
       geom_point(aes(x=temperature, y=pressure))
```

A **ggplot2** plot is built up like this by creating plot components, or *layers*, and combining them using the + operator.

Figure 5.3
A diagram showing how data is mapped to features of a geom (geometric shape) via aesthetics in **ggplot2**.

The following sections describe these ideas of geoms and aesthetics in more detail and go on to look at several other important components that allow for more complex plots that contain multiple groups, legends, facetting (similar to **lattice**'s multipanel conditioning), and more.

5.2.1   Why another graphics system?

Many of the plots that can be produced with **ggplot2** are very similar to the output of the traditional graphics system or the **lattice** graphics system, but there are several reasons for using **ggplot2** over the others:

- The default appearance of plots has been carefully chosen with visual perception in mind, like the defaults for **lattice** plots. The **ggplot2** style may be more appealing to some people than the **lattice** style.

- The arrangement of plot components and the inclusion of legends is automated. This is also like **lattice**, but the **ggplot2** facility is more comprehensive and sophisticated.

- Although the conceptual framework in **ggplot2** can take a little getting used to, once mastered, it provides a very powerful language for concisely expressing a wide variety of plots.

- The **ggplot2** package uses **grid** for rendering, which provides a lot of flexibility available for annotating, editing, and embedding **ggplot2** output (see Sections 6.9 and 7.8).

5.2.2   An example data set

The examples throughout this section will make use of the `mtcars2` data set. This data set is based on the `mtcars` data set from the **datasets** package and contains information on 32 different car models, including the size of the car engine (`disp`), its fuel efficiency (`mpg`), type of transmission (`trans`), number of forward gears (`gear`), and number of cylinders (`cyl`). The first few lines of the data set are shown below.

```
> head(mtcars2)
```

	mpg	cyl	disp	gear	trans
Mazda RX4	21.0	6	160	4	manual
Mazda RX4 Wag	21.0	6	160	4	manual
Datsun 710	22.8	4	108	4	manual
Hornet 4 Drive	21.4	6	258	3	automatic
Hornet Sportabout	18.7	8	360	3	automatic
Valiant	18.1	6	225	3	automatic

5.3   Data

The starting point for a plot is a set of data to visualize. The following call to the `ggplot()` function creates a new plot for the `mtcars` data set. The data for a plot must always be a data frame.

```
> p <- ggplot(mtcars2)
```

There is no information yet about how to display these data, so nothing is drawn. However, the result, a `"ggplot"` object, is assigned to the symbol p so that we can add more components to the plot in later examples.

5.4   Geoms and aesthetics

The next step in creating a plot is to specify what sort of shape will be used in the plot, for example, data symbols for a scatterplot or bars for a barplot. This step also involves deciding which variables in the data set will be used to control features of the shapes, for example, which variables will be used for the (x, y) positions of the data symbols in a scatterplot.

The following code adds this information to the plot that was created in the last section. This code produces a new `"ggplot"` object by adding information that says to draw data symbols, using the `geom_point()` function, and that the `disp` variable should be used for the x location and the and `mpg` variable should be used for the y location of the data symbols; these variables are mapped to the x and y aesthetics of the point geom, using the `aes()` function. The result is a scatterplot of fuel efficiency versus engine size (see Figure 5.4).

```
> p + geom_point(aes(x=disp, y=mpg))
```

Depending on what geom is being used to display the data, various other aesthetics are available. Another aesthetic that can be used with point geoms is the **shape** aesthetic. In the following code, the **gear** variable is associated with the data symbol shape so that cars with different numbers of forward gears are drawn with different data symbols (see Figure 5.4). Table 5.1 lists some of the common aesthetics for some common geoms.

```
> p + geom_point(aes(x=disp, y=mpg, shape=gear),
                 size=4)
```

This example also demonstrates the difference between *setting* an aesthetic and *mapping* an aesthetic. The **gear** variable is *mapped* to the **shape** aesthetic, using the `aes()` function, which means that the shapes of the data symbols are taken from the value of the variable and different data symbols will get different shapes. By contrast, the **size** aesthetic is *set* to the constant value of **4** (it is not part of the call to `aes()`), so all data symbols get this size. This is the reason for the use of the `I()` function on page 148; that is how to set an aesthetic when using `qplot()`.

The **ggplot2** package provides a range of geometric shapes that can be used to produce different sorts of plots. Other geoms include the standard graphical primitives, such as lines, text, and polygons, plus several more complex graphical shapes such as bars, contours, and boxplots (see later examples). Table 5.1 lists some of the common geoms that are available. As an example of a different sort of geom, the following code uses text labels rather than data symbols to plot the relationship between engine displacement and miles per gallon (see Figure 5.4). The locations of the the text are the same as the locations of the data symbols from before, but the text drawn at each location is based on the value of the **gear** variable. This example also demonstrates another aesthetic, **label**, which is relevant for text geoms.

```
> p + geom_text(aes(x=disp, y=mpg, label=gear))
```

A plot can be made up of multiple geoms by simply adding further geoms to the plot description. The following code draws a plot consisting of both data symbols and a straight line that is based on a linear model fit to the data (see Figure 5.4). The line is defined by its **intercept** and **slope** aesthetics.

Table 5.1

Some of the common geoms and their common aesthetics that are available in the **ggplot2** graphics system. All geoms have `color`, `size`, and `group` aesthetics. The `size` aesthetic means size of shape for points, height for text, and width for lines and it is in units of millimeters.

Geom	Description	Aesthetics
geom_point()	Data symbols	x, y, shape, fill
geom_line()	Line (ordered on x)	x, y, linetype
geom_path()	Line (original order)	x, y, linetype
geom_text()	Text labels	x, y, label, angle, hjust, vjust
geom_rect()	Rectangles	xmin, xmax, ymin, ymax, fill, linetype
geom_polygon()	Polygons	x, y, fill, linetype
geom_segment()	Line segments	x, y, xend, yend, linetype
geom_bar()	Bars	x, fill, linetype, weight
geom_histogram()	Histogram	x, fill, linetype, weight
geom_boxplot()	Boxplots	x, y, fill, weight
geom_density()	Density	x, y, fill, linetype
geom_contour()	Contour lines	x, y, fill, linetype
geom_smooth()	Smoothed line	x, y, fill, linetype
ALL		color, size, group

```
> lmcoef <- coef(lm(mpg ~ disp, mtcars2))

> p + geom_point(aes(x=disp, y=mpg)) +
      geom_abline(intercept=lmcoef[1], slope=lmcoef[2])
```

Specifying geoms and aesthetics provides the basis for creating a wide variety of plots with **ggplot2**. The remaining sections of this chapter introduce a number of other plot components within the **ggplot2** system, which are re-

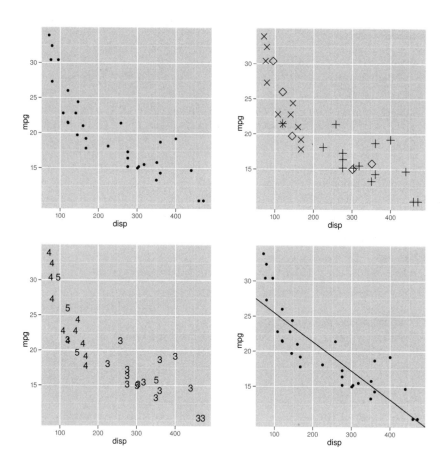

Figure 5.4

Variations on a scatterplot that shows the relationship between miles per gallon
(`mpg`) and engine displacement (`disp`): at top-left, a points geom is used to plot
data symbols; at top-right, the **shape** aesthetic of the points geom is used to plot
different data symbols for cars with different numbers of forward gears; at bottom-
left, a text geom is used to plot labels rather than data symbols; and at bottom-right,
both a points geom and an abline geom are used on the same plot to draw both
data symbols and a straight line (of best fit).

quired to control the details of plots and which extend the range of plots even further.

5.5   Scales

Another important type of component that has not yet been mentioned is the *scale* component. In **ggplot2** this encompasses the ideas of both axes and legends on plots.

Scales have not been mentioned to this point because **ggplot2** will often automatically generate appropriate scales for plots. For example, the x-axes and y-axes on the previous plots in this section are actually scale components that have been automatically generated by **ggplot2**.

One reason for explicitly adding a scale component to a plot is to override the detail of the scale that **ggplot2** creates. For example, the following code explicitly sets the axis labels using the `scale_x_continuous()` and `scale_y_continuous()` functions (see Figure 5.5).

```
> p + geom_point(aes(x=disp, y=mpg)) +
      scale_y_continuous(name="miles per gallon") +
      scale_x_continuous(name="displacement (cu.in.)")
```

It is also possible to control features such as the limits of the axis, where the tick marks should go, and what the tick labels should look like. Table 5.2 shows some of the common scale functions and their arguments. In the following code, the limits of the y-axis are widened to include zero (see Figure 5.5).

```
> p + geom_point(aes(x=disp, y=mpg)) +
      scale_y_continuous(limits=c(0, 40))
```

The **ggplot2** package also automatically creates legends when it is appropriate to do so. For example, in the following code, the `color` aesthetic is mapped to the `trans` variable in the `mtcars` data frame, so that the data symbols are colored according to what sort of transmission a car has. This automatically produces a legend to display the mapping between type of transmission and color.

```
> p + geom_point(aes(x=disp, y=mpg,
                   color=trans), size=4)
```

The plot resulting from the above code is not shown because this example demonstrates another important role that scales play in the **ggplot2** system.

When the `aes()` function is used to set up a mapping, the values of a variable are used to generate values of an aesthetic. Sometimes this is very straightforard. For example, when the variable `disp` is mapped to the aesthetic x for a points geom, the numeric values of `disp` are used directly as x locations for the points.

However, in other cases, the mapping is less obvious. For example, when the variable `trans`, with values `"manual"` and `"automatic"`, is mapped to the aesthetic `color` for a points geom, what color does the value `"manual"` correspond to?

As usual, **ggplot2** provides a reasonable answer to this question by default, but a second reason for explicitly adding a scale component to a plot is to explicitly control this mapping of variable values to aesthetic values (see Figure 5.6). For example, the following code uses the `scale_color_manual()` function to specify the two colors (shades of gray) that will correspond to the two values of the `trans` variable (see Figure 5.5).

```
> p + geom_point(aes(x=disp, y=mpg,
                  color=trans), size=4) +
    scale_color_manual(values=c(automatic=gray(2/3),
                        manual=gray(1/3)))
```

5.6   Statistical transformations

In the examples so far, data values have been mapped directly to aesthetic settings. For example, the numeric `disp` values have been used as x-locations for data symbols and the levels of the `trans` factor have been associated with different symbol colors.

Some geoms do not use the raw data values like this. Instead, the data values undergo some form of statistical transformation, or *stat*, and the transformed values are mapped to aesthetics (see Figure 5.7).

A good example of this sort of thing is the bar geom. This geom bins the raw values and uses the counts in each bin as the data to plot. For example, in the following code, the `trans` variable is mapped to the x aesthetic in the `geom_bar()` call. This establishes that the x-locations of the bars should be

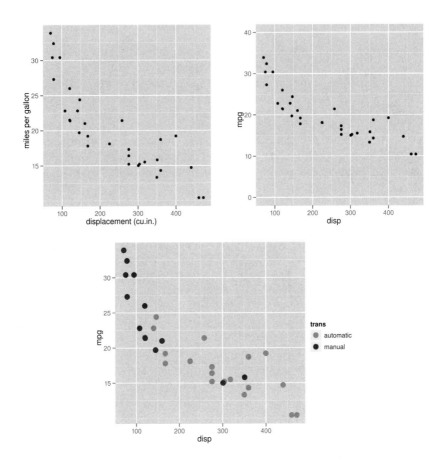

Figure 5.5
Scatterplots that have explicit scale components to control the labeling of axes or
the mapping from variable values to colors: at top-left, the x-axis and y-axis labels
are specified explicitly; at top-right, the y-axis range has been expanded; and the
bottom plot has an explicit mapping between transmission type and shades of gray.

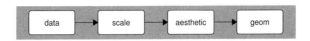

Figure 5.6
A diagram showing how the mapping of data to the features of geometric shapes is
controlled by a scale. The scale specifies how data values are mapped to aesthetic
values.

Table 5.2

Some of the common scales that are available in the **ggplot2** graphics system. All scales have `name`, `breaks`, `labels`, `limits` parameters. For every x-axis scale there is a corresponding y-axis scale.

Scale	Description	Parameters
`scale_x_continuous()`	Continuous axis	`expand, trans`
`scale_x_discrete()`	Categorical axis	
`scale_x_date()`	Date axis	`major, minor, format`
`scale_shape()`	Symbol shape legend	
`scale_linetype()`	Line pattern legend	
`scale_color_manual()`	Symbol/line color legend	`values`
`scale_fill_manual()`	Symbol/bar fill legend	`values`
`scale_size()`	Symbol size legend	`trans, to`
ALL		`name, breaks,`
		`labels, limits`

Figure 5.7

A diagram showing how the scaled data may be undergo a statistical transformation before being mapped to the values of an aesthetic.

the levels of `trans`, but heights of the bars (the `y` aesthetic) is automatically generated from the counts of each level of `trans` to produce a bar plot (see Figure 5.8).

```
> p + geom_bar(aes(x=trans))
```

The stat that is used in this case is a binning stat. Another option is an identity stat, which does not transform the data at all. The following code shows how to explicitly set the stat for a geom by creating the same bar plot from data that have already been binned.

```
> transCounts <- as.data.frame(table(mtcars2$trans))
> transCounts
```

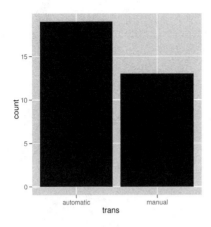 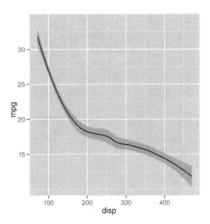

Figure 5.8
Examples of geoms with stat components: a bar geom, which uses a binning stat, and a smooth geom, which uses a smoother stat.

```
       Var1 Freq
1 automatic   19
2    manual   13
```

Now, both the x and the y aesthetics are set explicitly for the bar geom and the stat is set to "identity" to tell the geom not to bin again.

```
> ggplot(transCounts) +
      geom_bar(aes(x=Var1, y=Freq), stat="identity")
```

The following code presents another common transformation, which involves *smoothing* the original values. In this code, a smooth geom is added to the original empty plot. Rather than drawing a line through the original (x, y) values, this geom draws a smoothed line (plus a confidence band; see Figure 5.8).

```
> p + geom_smooth(aes(x=disp, y=mpg))
```

A similar result (without the confidence band) can be obtained using a line geom and explicitly specifying a "smooth" stat, as shown below.

```
> p + geom_line(aes(x=disp, y=mpg), stat="smooth")
```

Table 5.3

Some of the common stats that are available in the **ggplot2** graphics system.

Stat	Description	Parameters
stat_identity()	No transformation	–
stat_bin()	Binning	binwidth, origin
stat_smooth()	Smoother	method, se, n
stat_boxplot()	Boxplot statistics	width
stat_contour()	Contours	breaks

Yet another alternative is to add an explicit stat component, as in the following code. This works because stat components automatically have a geom associated with them, just as geoms automatically have a stat associated with them. The default geom for a smoother stat is a line.

```
> p + stat_smooth(aes(x=disp, y=mpg))
```

Similarly, the bar plot in Figure 5.8 could be created with an explicit binning stat component, as shown below. The default geom for a binning stat is a bar.

```
> p + stat_bin(aes(x=trans))
```

One advantage of this approach is that parameters of the stat, such as binwidths for binning data, can be specified clearly as part of the stat. For example, the following code controls the `method` for the smooth stat to get a straight line (the result is similar to the line in Figure 5.4).

```
> p + stat_smooth(aes(x=disp, y=mpg), method="lm")
```

Table 5.3 shows some common **ggplot2** stats and their parameters.

5.7 The group aesthetic

Previous examples have demonstrated that **ggplot2** automatically handles plotting multiple groups of data on a plot. For example, in the following

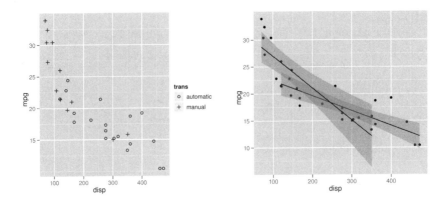

Figure 5.9
The **group** aesthetic in **ggplot2**. At left, mapping the **shape** aesthetic for point
geoms automatically generates a legend. At right, mapping the **group** aesthetic for
a smoother stat generates separate smoothed lines for different groups.

code, by introducing the **trans** variable as an aesthetic that controls shape,
two groups of data symbols are generated on the plot and a legend is produced
(the **scale_shape_manual()** function is used to control the mapping from
trans to data symbol **shape**; see Figure 5.9).

```
> p + geom_point(aes(x=disp, y=mpg, shape=trans)) +
      scale_shape_manual(values=c(1, 3))
```

It is also useful to be able to explicitly force a grouping for a plot and this can
be achieved via the **group** aesthetic. For example, the following code adds
a smoother stat to a scatterplot where the data symbols are all the same,
but there are separate smoothed lines for separate types of transmissions; the
group aesthetic is set for the smoother stat. The **method** parameter is also
set for the smoother stat so that the result is a straight line of best fit (see
Figure 5.9).

```
> ggplot(mtcars2, aes(x=disp, y=mpg)) +
      geom_point() +
      stat_smooth(aes(group=trans),
                  method="lm")
```

Notice that in the code above, aesthetic mappings have been specified in the
call to **ggplot()**. This is more efficient when several components in a plot
share the same aesthetic settings.

5.8 Position adjustments

Another detail that **ggplot2** often handles automatically is the problem of how to arrange geoms that overlap with each other. For example, the following code produces a bar plot of the number of cars with different transmissions, but also with the number of cylinders, `cyl`, mapped to the fill color for the bars (see Figure 5.10). The `color` aesthetic for the bars is set to `"black"` to provide borders for the bars and the fill color scale is explicitly set to three shades of gray.

```
> p + geom_bar(aes(x=trans, fill=factor(cyl)),
            color="black") +
    scale_fill_manual(values=gray(1:3/3))
```

There are three bars in this plot for automatic transmission cars (i.e., three bars share the same x-location). Rather than draw these bars over the top of each other, **ggplot2** has automatically stacked them up. This is an example of *position adjustment*.

An alternative is to use a dodge position adjustment, which places the bars side-by-side. This is shown in the following code and the result is shown in Figure 5.10.

```
> p + geom_bar(aes(x=trans, fill=factor(cyl)),
            color="black",
            position="dodge") +
    scale_fill_manual(values=gray(1:3/3))
```

Another option is a fill position adjustment. This expands the bars to fill the available space to produce a spine plot (see Figure 5.10).

```
> p + geom_bar(aes(x=trans, fill=factor(cyl)),
            color="black",
            position="fill") +
    scale_fill_manual(values=gray(1:3/3))
```

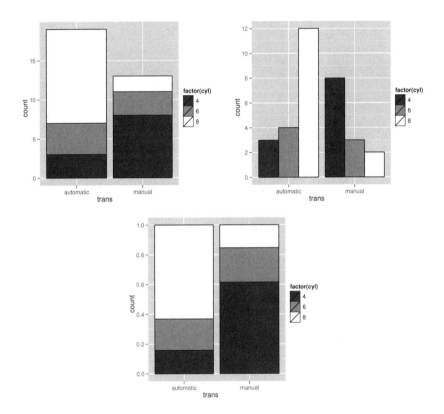

Figure 5.10
Examples of position adjustments in **ggplot2**: at top-left, the bars are `"stacked"`; at top-right, the bar position is `"dodge"` so the bars are side-by-side; and at the bottom, the position is `"fill"`, so the bars are scaled to fill the available (vertical) space.

5.9   Coordinate transformations

Section 5.5 described how scale components can be used to control the mapping between data values and the values of an aesthetic (e.g., map the `trans` value `"automatic"` to the `color` value `gray(2/3)`).

Another way to view this feature is as a *transformation* of the data values into the aesthetic domain. Another example of a transformation of data values is to use log axes on a plot. The following code does this for the plot of engine displacement versus miles per gallon via the `trans` argument of the `scale_x_continuous()` function. The result is shown in Figure 5.11.

```
> p + geom_point(aes(x=disp, y=mpg)) +
    scale_x_continuous(trans="log") +
    scale_y_continuous(trans="log") +
    geom_line(aes(x=disp, y=mpg), stat="smooth",
              method="lm")
```

This is another reason for using an explicit scale component in a plot. Notice that the data are transformed by the scale *before* any stat components are applied (see Figure 5.7), so the line is fitted to the log transformed data.

Another type of transformation is also possible in **ggplot2**. There is a coordinate system component, or *coord*, which by default is simple linear cartesian coordinates, but this can be explicitly set to something else.

For example, the following code adds a coordinate system component to the previous plot, using the `coord_trans()` function. This transformation says that both dimensions should be exponential.

```
> p + geom_point(aes(x=disp, y=mpg)) +
    scale_x_continuous(trans="log") +
    scale_y_continuous(trans="log") +
    geom_line(aes(x=disp, y=mpg), stat="smooth",
              method="lm") +
    coord_trans(x="exp", y="exp")
```

This sort of transformation occurs *after* the plot geoms have been created and controls how the graphical shapes are drawn on the page or screen (see Figure 5.12). In this case, the effect is to reverse the transformation of the data, so that the data points are back in their familiar arrangement and the line of

best fit, which was fitted to the logged data, has become a curve (see Figure
5.11).

Another example of a coordinate system in **ggplot2** is polar coordinates,
where the x- and y-values are treated as angle and radius values. The following
code creates a normal, cartesian coordinate system, stacked barplot showing
the number of cars with automatic versus manual transmissions (see Figure
5.11).

```
> p + geom_bar(aes(x="", fill=trans)) +
      scale_fill_manual(values=gray(1:2/3))
```

This next code sets the coordinate system to be polar, so that the y-values
(the heights of the bars) are treated as angles and x-values (the width of the
bar) is a (constant) radius. The result is a pie chart (see Figure 5.11).

```
> p + geom_bar(aes(x="", fill=trans)) +
      scale_fill_manual(values=gray(1:2/3)) +
      coord_polar(theta="y")
```

5.10   Facets

Facetting means breaking the data into several subsets and producing a separate plot for each subset on a single page. This is similar to **lattice**'s idea of
multipanel conditioning and is also known as producing *small multiples*.

The `facet_wrap()` function can be used to add facetting to a plot. The main
argument to this function is a formula that describes the variable to use for
subsetting the data. For example, in the following code a separate scatterplot
is produced for each value of `gear`. The `nrow` argument is used here to ensure
a single row of plots is produced.

```
> p + geom_point(aes(x=disp, y=mpg)) +
      facet_wrap(~ gear, nrow=1)
```

There is also a `facet_grid()` function for producing plots arranged on a grid.
The main difference is that the formula argument is of the form `y ~ x` and a
separate row of plots is produced for each level of `y` and a separate column of
plots is produced for each level of `x`.

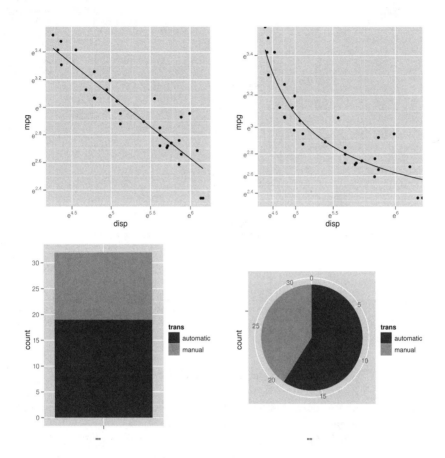

Figure 5.11
Examples of coordinate system transformations in **ggplot2**: at top-left is a cartesian plot of logged data with linear axes; at top-right is a cartesian plot of logged data with exponential axes; at bottom-left is a cartesian stacked barplot; and at bottom-right is a polar stacked barplot (a pie chart).

Figure 5.12
A diagram showing how geometric shapes may be transformed by a coordinate system before they are drawn on the page or screen.

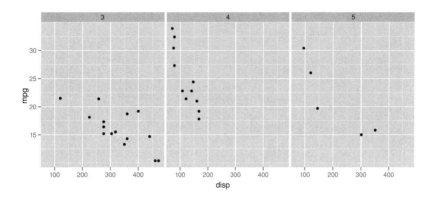

Figure 5.13
A facetted **ggplot2** scatterplot. A separate panel is produced for each level of a facetting variable, `gear`.

5.11   Themes

The **ggplot2** package takes a different approach to controlling the appearance of graphical objects, by separating output into data and non-data elements. Geoms represent the data-related elements of a plot and aesthetics are used to control the appearance of a geom, as was described in Section 5.4. This section looks at how to control the non-data elements of a plot, such as the labels and lines used to create the axes and legends.

The collection of graphical parameters that control non-data elements is called a *theme* in **ggplot2**. A theme can be added as another component to a plot in the now-familiar way. For example, the following code creates a basic scatterplot, but changes the basic color settings for the plot using the function `theme_bw()`. Instead of the standard gray background with white grid lines, this plot has a white background with gray gridlines (see Figure 5.14).

```
> p + geom_point(aes(x=disp, y=mpg)) +
      theme_bw()
```

It is also possible to set just specific *theme elements* of the overall theme for a plot. This requires the `opts()` function and one of the *element functions* to specify the new setting. For example, the following code uses the `theme_text()` function to make the y-axis label horizontal (see Figure 5.14).

Table 5.4

Some of the common plot elements in the **ggplot2** graphics system. The type implies which element function should be used to provide graphical parameter settings (e.g., text implies `theme_text()`).

Element	Type	Description
`axis.text.x`	text	X-axis tick labels
`legend.text`	text	Legend labels
`panel.background`	rect	Background of panel
`panel.grid.major`	line	Major grid lines
`panel.grid.minor`	line	Minor grid lines
`plot.title`	text	Plot title
`strip.background`	rect	Background of facet labels
`strip.text.x`	text	Text for horizontal strips

This example sets the text angle of rotation; it is also possible to set other parameters such as text font, color, and justification.

```
> p + geom_point(aes(x=disp, y=mpg)) +
      opts(axis.title.y=theme_text(angle=0))
```

There are other functions for setting graphical parameters for lines, segments, and rectangles, plus a `theme_blank()`, which removes the relevant plot element completely (see Figure 5.14).

```
> p + geom_point(aes(x=disp, y=mpg)) +
      opts(axis.title.y=theme_blank())
```

Table 5.4 shows some of the plot elements that can be controlled in this way.

The `opts()` function can also be used to control other features of the plot. For example, the following code specifies an overall title for a scatterplot (see Figure 5.14).

```
> p + geom_point(aes(x=disp, y=mpg)) +
      opts(title="Vehicle Fuel Efficiency")
```

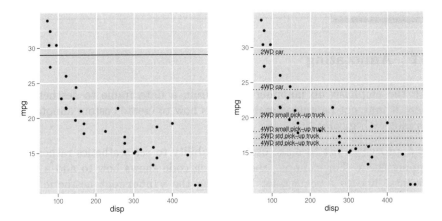

Figure 5.15
Some examples of annotation in **ggplot2**: at left, a single horizontal line has been
added by setting a geom aesthetic (rather than mapping the aesthetic) and, at right,
several horizontal lines and text labels have been added by using a completely new
data set for the relevant geoms.

```
> p + geom_point(aes(x=disp, y=mpg)) +
    geom_hline(data=gcLimits,
               aes(yintercept=limit),
               linetype="dotted") +
    geom_text(data=gcLimits,
              aes(y=limit + .1, label=category),
              x=70, hjust=0, vjust=0, size=3)
```

5.13   Extending ggplot2

Because **ggplot2** is based on a set of plot components that are combined
to form plots, developing a new type of plot is usually simply a matter of
combining the existing components in a new way.

Hadley Wickham's **ggplot2** book provides further discussion, including advice
on how to write a high-level function for producing a plot from **ggplot2**
functions.

Chapter summary

The **ggplot2** package implements and extends the Grammar of Graphics paradigm for statistical plots. The `qplot()` function works like `plot()` in very simple cases. Otherwise, a plot is created from basic components: a data frame, plus a set of geometric shapes (geoms), with a set of mappings from data values to properties of the shapes (aesthetics). Legends and axes are generated automatically, but the detailed appearance of all aspects of a plot can still be controlled. Multipanel plots are also possible.

6

The grid Graphics Model

Chapter preview

This chapter describes the fundamental tools that **grid** provides for drawing graphical scenes (including plots). There are basic features such as functions for drawing lines, rectangles, and text, together with more sophisticated and powerful concepts such as viewports, layouts, and units, which allow basic output to be located and sized in very flexible ways.

This chapter is useful for drawing a wide variety of pictures, including statistical plots from scratch, and for adding output to plots created by **lattice** or **ggplot2**.

The functions that make up the **grid** graphics system are provided in an extension package called `grid`. The **grid** system is loaded into R as follows.

```
> library(grid)
```

In addition to the standard on-line documentation available via the `help()` function, **grid** provides both broader and more in-depth on-line documentation in a series of vignettes, which are available via the `vignette()` function.

The **grid** graphics system only provides low-level graphics functions. There are no high-level functions for producing complete plots. Section 6.1 briefly introduces the concepts underlying the **grid** system, but this only provides an indication of how to work with **grid** and some of the things that are possible. An effective direct use of **grid** functions requires a deeper understanding of the **grid** system (see later sections of this chapter and Chapter 7).

```
> grid.points(pressure$temperature, pressure$pressure,
              name="dataSymbols")
> grid.rect()
> grid.xaxis()
> grid.yaxis()
```

Adding labels to the axes demonstrates the use of the different coordinate systems that are available in **grid**. The label text is drawn outside the edges of the plot region and is positioned in terms of a number of lines of text (i.e., the height that a line of text would occupy).

```
> grid.text("temperature", y=unit(-3, "line"))
> grid.text("pressure", x=unit(-3, "line"), rot=90)
```

The obvious result of running the above code is the graphical output (see the top-left image in Figure 6.1). Less obvious is the fact that several objects have been created. There are objects representing the viewport regions and there are objects representing the graphical output. The following code makes use of this fact to modify the plotting symbol from a circle to a triangle (see the top-right image in Figure 6.1). The object representing the data symbols was named "dataSymbols" (see the code above) and this name is used to find that object and modify it using the grid.edit() function.

```
> grid.edit("dataSymbols", pch=2)
```

The next piece of code makes use of the objects representing the viewports. The upViewport() and downViewport() functions are used to navigate between the different viewport regions to perform some extra annotations. First of all, a call to the upViewport() function is used to go back to working within the entire page so that a dashed rectangle can be drawn around the complete plot.

```
> upViewport(2)
> grid.rect(gp=gpar(lty="dashed"))
```

Next, the downViewport() function is used to return to the plot region to add a text annotation that is positioned relative to the scale on the axes of the plot (see bottom-right image in Figure 6.1).

```
> downViewport("plotRegion")
> grid.text("Pressure (mm Hg)\nversus\nTemperature (Celsius)",
            x=unit(150, "native"), y=unit(600, "native"))
```

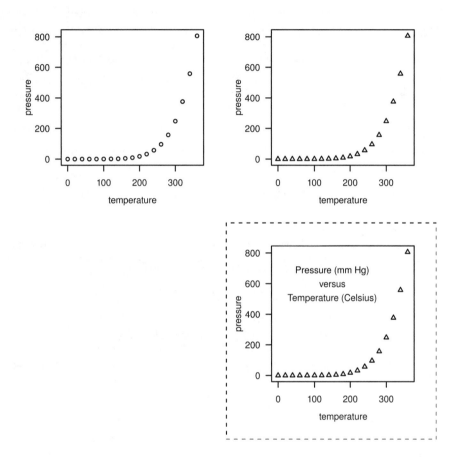

Figure 6.1

A simple scatterplot produced using **grid**. The top-left plot was constructed from a series of calls to primitive **grid** functions that produce graphical output. The top-right plot shows the result of calling the `grid.edit()` function to interactively modify the plotting symbol. The bottom-right plot was created by making calls to `upViewport()` and `downViewport()` to navigate between different drawing regions and adding further output (a dashed border and text within the plot).

The final scatterplot is still quite simple in this example, but the techniques that were used to produce it are very general and powerful. It is possible to produce a very complex plot, yet still have complete access to modify and add to any part of the plot.

In the remaining sections of this chapter, the basic **grid** concepts of viewports and units are discussed in full detail. A complete understanding of the **grid** system will be useful in two ways: it will allow the user to produce very complex images from scratch and it will allow the user to work effectively with complex **grid** output that is produced by other people's code, for example plots that are produced using **lattice** or **ggplot2**.

6.2   Graphical primitives

The most simple **grid** functions to understand are those that draw something. There are a set of **grid** functions for producing basic graphical output such as lines, circles, and text.* Table 6.1 lists the full set of these functions.

The first arguments to most of these functions is a set of locations and dimensions for the graphical object to draw. For example, grid.rect() has arguments x, y, width, and height for specifying the locations and sizes of the rectangles to draw. An important exception is the grid.text() function, which requires the text to draw as its first argument. The text to draw may be a character vector or an R expression (to produce special symbols and formatting; see Section 10.5).

In most cases, multiple locations and sizes can be specified and multiple primitives will be produced in response. For example, the following function call produces 100 circles because 100 locations and radii are specified (see Figure 6.2).

```
> grid.circle(x=seq(0.1, 0.9, length=100),
              y=0.5 + 0.4*sin(seq(0, 2*pi, length=100)),
              r=abs(0.1*cos(seq(0, 2*pi, length=100)))))
```

All of these functions are of the form grid.() and, for each one, there is a corresponding *Grob() function that creates an object containing a description of primitive graphical output, but does not draw anything. The *Grob() versions are addressed fully in Chapter 7.

Table 6.1

Graphical primitives in **grid**. This is the complete set of low-level functions that produce graphical output. For each function that produces graphical output (left-most column), there is a corresponding function that returns a graphical object containing a description of graphical output instead of producing graphical output (right-most column). The latter set of functions is described further in Chapter 7.

Function to Produce Output	Description	Function to Produce Object
grid.move.to()	Set the current location.	moveToGrob()
grid.line.to()	Draw a line from the current location to a new location and reset the current location.	lineToGrob()
grid.lines()	Draw a single line through multiple locations in sequence.	linesGrob()
grid.polyline()	Draw multiple lines through multiple locations in sequence.	polylineGrob()
grid.segments()	Draw multiple lines between pairs of locations.	segmentsGrob()
grid.xspline()	Draw smooth curve relative to control points.	xsplineGrob()
grid.rect()	Draw rectangles given locations and sizes.	rectGrob()
grid.roundrect()	Draw rectangles with rounded corners, given locations and sizes.	roundrectGrob()
grid.circle()	Draw circles given locations and radii.	circleGrob()
grid.polygon()	Draw polygons given vertexes.	polygonGrob()
grid.path()	Draw single polygon consisting of multiple paths.	pathGrob()
grid.text()	Draw text given strings, locations and rotations.	textGrob()
grid.raster()	Draw bitmap image.	rasterGrob()
grid.curve()	Draw smooth curve between two end points.	curveGrob()
grid.points()	Draw data symbols given locations.	pointsGrob()
grid.xaxis()	Draw x-axis.	xaxisGrob()
grid.yaxis()	Draw y-axis.	yaxisGrob()

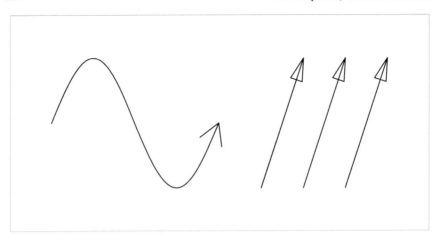

Figure 6.4
Drawing arrows using line-drawing functions. Arrows can be added to the output from `grid.lines()`, `grid.polyline()`, `grid.segments()`, `grid.line.to()`,
`grid.xspline()`, and `grid.curve()`. Examples are shown for `grid.lines()` (the
sine curve in the left half of the figure) and `grid.segments()` (the three straight
lines in the right half of the figure).

In simple usage, the `grid.polygon()` function draws a single polygon through
the specified x- and y-locations, automatically joining the last location to
the first to close the polygon. It is possible to produce multiple polygons
from a single call if the `id` argument is specified. In this case, a polygon is
drawn for each set of x- and y-locations corresponding to a different value of
`id`. The following code demonstrates both usages (see Figure 6.5). The two
`grid.polygon()` calls use the same x- and y-locations, but the second call
splits the locations into three separate polygons using the `id` argument.

```
> angle <- seq(0, 2*pi, length=10)[-10]
> grid.polygon(x=0.25 + 0.15*cos(angle), y=0.5 + 0.3*sin(angle),
               gp=gpar(fill="gray"))
> grid.polygon(x=0.75 + 0.15*cos(angle), y=0.5 + 0.3*sin(angle),
               id=rep(1:3, each=3),
               gp=gpar(fill="gray"))
```

The `grid.path()` function also has an `id` argument, but instead of producing
multiple polygons, the result is a single polygon consisting of multiple paths.
This can be used to create a shape with an internal hole. The following code
shows an example where a polygon shape is created with a rectangular hole
in the middle (see Figure 6.6).

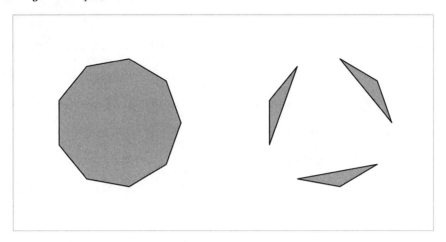

Figure 6.5
Drawing polygons using the `grid.polygon()` function. By default, a single polygon
is produced from multiple (x, y) locations (the nonagon on the left), but it is
possible to associate subsets of the locations with separate polygons using the `id`
argument (the three triangles on the right).

```
> angle <- seq(0, 2*pi, length=10)[-10]
> grid.path(x=0.25 + 0.15*cos(angle), y=0.5 + 0.3*sin(angle),
            gp=gpar(fill="gray"))
> grid.path(x=c(0.75 + 0.15*cos(angle), .7, .7, .8, .8),
            y=c(0.5 + 0.3*sin(angle),   .4, .6, .6, .4),
            id=rep(1:2, c(9, 4)),
            gp=gpar(fill="gray"))
```

The `grid.points()` function draws small shapes as data symbols at the (x,
y) locations. The `pch` argument specifies the data symbol shape as an integer
(e.g., 0 means an open square and 1 means an open circle) or as a single
character (see Section 10.3).

The `grid.raster()` function draws a bitmap image. The bitmap image can
be specified as a vector, matrix, or array. Chapter 18 describes ways to source
an image from an external file.

The `grid.xaxis()` and `grid.yaxis()` functions are not really graphical primitives as they produce relatively complex output consisting of both lines and
text. They are included here because they complete the set of **grid** functions
that produce graphical output. The main argument to these functions is the
`at` argument. This is used to specify where tick marks should be placed. If the
argument is not specified, sensible tick marks are drawn based on the current
scales in effect (see Section 6.5 for information about viewport scales). The

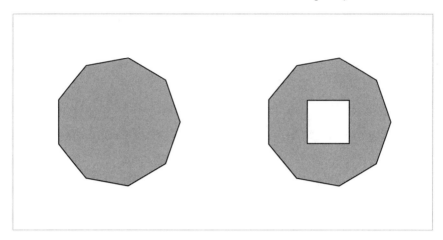

Figure 6.6
Drawing paths using the `grid.path()` function. In simple cases, a single polygon is
produced, from multiple `(x, y)` locations (the nonagon on the left), but it is also
possible to associate subsets of the locations with separate subpaths using the `id`
argument, which can be used to create holes in the polygon (the shape on the right).

values specified for the `at` argument are always relative to the current scales
(see the concept of the `"native"` coordinate system in Section 6.3). These
functions are much less flexible and general than the traditional `axis()` function. For example, they do not provide automatic support for generating
labels from time-based or date-based `at` locations.

6.2.1   Standard arguments

All primitive graphics functions accept a `gp` argument that allows control over
aspects such as the color and line type of the relevant output. For example, the
following code specifies that the boundary of the rectangle should be dashed
and colored red.

```
> grid.rect(gp=gpar(col="red", lty="dashed"))
```

Section 6.4 provides more information about setting graphical parameters.

All primitive graphics functions also accept a `vp` argument that can be used
to specify a viewport in which to draw the relevant output. The following
code shows a simple example of the syntax (the result is a rectangle drawn in
the left half of the page); Section 6.5 describes viewports and the use of `vp`
arguments in full detail.

```
> grid.rect(vp=viewport(x=0, width=0.5, just="left"))
```

Finally, all primitive graphics functions also accept a **name** argument. This can be used to identify the graphical object produced by the function. It is useful for editing graphical output and when working with graphical objects (see Chapter 7). The following code demonstrates how to associate a name with a rectangle.

```
> grid.rect(name="myrect")
```

6.2.2 Clipping

The `grid.clip()` function is not really a graphical primitive because it does not draw anything. Instead, this function specifies a clipping rectangle. After this function has been called, any subsequent drawing will only be visible if it occurs inside the clipping rectangle.

The clipping rectangle can be reset by calling `grid.clip()` again or by changing the drawing viewport (see Section 6.5, especially Section 6.5.2).

6.3 Coordinate systems

When drawing in **grid**, there are always a large number of coordinate systems available for specifying the locations and sizes of graphical output. For example, it is possible to specify an x-location as a proportion of the width of the drawing region, or as a number of inches (or centimeters, or millimeters) from the left-hand edge of the drawing region, or relative to the current x-axis scale. The full set of coordinate systems available is shown in Table 6.2. The meaning of some of these will only become clear with an understanding of viewports (Section 6.5) and graphical objects (Chapter 7).*

With so many coordinate systems available, it is necessary to specify which coordinate system a location or size refers to. This is the purpose of the `unit()` function. This function creates an object of class `"unit"` (hereafter referred to simply as a *unit*), which acts very much like a normal `numeric`

---

*Absolute units, such as inches, may not be rendered with full accuracy in all output formats (see the footnote on page 95).

Table 6.2
The full set of coordinate systems available in **grid**.

Coordinate System Name	Description
`"native"`	Locations and sizes are relative to the x- and y-scales for the current viewport.
`"npc"`	Normalized Parent Coordinates. Treats the bottom-left corner of the current viewport as the location $(0, 0)$ and the top-right corner as $(1, 1)$.
`"snpc"`	Square Normalized Parent Coordinates. Locations and sizes are expressed as a proportion of the *smaller* of the width and height of the current viewport.
`"in"`	Locations and sizes are in terms of physical inches. For locations, $(0, 0)$ is at the bottom-left of the viewport.
`"cm"`	Same as `"in"`, except in centimeters.
`"mm"`	Millimeters.
`"pt"`	Points. There are 72.27 points per inch.
`"bigpts"`	Big points. There are 72 big points per inch.
`"picas"`	Picas. There are 12 points per pica.
`"dida"`	Dida. 1157 dida equals 1238 points.
`"cicero"`	Cicero. There are 12 dida per cicero.
`"scaledpts"`	Scaled points. There are 65536 scaled points per point.
`"char"`	Locations and sizes are specified in terms of multiples of the current nominal font size (dependent on the current `fontsize` and `cex`).
`"line"`	Locations and sizes are specified in terms of multiples of the height of a line of text (dependent on the current `fontsize`, `cex`, and `lineheight`).
`"strwidth"` `"strheight"`	Locations and sizes are expressed as multiples of the width (or height) of a given string (dependent on the string and the current `fontsize`, `cex`, `fontfamily`, and `fontface`).
`"grobx"` `"groby"`	Locations and sizes are expressed as multiples of the x- or y-location on the boundary of a given graphical object (dependent on the type, location, and graphical settings of the graphical object).
`"grobwidth"` `"grobheight"`	Locations and sizes are expressed as multiples of the width (or height) of a given graphical object (dependent on the type, location, and graphical settings of the graphical object).

object — it is possible to perform basic operations such as subsetting units, and adding and subtracting units.

Each value in a unit can be associated with a different coordinate system and each location and dimension of a graphical object is a separate unit so, for example, a rectangle can have its x-location, y-location, width, and height all specified relative to different coordinate systems.

The following pieces of code demonstrate some of the flexibility of **grid** units. The first code examples show some different uses of the unit() function: a single value is associated with a coordinate system, then several values are associated with a coordinate system (notice the recycling of the coordinate system), then several values are associated with different coordinate systems.

```
> unit(1, "mm")
```

```
[1] 1mm
```

```
> unit(1:4, "mm")
```

```
[1] 1mm 2mm 3mm 4mm
```

```
> unit(1:4, c("npc", "mm", "native", "line"))
```

```
[1] 1npc      2mm      3native 4line
```

The next code examples show how units can be manipulated in many of the ways that normal numeric vectors can: firstly by subsetting, then simple arithmetic (again notice the recycling), then finally the use of a summary function (max() in this case).

```
> unit(1:4, "mm")[2:3]
```

```
[1] 2mm 3mm
```

```
> unit(1, "npc") - unit(1:4, "mm")
```

```
[1] 1npc-1mm 1npc-2mm 1npc-3mm 1npc-4mm
```

```
> max(unit(1:4, c("npc", "mm", "native", "line")))
```

[1] max(1npc, 2mm, 3native, 4line)

Some operations on units are not as straightforward as with numeric vectors, but require the use of functions written specifically for units. For example, units must be concatenated (in the sense of the c() function) using unit.c().

The following code provides an example of using units to locate and size a rectangle. The rectangle is at a location 40% of the way across the drawing region and 1 inch from the bottom of the drawing region. It is as wide as the text "very snug", and it is one line of text high (see Figure 6.7).

```
> grid.rect(x=unit(0.4, "npc"), y=unit(1, "in"),
            width=stringWidth("very snug"),
            height=unit(1, "line"),
            just=c("left", "bottom"))
```

6.3.1   Conversion functions

As demonstrated in the previous section, a unit is not simply a numeric value. Units only reduce to a simple numeric value (a physical location on a graphics device) when drawing occurs. A consequence of this is that a unit can mean very different things, depending on when it gets drawn (this should become more apparent with an understanding of graphical parameters in Section 6.4 and viewports in Section 6.5).

In some cases, it can be useful to convert a unit to a simple numeric value. For example, it is sometimes necessary to know the current scale limits for numerical calculations. There are several functions that can assist with this problem: convertUnit(), convertX(), convertY(), convertWidth(), and convertHeight(). The following code shows a calculation of the current page height in inches.

```
> convertHeight(unit(1, "npc"), "in")
```

[1] 7in

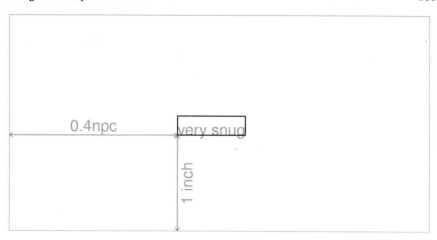

Figure 6.7
A demonstration of **grid** units. A diagram demonstrating how graphical output can be located and sized using **grid** units to associate numeric values with different coordinate systems. The gray border represents the current viewport. A black rectangle has been drawn with its bottom-left corner 40% of the way across the current viewport and 1 inch above the bottom of the current viewport. The rectangle is 1 line of text high and as wide as the text "very snug" (as it would be drawn in the current font).

WARNING: These conversion functions must be used with care. The output from these functions is only valid for the current page or screen size. If, for example, a window on screen is resized, or output is copied from the screen to a file format with a different physical size, these calculations may no longer be correct. In other words, only rely on these functions when it is known that the size of the screen will not change. See Chapter 19 for more information on this topic and for a way to be able to use these functions when the screen may be resized. The discussion on the use of these functions in `drawDetails()` methods and the function `grid.record()` is also relevant (see "Calculations during drawing" in Section 8.3.11).

6.3.2   Complex units

A number of coordinate systems in **grid** are *relative* in the sense that a value is interpreted as a multiple of the location or size of some other object. There are two peculiarities of these sorts of coordinate systems, `"strwidth"`, `"strheight"`, `"grobx"`, `"groby"`, `"grobwidth"`, and `"grobheight"`, that require further explanation. In the first two cases, the other object is just a text string (e.g., `"a label"`), but in the latter four cases, the other object can be any graphical object (see Chapter 7). It is necessary to specify the other object when generating a unit for these coordinate systems and this is achieved via the `data` argument. The following code shows some simple examples.

```
> unit(1, "strwidth", "some text")
```

[1] 1strwidth

```
> unit(1, "grobwidth", textGrob("some text"))
```

[1] 1grobwidth

A more convenient interface for generating units, when all values are relative to a single coordinate system, is also available via the `stringWidth()`, `stringHeight()`, `grobX()`, `grobY()`, `grobWidth()`, and `grobHeight()` functions. The following code is equivalent to the previous example.

```
> stringWidth("some text")
```

[1] 1strwidth

```
> grobWidth(textGrob("some text"))
```

[1] 1grobwidth

In this particular example, the `"strwidth"` and `"grobwidth"` units will be identical as they are based on identical pieces of text. The difference is that a graphical object can contain not only the text to draw, but also other information that may affect the size of the text, such as the font family and size.

In the following code, the two units are no longer identical because the `text` grob represents text drawn at font size of 18, whereas the simple string represents text at the default size of 10. The `convertWidth()` function is used to demonstrate the difference.

```
> convertWidth(stringWidth("some text"), "in")
```

[1] 0.715666666666667in

```
> convertWidth(grobWidth(textGrob("some text",
                                gp=gpar(fontsize=18))),
          "in")
```

[1] 1.0735in

For units that contain multiple values, there must be an object specified for every "strwidth", "strheight", "grobx", "groby", "grobwidth", and "grobheight" value. Where there is a mixture of coordinate systems within a unit, a value of NULL can be supplied for the coordinate systems that do not require data. The following code demonstrates this.

```
> unit(rep(1, 3), "strwidth", list("one", "two", "three"))
```

[1] 1strwidth 1strwidth 1strwidth

```
> unit(rep(1, 3),
       c("npc", "strwidth", "grobwidth"),
       list(NULL, "two", textGrob("three")))
```

[1] 1npc   1strwidth 1grobwidth

Again, there is a simpler interface for straightforward situations.

```
> stringWidth(c("one", "two", "three"))
```

[1] 1strwidth 1strwidth 1strwidth

For "grobx", "groby", "grobwidth", and "grobheight" units, it is also possible to specify the name of a graphical object rather than the graphical object itself. This can be useful for establishing a reference to a graphical object, so that when the named graphical object is modified, the unit is updated for the change. The following code demonstrates this idea. First of all, a text grob is drawn with the name "tgrob".

```
> grid.text("some text", name="tgrob")
```

Next, a unit is created that is based on the width of the grob called "tgrob".

```
> theUnit <- grobWidth("tgrob")
```

The `convertWidth()` function can be used to show the current value of the unit.

```
> convertWidth(theUnit, "in")
```

[1] 0.715666666666666in

The following code modifies the grob named `"tgrob"` and `convertWidth()` is used to show that the value of the unit reflects the new width of the `text` grob.

```
> grid.edit("tgrob", gp=gpar(fontsize=18))
> convertWidth(theUnit, "in")
```

[1] 1.0735in

See Section 7.5 for more examples of calculating the sizes of graphical objects.

6.4   Controlling the appearance of output

All graphical primitives functions (and the `viewport()` function; see Section 6.5) have a `gp` argument that can be used to provide a set of graphical parameters to control the appearance of the graphical output. There is a fixed set of graphical parameters (see Table 6.3), all of which can be specified for all types of graphical output.

The value supplied for the `gp` argument must be an object of class `"gpar"`, which is produced using the `gpar()` function. For example, the following code produces a `gpar` object containing graphical parameter settings controlling color and line type.

```
> gpar(col="red", lty="dashed")
```

$col
[1] "red"

$lty
[1] "dashed"

Table 6.3
The full set of graphical parameters available in **grid**.

Parameter	Description
col	Color of lines, text, rectangle borders, ...
fill	Color for filling rectangles, circles, polygons, ...
alpha	Alpha blending coefficient for transparency
lwd	Line width
lex	Line width expansion multiplier applied to lwd to obtain final line width
lty	Line type
lineend	Line end style (round, butt, square)
linejoin	Line join style (round, miter, bevel)
linemitre	Line miter limit
cex	Character expansion multiplier applied to fontsize to obtain final font size
fontsize	Size of text (in points)
fontface	Font face (bold, italic, ...)
fontfamily	Font family
lineheight	Multiplier applied to final font size to obtain the height of a line

The function `get.gpar()` can be used to obtain current graphical parameter settings. The following code shows how to query the current line type and fill color. When called with no arguments, the function returns a complete list of current settings.

```
> get.gpar(c("lty", "fill"))
```

```
$lty
[1] "solid"
```

```
$fill
[1] NA
```

A `gpar` object represents an *explicit graphical context* — settings for a small number of specific graphical parameters. The example above produces a graphical context that ensures that the color setting is `"red"` and the line-type setting is `"dashed"`. There is always an *implicit graphical context* consisting of default settings for all graphical parameters. The implicit graphical context is initialized automatically by **grid** device and can be modified by viewports (see Section 6.5.5) or by gTrees (see Section 7.2.1).*

A graphical primitive will be drawn with graphical parameter settings taken from the implicit graphical context, except where there are explicit graphical parameter settings from the graphical primitive's `gp` argument. For graphical primitives, the explicit graphical context is only in effect for the duration of the drawing of the graphical primitive. The following code example demonstrates these rules.

The default initial implicit graphical context includes settings such as `lty="solid"` and `fill="transparent"`. The first rectangle has an explicit setting `fill="black"` so it only uses the implicit setting `lty="solid"`. The second rectangle has no explicit graphical parameter settings so it uses all of the implicit graphical parameter settings. In particular, it is not at all affected by the explicit settings of the first rectangle (see Figure 6.8).

```
> grid.rect(x=0.33, height=0.7, width=0.2,
            gp=gpar(fill="black"))
> grid.rect(x=0.66, height=0.7, width=0.2)
```

---

*The ideas of implicit and explicit graphical contexts are similar to the specification of settings in Cascading Style Sheets and the graphics state in PostScript.

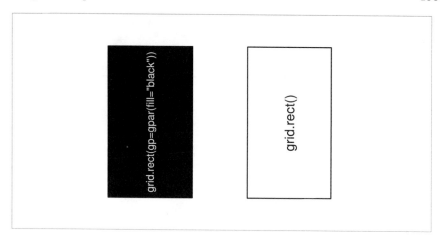

Figure 6.8
Graphical parameters for graphical primitives. The gray rectangle represents the current viewport. The right-hand rectangle has been drawn with no specific graphical parameters so it inherits the defaults for the current viewport (which in this case are a black border and no fill color). The left-hand rectangle has been drawn with a specific fill color of black (it is still drawn with the inherited black border). The graphical parameter settings for one rectangle have no effect on the other rectangle.

6.4.1 Specifying graphical parameter settings

The values that can be specified for colors, line types, line widths, line ends, line joins, and fonts are mostly the same as for the traditional graphics system. For example, colors can be specified by names such as "red". Chapter 10 describes the specification of graphical parameters in R in complete detail.

One peculiarity to **grid** is that the fontface value can be a name instead of an integer. Table 6.4 shows the possible values.

Many of the parameter names in **grid** are also the same as those in traditional graphics, though several of the **grid** names are slightly more verbose (e.g., lineend and fontfamily).

In **grid**, the cex value is cumulative. This means that it is multiplied by the previous cex value to obtain a current cex value. The following code shows a simple example. A viewport is pushed with cex=0.5. This means that text will be half size. Next, some text is drawn, also with cex=0.5. This text is drawn quarter size because cex was already 0.5 from the viewport (0.5*0.5 = 0.25).

```
> pushViewport(viewport(gp=gpar(cex=0.5)))
> grid.text("How small do you think?", gp=gpar(cex=0.5))
```

Table 6.4
Possible font face specifications in **grid**.

Integer	Name	Description
1	`"plain"`	Roman or upright face
2	`"bold"`	Bold face
3	`"italic"` or `"oblique"`	Slanted face
4	`"bold.italic"`	Bold and slanted face

The `lex` parameter, which is a multiplier that affects line width, is similarly cumulative.

The `alpha` graphical parameter provides a general alpha-transparency setting. It is a value between 1 (fully opaque) and 0 (fully transparent). The `alpha` value is combined with the alpha channel of colors by multiplying the two and this setting is cumulative like the `cex` setting. The following code shows a simple example. A viewport is pushed with `alpha=0.5`, then a rectangle is drawn using a semitransparent red fill color (alpha channel set to 0.5). The final alpha channel for the fill color is 0.25 (0.5*0.5 = 0.25).

```
> pushViewport(viewport(gp=gpar(alpha=0.5)))
> grid.rect(width=0.5, height=0.5,
            gp=gpar(fill=rgb(1, 0, 0, 0.5)))
```

The **grid** system does not provide any support for fill gradients or patterns, but some effects are possible through judicious use of raster images, graphical primitives, and clipping. Section 11.2 describes some functions in external packages that implement this idea.

6.4.2   Vectorized graphical parameter settings

All graphical parameter settings may be vector values. Many graphical primitive functions produce multiple primitives as output and graphical parameter settings will be recycled over those primitives. The following code produces 100 circles, cycling through 50 different shades of gray for the circles (see Figure 6.9).

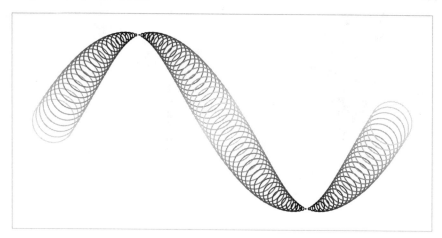

Figure 6.9

Recycling graphical parameters. The 100 circles are drawn by a single function call with 50 different grays specified for the border color (from a very light gray to a very dark gray and back to a very light gray). The 50 colors are recycled over the 100 circles so circle i gets the same color as circle $i + 50$.

```
> levels <- round(seq(90, 10, length=25))
> grays <- paste("gray", c(levels, rev(levels)), sep="")
> grid.circle(x=seq(0.1, 0.9, length=100),
              y=0.5 + 0.4*sin(seq(0, 2*pi, length=100)),
              r=abs(0.1*cos(seq(0, 2*pi, length=100))),
              gp=gpar(col=grays))
```

The grid.polygon() function is a slightly complex case. There are two ways in which this function will produce multiple polygons: when the id argument is specified *and* when there are NA values in the x- or y-locations (see Section 6.6). For grid.polygon(), a different graphical parameter will only be applied to each polygon identified by a different id. When a single polygon (as identified by a single id value) is split into multiple subpolygons by NA values, all subpolygons receive the same graphical parameter settings. The following code demonstrates these rules (see Figure 6.10). The first call to grid.polygon() draws two polygons as specified by the id argument. The fill graphical parameter setting contains two colors so the first polygon gets the first color (gray) and the second polygon gets the second color (white). In the second call, all that has changed is that an NA value has been introduced. This means that the first polygon as specified by the id argument is split into two separate polygons, but both of these polygons use the same fill setting because they both correspond to an id of 1. Both of these polygons get the first color (gray).

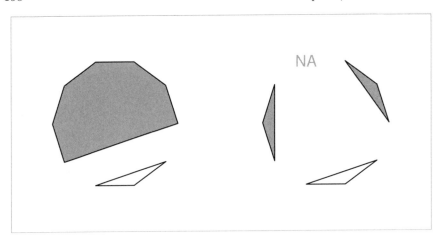

Figure 6.10
Recycling graphical parameters for polygons. On the left, a single function call
produces two polygons with different fill colors by specifying an `id` argument and
two fill colors. On the right, there are three polygons because an `NA value` has been
introduced in the `(x, y)` locations for the polygon, but there are still only two colors
specified. The colors are allocated to polygons using the `id` argument and ignoring
any `NA` values.

```
> angle <- seq(0, 2*pi, length=11)[-11]
> grid.polygon(x=0.25 + 0.15*cos(angle), y=0.5 + 0.3*sin(angle),
               id=rep(1:2, c(7, 3)),
               gp=gpar(fill=c("gray", "white")))
> angle[4] <- NA
> grid.polygon(x=0.75 + 0.15*cos(angle), y=0.5 + 0.3*sin(angle),
               id=rep(1:2, c(7, 3)),
               gp=gpar(fill=c("gray", "white")))
```

Other functions with an `id` argument, for example, `grid.polyline()` and
`grid.xspline()`, obey similar rules. On the other hand, the `grid.path()`
function is an exception to the exception because it (conceptually) only ever
draws a single shape.

All graphical primitives have a `gp` component, so it is possible to specify any
graphical parameter setting for any graphical primitive. This may seem inef-
ficient, and indeed in some cases the values are completely ignored (e.g., text
drawing ignores the `lty` setting), but in many cases the values are potentially
useful. For example, even when there is no text being drawn, the settings for
`fontsize`, `cex`, and `lineheight` are always used to calculate the meaning of
`"line"` and `"char"` coordinates. For example, the rectangles produced by the
following code are different heights.

```
> grid.rect(height=unit(1, "lines"))
> grid.rect(height=unit(1, "lines"),
             gp=gpar(lineheight=2))
```

6.5   Viewports

A *viewport* is a rectangular region that provides a context for drawing.

A viewport provides a *drawing context* consisting of both a *geometric context* and a *graphical context*. A geometric context consists of a set of coordinate systems for locating and sizing output and all of the coordinate systems described in Section 6.3 are available within every viewport. A graphical context consists of explicit graphical parameter settings for controlling the appearance of output. This is specified as a **gpar** object via the **gp** argument.

By default, **grid** creates a *root* viewport that corresponds to the entire page and, until another viewport is created, drawing occurs within the full extent of the page and using the default graphical parameter settings.[*]

A new viewport is created using the **viewport()** function. A viewport has a location (given by x and y), a size (given by width and height), and it is justified relative to its location (according to the value of the **just** argument). The location and size of a viewport are specified in units, so a viewport can be positioned and sized within another viewport in a very flexible manner. The following code creates a viewport that is left-justified at an x-location 0.4 of the way across the drawing region, and bottom-justified 1 centimeter from the bottom of the drawing region. It is as wide as the text "very very snug indeed", and it is six lines of text high. Figure 6.11 shows a diagram representing this viewport.

```
> viewport(x=unit(0.4, "npc"), y=unit(1, "cm"),
           width=stringWidth("very very snug indeed"),
           height=unit(6, "line"),
           just=c("left", "bottom"))
```

viewport[GRID.VP.24]

---

[*]Warning: some default parameter settings vary between different graphics formats. For example, the **fill** parameter is usually **"transparent"**, but for PNG output it is **"white"**.

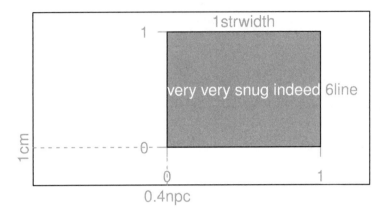

Figure 6.11
A diagram of a simple viewport. A viewport is a rectangular region specified by an (x, y) location, a (width, height) size, and a justification (and possibly a rotation). This diagram shows a viewport that is left-bottom justified 1 centimeter off the bottom of the page and 0.4 of the way across the page. It is six lines of text high and as wide as the text "very very snug indeed."

An important thing to notice in the above example is that the result of the viewport() function is an object of class "viewport". No region has actually been created on the page. In order to create regions on the page, a viewport object must be *pushed*, as described in the next section.

6.5.1   Pushing, popping, and navigating between viewports

The pushViewport() function takes a viewport object and uses it to create a region on the graphics device. This region becomes the drawing context for all subsequent graphical output, until the region is removed or another region is defined.

The following code demonstrates this idea (see Figure 6.12). To start with, the entire page, and the default graphical parameter settings, provide the drawing context. Within this context, the grid.text() call draws some text at the top-left corner of the device. A viewport is then pushed, which creates a region 80% as wide as the page, half the height of the page, and rotated at an angle of 10 degrees.* The viewport is given a name, "vp1", which will help

---

*It is not often very useful to rotate a viewport, but it helps in this case to dramatize the difference between the drawing regions.

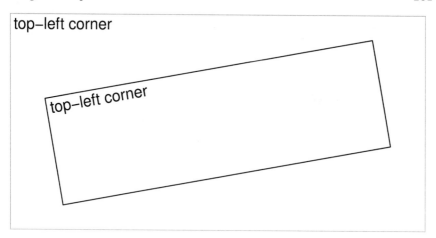

Figure 6.12

Pushing a viewport. Drawing occurs relative to the entire device until a viewport is pushed. For example, some text has been drawn in the top-left corner of the device. Once a viewport has been pushed, output is drawn relative to that viewport. The black rectangle represents a viewport that has been pushed and text has been drawn in the top-left corner of that viewport.

us to navigate back to this viewport from another viewport later.

Within the new drawing context defined by the viewport that has been pushed, *exactly the same* `grid.text()` call produces some text at the top-left corner of the viewport. A rectangle is also drawn to make the extent of the new viewport clear.

```
> grid.text("top-left corner", x=unit(1, "mm"),
            y=unit(1, "npc") - unit(1, "mm"),
            just=c("left", "top"))
> pushViewport(viewport(width=0.8, height=0.5, angle=10,
              name="vp1"))
> grid.rect()
> grid.text("top-left corner", x=unit(1, "mm"),
            y=unit(1, "npc") - unit(1, "mm"),
            just=c("left", "top"))
```

The pushing of viewports is entirely general. A viewport is pushed relative to the current drawing context. The following code slightly extends the previous example by pushing a further viewport, exactly like the first, and again drawing text at the top-left corner (see Figure 6.13). The location, size, and rotation of this second viewport are all relative to the context provided by the first viewport. Viewports can be nested like this to any depth.

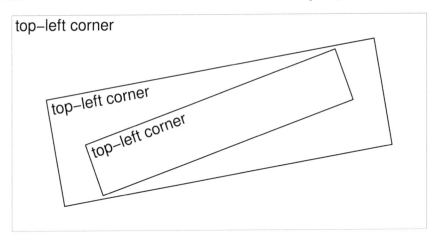

Figure 6.13
Pushing several viewports. Viewports are pushed relative to the current viewport.
Here, a second viewport has been pushed relative to the viewport that was pushed
in Figure 6.12. Again, text has been drawn in the top-left corner.

```
> pushViewport(viewport(width=0.8, height=0.5, angle=10,
             name="vp2"))
> grid.rect()
> grid.text("top-left corner", x=unit(1, "mm"),
           y=unit(1, "npc") - unit(1, "mm"),
           just=c("left", "top"))
```

In **grid**, drawing is always within the context of the current viewport. One
way to change the current viewport is to push a viewport (as in the previous
examples), but there are other ways too. For a start, it is possible to *pop* a
viewport using the popViewport() function. This removes the current viewport and the drawing context reverts to whatever it was before the current
viewport was pushed. It is illegal to pop the top-most viewport that represents the entire page and the default graphical parameter settings and trying
to do so will result in an error.

The following code demonstrates popping viewports (see Figure 6.14). The
call to popViewport() removes the last viewport that was created on the page.
Text is drawn at the bottom-right of the resulting drawing region (which has
reverted back to being the first viewport that was pushed).

```
> popViewport()
> grid.text("bottom-right corner",
           x=unit(1, "npc") - unit(1, "mm"),
           y=unit(1, "mm"), just=c("right", "bottom"))
```

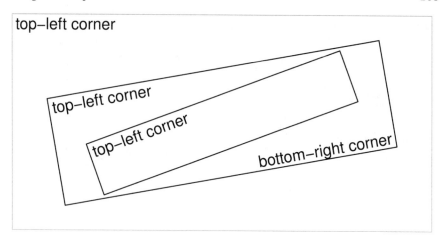

Figure 6.14
Popping a viewport. When a viewport is popped, the drawing context reverts to
the parent viewport. In this figure, the second viewport (pushed in Figure 6.13) has
been popped to go back to the first viewport (pushed in Figure 6.12). This time
text has been drawn in the bottom-right corner.

The popViewport() function has an integer argument n that specifies how
many viewports to pop. The default is 1, but several viewports can be popped
at once by specifying a larger value. The special value of 0 means that all
viewports should be popped. In other words, the drawing context should
revert to the entire device and the default graphical parameter settings.

Another way to change the current viewport is by using the upViewport()
and downViewport() functions. The upViewport() function is similar to
popViewport() in that the drawing context reverts to whatever it was prior to
the current viewport being pushed. The difference is that upViewport() does
not remove the current viewport from the page. This difference is significant
because it means that that a viewport can be revisited without having to
push it again. Revisiting a viewport is faster than pushing a viewport and it
allows the creation of viewport regions to be separated from the production
of output (see "viewport paths" in Section 6.5.3 and Chapter 8).

A viewport can be revisited using the downViewport() function. This function
has an argument name that can be used to specify the name of an existing
viewport. The result of downViewport() is to make the named viewport
the current drawing context. The following code demonstrates the use of
upViewport() and downViewport() (see Figure 6.15).

A call to upViewport() is made, which reverts the drawing context to the
entire page (recall that prior to this navigation the current viewport was the
first viewport that was pushed) and text is drawn in the bottom-right corner.

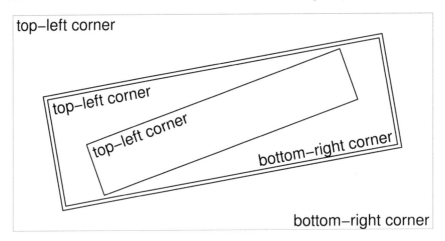

Figure 6.15
Navigating between viewports. Rather than popping a viewport, it is possible to
navigate up from a viewport (and leave the viewport on the device). Here navigation
has occurred from the first viewport to revert the drawing context to the entire
device and text has been drawn in the bottom-right corner. Next, there has been
a navigation down to the first viewport again and a second border has been drawn
around the outside of the viewport.

The `downViewport()` function is then used to navigate back down to the
viewport that was first pushed and a second border is drawn around this
viewport. The viewport to navigate down to is specified by its name, `"vp1"`.

```
> upViewport()
> grid.text("bottom-right corner",
            x=unit(1, "npc") - unit(1, "mm"),
            y=unit(1, "mm"), just=c("right", "bottom"))
> downViewport("vp1")
> grid.rect(width=unit(1, "npc") + unit(2, "mm"),
            height=unit(1, "npc") + unit(2, "mm"))
```

There is also a `seekViewport()` function that can be used to travel across
the viewport tree. This can be convenient for interactive use, but the result is
less predictable, so it is less suitable for use in writing **grid** functions for others to use. The call `seekViewport("avp")` is equivalent to `upViewport(0)`;
`downViewport("avp")`.

Drawing between viewports

Sometimes it is useful to be able to locate graphical output relative to more than one viewport. One way to do this in **grid** is via the `grid.move.to()` and `grid.line.to()` functions. It is possible to call `grid.move.to()` within one viewport, change viewports, and call `grid.line.to()`.

Another approach is to use the `grid.null()` function. This is a special graphical primitive that does not draw anything, but it draws nothing at a very specific location. Through the use of the functions `grobX()` and `grobY()` this makes it possible to perform drawing relative to one or more invisible locations, represented by one or more "null" grobs, which can be located in one or more different viewports. Section 7.5 has an example of this approach.

6.5.2 Clipping to viewports

Drawing can be restricted to only the interior of the current viewport (*clipped* to the viewport) by specifying the `clip` argument to the `viewport()` function. This argument has three values: `"on"` indicates that output should be clipped to the current viewport; `"off"` indicates that output should not be clipped at all; `"inherit"` means that the clipping region of the previous viewport should be used (this may not have been set by the previous viewport if that viewport's `clip` argument was also `"inherit"`). The following code provides a simple example (see Figure 6.16). A viewport is pushed with clipping on and a circle with a very thick black border is drawn relative to the viewport. A rectangle is also drawn to show the extent of the viewport. The circle partially extends beyond the limits of the viewport, so only those parts of the circle that lie within the viewport are drawn.

```
> pushViewport(viewport(w=.5, h=.5, clip="on"))
> grid.rect()
> grid.circle(r=.7, gp=gpar(lwd=20))
```

Next, another viewport is pushed and this viewport just inherits the clipping region from the first viewport. Another circle is drawn, this time with a gray and slightly thinner border and again the circle is clipped to the viewport.

```
> pushViewport(viewport(clip="inherit"))
> grid.circle(r=.7, gp=gpar(lwd=10, col="gray"))
```

Finally, a third viewport is pushed with clipping turned off. Now, when a third circle is drawn (with a thin, black border) all of the circle is drawn, even though parts of the circle extend beyond the viewport.

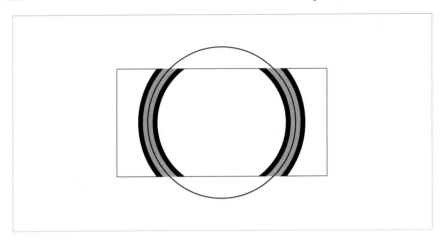

Figure 6.16
Clipping output in viewports. When a viewport is pushed, output can be clipped to
that viewport, or the clipping region can be left in its current state, or clipping can
be turned off entirely. In this figure, a viewport is pushed (the black rectangle) with
clipping on. A circle is drawn with a very thick black border and it gets clipped.
Next, another viewport is pushed (in the same location) with clipping left as it was.
A second circle is drawn with a slightly thinner gray border and it is also clipped.
Finally, a third viewport is pushed, which turns clipping off. A circle is drawn with
a thin black border and this circle is not clipped.

```
> pushViewport(viewport(clip="off"))
> grid.circle(r=.7)
> popViewport(3)
```

6.5.3   Viewport lists, stacks, and trees

It can be convenient to work with several viewports at once and there are
several facilities for doing this in **grid**. The `pushViewport()` function will
accept multiple arguments and will push the specified viewports one after
another. For example, the fourth expression below is a shorter equivalent
version of the first three expressions.

```
> pushViewport(vp1)
> pushViewport(vp2)
> pushViewport(vp3)
```

```
> pushViewport(vp1, vp2, vp3)
```

The pushViewport() function will also accept objects that contain several viewports: viewport lists, viewport stacks, and viewport trees. The function vpList() creates a list of viewports and these are pushed "in parallel." The first viewport in the list is pushed, then **grid** navigates back up before the next viewport in the list is pushed. The vpStack() function creates a stack of viewports and these are pushed "in series." Pushing a stack of viewports is exactly the same as specifying the viewports as multiple arguments to pushViewport(). The vpTree() function creates a tree of viewports that consists of a parent viewport and any number of child viewports. The parent viewport is pushed first, then the child viewports are pushed in parallel within the parent.

The current set of viewports that have been pushed on the current device constitute a viewport tree and the current.vpTree() function prints out a representation of the current viewport tree. The following code demonstrates the output from current.vpTree() and the difference between lists, stacks, and trees of viewports. First of all, some (trivial) viewports are created to work with.

```
> vp1 <- viewport(name="A")
> vp2 <- viewport(name="B")
> vp3 <- viewport(name="C")
```

The next piece of code shows these three viewports pushed as a list. The output of current.vpTree() shows the root viewport (which represents the entire device) and then all three viewports as children of the root viewport. A graph of the resulting viewport tree is shown in Figure 6.17 (top-left).

```
> pushViewport(vpList(vp1, vp2, vp3))
> current.vpTree()
```

viewport[ROOT]->(viewport[A], viewport[B], viewport[C])

This next code pushes the three viewports as a stack. The viewport vp1 is now the only child of the root viewport with vp2 a child of vp1, and vp3 a child of vp2. A graph of the resulting viewport tree is shown in Figure 6.17 (top-right).

```
> grid.newpage()
> pushViewport(vpStack(vp1, vp2, vp3))
> current.vpTree()
```

viewport[ROOT]->(viewport[A]->(viewport[B]->(viewport[C])))

Finally, the three viewports are pushed as a tree, with **vp1** as the parent and **vp2** and **vp3** as its children. A graph of the resulting viewport tree is shown in Figure 6.17 (bottom-left).

```
> grid.newpage()
> pushViewport(vpTree(vp1, vpList(vp2, vp3)))
> current.vpTree()
```

viewport[ROOT]->(viewport[A]->(viewport[B], viewport[C]))

As with single viewports, viewport lists, stacks, and trees can be provided as the **vp** argument for graphical functions (see Section 6.5.4).

Viewport paths

The **downViewport()** function, by default, searches down the current viewport tree as far as is necessary to find a given viewport name. This is convenient for interactive use, but can be ambiguous if there is more than one viewport with the same name in the viewport tree.

The **grid** system provides the concept of a *viewport path* to resolve such ambiguity. A viewport path is an ordered list of viewport names, which specify a series of parent-child relations. A viewport path is created using the **vpPath()** function. For example, the following code produces a viewport path that specifies a viewport called "C" with a parent called "B", which in turn has a parent called "A".

```
> vpPath("A", "B", "C")
```

A::B::C

For convenience in interactive use, a viewport path may be specified directly as a string. For example, the previous viewport path could be specified simply as "A::B::C". The **vpPath()** function should be used when writing graphics functions for others to use.

The **name** argument to the **downViewport()** function will accept a viewport path, in which case it searches for a viewport that matches the entire path. The **strict** argument to **downViewport()** ensures that a viewport will only be found if the full viewport path is found, *starting from the current location in the viewport tree.*

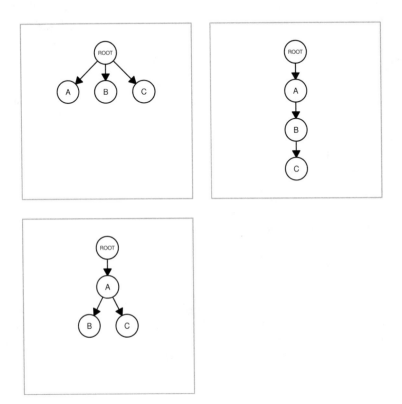

Figure 6.17

Viewport lists, stacks, and trees. There is always a ROOT viewport. At top-left, a *list* of three viewports has been pushed. At top-right, a *stack* of three viewports has been pushed. At bottom-left, a *tree* of three viewports has been pushed (where the tree consists of a parent with two children).

6.5.4   Viewports as arguments to graphical primitives

As mentioned in Section 6.2.1, a viewport may be specified as an argument to functions that produce graphical output (via an argument called vp). When a viewport is specified in this way, the viewport gets pushed before the graphical output is produced and the viewport is popped again afterward. To make this completely clear, the following two code segments are identical. First of all, a simple viewport is defined.

```
> vp1 <- viewport(width=0.5, height=0.5, name="vp1")
```

The next code explicitly pushes the viewport, draws some text, then pops the viewport.

```
> pushViewport(vp1)
> grid.text("Text drawn in a viewport")
> popViewport()
```

This next piece of code does the same thing in a single call.

```
> grid.text("Text drawn in a viewport", vp=vp1)
```

It is also possible to specify the name of a viewport (or a viewport path) for a vp argument. In this case, the name (or path) is used to navigate down to the viewport, via a call to downViewport(), and then back up again afterward, via a call to upViewport(). This promotes the practice of pushing viewports once, then specifying where to draw different output by simply naming the appropriate viewport. The following code does the same thing as the previous example, but leaves the viewport intact (so that it can be used for further drawing).

```
> pushViewport(vp1)
> upViewport()
> grid.text("Text drawn in a viewport", vp="vp1")
```

This feature is also very useful when annotating a plot produced by a high-level graphics function. As long as the graphics function names the viewports that it creates and does not pop them, it is possible to revisit the viewports to add further output. This is what both **lattice** and **ggplot2** do and examples of this sort of annotation are given in Section 6.8. This approach to writing high-level **grid** functions is discussed further in Chapter 8.

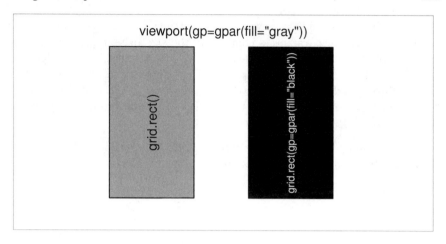

Figure 6.18
The inheritance of viewport graphical parameters. A diagram demonstrating how viewport graphical parameter settings are inherited by graphical output within the viewport. The viewport sets the default fill color to gray. The left-hand rectangle specifies no fill color itself so it is filled with gray. The right-hand rectangle specifies a black fill color that overrides the viewport setting.

6.5.5 Graphical parameter settings in viewports

A viewport can have graphical parameter settings associated with it via the gp argument to viewport(). When a viewport has graphical parameter settings, those settings affect all graphical objects drawn within the viewport, and all other viewports pushed within the viewport, unless the graphical objects or the other viewports specify their own graphical parameter setting. In other words, the graphical parameter settings for a viewport modify the implicit graphical context (see page 194).

The following code demonstrates this rule. A viewport is pushed that has a fill="gray" setting. A rectangle with no graphical parameter settings is drawn within that viewport and this rectangle "inherits" the fill="gray" setting. Another rectangle is drawn with its own fill setting so it does not inherit the viewport setting (see Figure 6.18).

```
> pushViewport(viewport(gp=gpar(fill="gray")))
> grid.rect(x=0.33, height=0.7, width=0.2)
> grid.rect(x=0.66, height=0.7, width=0.2,
            gp=gpar(fill="black"))
> popViewport()
```

The graphical parameter settings in a viewport only affect other viewports and graphical output within that viewport. The settings do not affect the viewport itself. For example, parameters controlling the size of text (`fontsize`, `cex`, etc.) do not affect the meaning of `"line"` units when determining the location and size of the viewport, but they will affect the location and size of other viewports or graphical output within the viewport. A layout (see Section 6.5.6) counts as being within the viewport (i.e., it is affected by the graphical parameter settings of the viewport).

If there are multiple values for a graphical parameter setting, only the first is used when determining the location and size of a viewport.

6.5.6 Layouts

A viewport can have a *layout* specified via the `layout` argument. A layout in **grid** is similar to the same concept in traditional graphics (see Section 3.3.2). It divides the viewport region into several columns and rows, where each column can have a different width and each row can have a different height. For several reasons, however, layouts are much more flexible in **grid**: there are many more coordinate systems for specifying the widths of columns and the heights of rows (see Section 6.3); viewports can occupy overlapping areas within the layout; and each viewport within the viewport tree can have a layout (layouts can be nested). There is also a `just` argument to justify the layout within a viewport when the layout does not occupy the entire viewport region.

Layouts provide a convenient way to position viewports using the standard set of coordinate systems, and provide an extra coordinate system, `"null"`, which is specific to layouts.

The basic idea is that a viewport can be created with a layout and then subsequent viewports can be positioned relative to that layout. In simple cases, this can be just a convenient way to position viewports in a regular grid, but in more complex cases, layouts are the only way to apportion regions. There are very many ways that layouts can be used in **grid**; the following sections attempt to provide a glimpse of the possibilities by demonstrating a series of example uses.

A **grid** layout is created using the function `grid.layout()` (*not* the traditional function `layout()`).

A simple layout

The following code produces a simple layout with three columns and three rows, where the central cell (row two, column two) is forced to always be square (using the `respect` argument).

```
> vplay <- grid.layout(3, 3,
                       respect=rbind(c(0, 0, 0),
                                     c(0, 1, 0),
                                     c(0, 0, 0)))
```

The next piece of code uses this layout in a viewport. Any subsequent view-ports may make use of the layout, or they can ignore it completely.

```
> pushViewport(viewport(layout=vplay))
```

In the next piece of code, two further viewports are pushed within the viewport with the layout. The `layout.pos.col` and `layout.pos.row` arguments are used to specify which cells within the layout each viewport should occupy. The first viewport occupies all of column two and the second viewport occupies all of row two. This demonstrates that viewports can occupy overlapping regions within a layout. A rectangle has been drawn within each viewport to show the region that the viewport occupies (see Figure 6.19).

```
> pushViewport(viewport(layout.pos.col=2, name="col2"))
> upViewport()
> pushViewport(viewport(layout.pos.row=2, name="row2"))
```

A layout with units

This section describes a layout that makes use of **grid** units. In the context of specifying the widths of columns and the heights of rows for a layout, there is an additional unit available, the `"null"` unit. All other units (`"cm"`, `"npc"`, etc.) are allocated first within a layout, then the `"null"` units are used to divide the remaining space proportionally (see Section 3.3.2). The following code creates a layout with three columns and three rows. The left column is one inch wide and the top row is three lines of text high. The remainder of the current region is divided into two rows of equal height and two columns with the right column twice as wide as the left column (see Figure 6.20).

```
> unitlay <-
    grid.layout(3, 3,
                widths=unit(c(1, 1, 2),
                            c("in", "null", "null")),
                heights=unit(c(3, 1, 1),
                             c("line", "null", "null")))
```

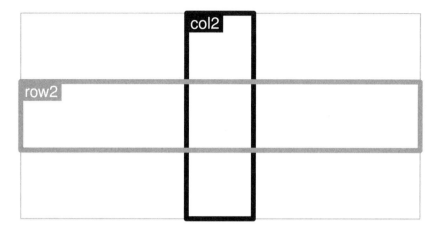

Figure 6.19
Layouts and viewports. Two viewports occupying overlapping regions within a
layout. Each viewport is represented by a rectangle with the viewport name at the
top-left corner. The layout has three columns and three rows with one viewport
occupying all of row two and the other viewport occupying all of column two.

With the use of "strwidth" and "grobwidth" units it is possible to produce
columns that are just wide enough to fit graphical output that will be drawn
in the column (and similarly for row heights — see Section 7.4).

A nested layout

This section demonstrates the nesting of layouts. The following code defines
a function that includes a trivial use of a layout consisting of two equal-width
columns to produce **grid** output.

```
> gridfun <- function() {
    pushViewport(viewport(layout=grid.layout(1, 2)))
    pushViewport(viewport(layout.pos.col=1))
    grid.rect()
    grid.text("black")
    grid.text("&", x=1)
    popViewport()
    pushViewport(viewport(layout.pos.col=2, clip="on"))
    grid.rect(gp=gpar(fill="black"))
    grid.text("white", gp=gpar(col="white"))
    grid.text("&", x=0, gp=gpar(col="white"))
    popViewport(2)
  }
```

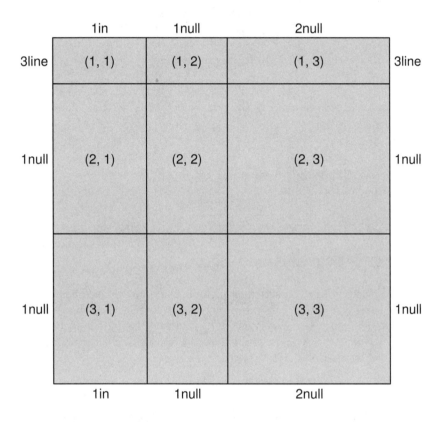

Figure 6.20
Layouts and units. A **grid** layout using a variety of coordinate systems to specify the widths of columns and the heights of rows.

The next piece of code creates a viewport with a layout and places the output from the above function within a particular cell of that layout (see Figure 6.21).

```
> pushViewport(
    viewport(
      layout=grid.layout(5, 5,
                        widths=unit(c(5, 1, 5, 2, 5),
                                    c("mm", "null", "mm",
                                      "null", "mm")),
                        heights=unit(c(5, 1, 5, 2, 5),
                                     c("mm", "null", "mm",
                                       "null", "mm")))))
> pushViewport(viewport(layout.pos.col=2, layout.pos.row=2))
> gridfun()
> popViewport()
```

The next piece of code calls the function again to draw the same output within a different cell of the layout.

```
> pushViewport(viewport(layout.pos.col=4, layout.pos.row=4))
> gridfun()
> popViewport(2)
```

Although the result of this particular example could be achieved using a single layout, what this shows is that it is possible to take **grid** code that makes use of a layout (and may have been written by someone else) and embed it within a layout of your own. A more sophisticated example of this involving lattice plots is given in Section 6.8.2.

6.6   Missing values and non-finite values

Non-finite values are not permitted in the location, size, or scales of a viewport. Viewport scales are checked when a viewport is created, but it is impossible to be certain that locations and sizes are not non-finite when the viewport is created, so this is only checked when the viewport is pushed. Non-finite values result in error messages.

The locations and sizes of graphical objects can be specified as missing values (NA, "NA") or non-finite values (NaN, Inf, -Inf). For most graphical primitives,

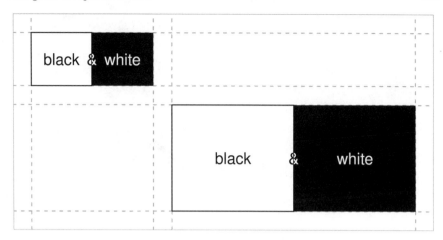

Figure 6.21
Nested layouts. An example of a layout nested within a layout. The black and white squares are drawn within a layout that has two equal-width columns. One instance of the black and white squares has been embedded within cell $(2, 2)$ of a layout consisting of five columns and five rows of varying widths and heights (as indicated by the dashed lines). Another instance has been embedded within cell $(4, 4)$.

non-finite values for locations or sizes result in the corresponding primitive not being drawn. For the `grid.line.to()` function, a line segment is only drawn if the previous location and the new location are both not non-finite. For `grid.polygon()`, a non-finite value breaks the polygon into two separate polygons. This break happens within the current polygon as specified by the `id` argument. All polygons with the same `id` receive the same `gp` settings. For line-drawing primitives that are supposed to draw arrowheads, an arrowhead is only drawn if the first or last line segment is drawn.

Figure 6.22 shows the behavior of these primitives where x- and y-locations are seven equally spaced locations around the perimeter of a circle. In the top-left figure, all locations are not non-finite. In each of the other figures, two locations have been made non-finite (indicated in each case by gray text).

6.7 Interactive graphics

The strength of the **grid** system is in the production of static graphics. Only very basic support for user interaction is provided via the `grid.locator()`

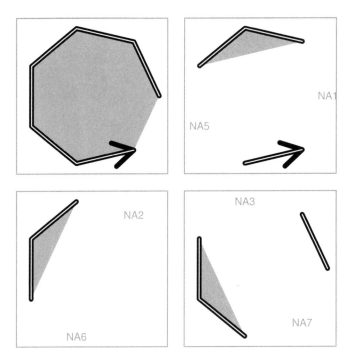

Figure 6.22
Non-finite values for line-tos, polygons, and arrows. The effect of non-finite values
for grid.line.to(), grid.polygon(), and grid.lines() (with an **arrow** specified).
In each panel, a single gray polygon, a single thick black line (with an arrow at the
end), and a series of thin white line-tos are drawn through the same set of seven
points. In some cases, certain locations have been set to NA (indicated by gray text),
which causes the polygon to become cropped, creates gaps in the lines, and can
cause the arrowhead to disappear. In the bottom-left panel, the seventh location is
not NA, but it produces no output.

function. This function returns the location of a single mouse click relative to the current viewport. The result is a list containing an x and a y unit. The unit argument can be used to specify the coordinate system that is to be used for the result.

The getGraphicsEvent() function provides additional capability (on Windows and X11) to respond to mouse movements, mouse ups, and key strokes. However, with this function, mouse activity is only reported relative to the native coordinate system of the device.

Chapter 17 describes more sophisticated interactive graphics solutions.

6.8 Customizing lattice plots

The **lattice** package described in Chapter 4 produces complete and very sophisticated plots using **grid**. It makes use of a sometimes large number of viewports to arrange the graphical output. A page of **lattice** output contains a top-level viewport with a quite complex layout that provides space for all of the panels and strips and margins used in the plot. Viewports are created for each panel and for each strip (among other things), and the plot is constructed from a large number of rectangles, lines, text, and data points.

In many cases, it is possible to use **lattice** without having to know anything about **grid**. However, a knowledge of **grid** provides a number of more advanced ways to work with **lattice** output (also see Section 7.7).

6.8.1 Adding grid output to lattice output

The functions that **lattice** provides for adding output to panels, for example, panel.text() and panel.points(), are restricted because they only allow output to be located and sized relative to the "native" coordinate system of the panel (i.e., relative to the panel axes). The low-level **grid** graphical primitives provide much more control over the location and size of additional panel output. It is even possible to create and push extra viewports within a panel if desired, although it is very important that they are popped again or **lattice** will get very confused.

In a similar vein, the **grid** functions upViewport() and downViewport() allow for more flexible navigation of a **lattice** plot compared to the trellis.focus() function.

The following code provides an example of `grid.text()` to add output within a **lattice** panel function. This produces a variation on Figure 4.5 with a text label in the top-right corner of each panel to indicate the number of data values in each panel (see Figure 6.23).*

```
> xyplot(mpg ~ disp | factor(gear), data=mtcars,
        panel=function(subscripts, ...) {
            grid.text(paste("n =", length(subscripts)),
                      unit(1, "npc") - unit(1, "mm"),
                      unit(1, "npc") - unit(1, "mm"),
                      just=c("right", "top"))
            panel.xyplot(subscripts=subscripts, ...)
        })
```

6.8.2 Adding lattice output to grid output

As well as the advantages of using **grid** functions to add further output to **lattice** plots, an understanding that **lattice** output is really **grid** output makes it possible to embed **lattice** output within **grid** output. The following code provides a simple example where two **lattice** plots are arranged together on a page by drawing them within **grid** viewports (see Figure 6.24).

```
> grid.newpage()
> pushViewport(viewport(x=0, width=.4, just="left"))
> print(barchart(table(mtcars$gear)),
        newpage=FALSE)
> popViewport()
> pushViewport(viewport(x=.4, width=.6, just="left"))
> print(xyplot(mpg ~ disp, data=mtcars,
              group=gear,
              auto.key=list(space="right")),
        newpage=FALSE)
> popViewport()
```

The viewports are set up using the standard **grid** functions, then the **lattice** plots are drawn within the viewports by explicitly calling `print()` and specifying `newpage=FALSE`.

---

*The data are from the `mtcars` data set (see page 126).

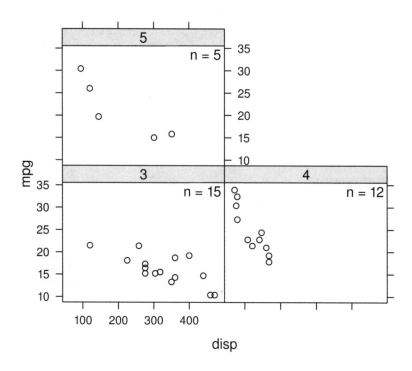

Figure 6.23
Adding **grid** output to a **lattice** plot (the **lattice** plot in Figure 4.5). The **grid**
function grid.text() is used within a **lattice** panel function to show the number
of points in each panel.

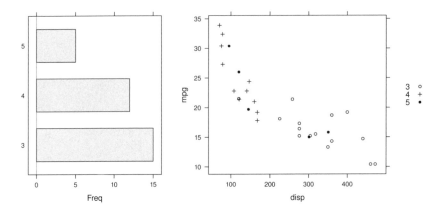

Figure 6.24
Embedding a **lattice** plot within **grid** output. Two **lattice** plots are arranged on a page by drawing them within **grid** viewports.

6.9   Customizing ggplot2 output

Like **lattice**, the **ggplot2** package uses **grid** to do its drawing, which involves creating a lot of viewports and drawing a lot of graphical primitives. This means that it is possible to use low-level **grid** functions to manipulate and add further drawing to **ggplot2** output.

6.9.1   Adding grid output to ggplot2 output

There is one obstacle to using **grid** functions to add further drawing to **ggplot2** output: the viewports created by **ggplot2** do not have any knowledge of the x-axis or y-axis scale on the plot, so it is not feasible to position extra output relative the plot scales.

Nevertheless, it is still possible to locate further drawing using any of the other **grid** coordinate systems. For example, the following code draws a **ggplot2** scatterplot and then navigates to the `"panel-3-3"` viewport to place a text label in the top-right corner of the plot (see Figure 6.25).

```
> ggplot(mtcars2, aes(x=disp, y=mpg)) +
      geom_point()
```

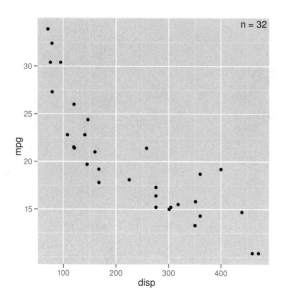

Figure 6.25
Adding **grid** output to **ggplot2**. A text label is added to a **ggplot2** scatterplot by navigating to the appropriate **ggplot2** viewport and calling `grid.text()`.

```
> downViewport("panel-3-3")
> grid.text(paste("n =", nrow(mtcars2)),
           x=unit(1, "npc") - unit(1, "mm"),
           y=unit(1, "npc") - unit(1, "mm"),
           just=c("right", "top"))
```

6.9.2 Adding ggplot2 output to grid output

The **ggplot2** functions create a `"ggplot"` object, which only produces output when it is printed. The printing can be controlled so that, for example, **ggplot2** does not start a new page for the plot. This makes it possible to set up **grid** viewports and draw **ggplot2** output within the viewports.

The following code demonstrates this idea by drawing a **ggplot2** barplot to the left of a **ggplot2** scatterplot (see Figure 6.26).

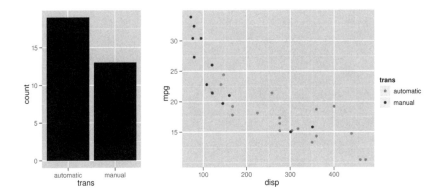

Figure 6.26
Embedding a **ggplot2** plot within **grid** output. Two **ggplot2** plots are drawn
within two **grid** viewports. This is how to get more than one **ggplot2** plot on the
same page.

```
> grid.newpage()
> pushViewport(viewport(x=0, width=1/3, just="left"))
> print(ggplot(mtcars2, aes(x=trans)) +
        geom_bar(),
        newpage=FALSE)
> popViewport()
> pushViewport(viewport(x=1/3, width=2/3, just="left"))
> print(ggplot(mtcars2, aes(x=disp, y=mpg)) +
        geom_point(aes(color=trans)) +
        scale_color_manual(values=gray(2:1/3)),
        newpage=FALSE)
> popViewport()
```

Chapter summary

The **grid** package provides a number of functions for producing basic graphical output such as lines, polygons, rectangles, and text, plus some functions for producing slightly more complex output such as data symbols, smooth curves, and axes. Graphical output can be located and sized relative to a large number of coordinate systems and there are a number of graphical parameter settings for controlling the appearance of output, such as colors, fonts, and line types.

Viewports can be created to provide contexts for drawing. A viewport defines a rectangular region on the device and all coordinate systems are available within all viewports. Viewports can be arranged using layouts and nested within one another to produce sophisticated arrangements of graphical output.

Because **lattice** and **ggplot2** output is **grid** output, **grid** functions can be used to add further output to a **ggplot2** or **lattice** plot and **grid** functions can also be used to control the size and placement of **ggplot2** and **lattice** plots.

7

The grid Graphics Object Model

Chapter preview

This chapter describes how to work with graphical objects (grobs). The main advantage of this approach is that it is possible to modify a scene that was produced using **grid** without having to modify the source code that produced the scene. Because **lattice** and **ggplot2** are built on **grid**, this means it is possible to modify a **ggplot2** or **lattice** plot.

There are also benefits from being able to do such things as ask a piece of graphical output how big it is. For example, this makes it easy to leave space for a legend beside a plot.

Graphical objects can be combined to form larger, hierarchical graphical objects (gTrees). This makes it possible to control the appearance and position of whole groups of graphical objects at once.

This chapter describes the **grid** concepts of grobs and gTrees as well as important functions for accessing, querying, and modifying these objects.

The previous chapter mostly dealt with using **grid** functions to produce graphical output. That knowledge is useful for annotating a plot produced using **grid** (such as a **lattice** plot), for producing one-off or customized plots for your own use, and for writing simple graphics functions.

This chapter on the other hand addresses **grid** functions for creating and manipulating graphical objects. This information is useful for querying or modifying graphical output and for writing graphical functions and objects for others to use (also see Chapter 8).

7.1   Working with graphical output

This section describes using **grid** to modify graphical output. Having called
functions to draw some output, there are functions to edit and delete elements
of the output.

All of the functions in Chapter 6 that produce graphical output also produce
graphical objects, or *grobs*, representing that output. For example, the following code produces a number of circles as output (see the left panel in Figure
7.1).

```
> grid.circle(name="circles", x=seq(0.1, 0.9, length=40),
              y=0.5 + 0.4*sin(seq(0, 2*pi, length=40)),
              r=abs(0.1*cos(seq(0, 2*pi, length=40)))))
```

As well as drawing the circles, this code produces a `circle` grob, an object
of class `"circle"`, which contains information describing the circles that have
been drawn. Importantly, this grob has been given a name, in this case
`"circles"`.

The **grid** system maintains a display list, a record of all viewports and grobs
drawn on the current page, and the object that `grid.circle()` created is
stored on this display list. This means that it can be accessed to obtain a
copy, to modify the output, or even to remove it altogether. The grob has
been given the name `"circles"` to make it easy to identify on the display list.

In the following code, the call to `grid.get()` obtains a copy of the `circle`
object. This can be useful for inspecting the elements of a scene.

```
> grid.get("circles")
```

```
circle[circles]
```

The following call to `grid.edit()` modifies the output by editing the `circle`
object to change the colors used for drawing the circles (see the middle panel
of Figure 7.1). In this case, the `gp` component of the `circle` grob is being
modified. Typically, most arguments that can be specified when first drawing
output can also be used when editing output.

```
> grid.edit("circles",
           gp=gpar(col=gray(c(1:20*0.04, 20:1*0.04))))
```

Figure 7.1

Modifying a `circle` grob. The left panel shows the output produced by a call to `grid.circle()`, the middle panel shows the result of using `grid.edit()` to modify the colors of the circles, and the right panel shows the result of using `grid.remove()` to delete the circles.

The next call below, to the `grid.remove()` function, deletes the output by removing the `circle` object from the display list (see the right panel of Figure 7.1).

```
> grid.remove("circles")
```

In each of these examples, the grob has been specified by giving its name (`"circles"`). Other standard arguments to these functions are discussed in the next section.

Any output produced by **grid** functions can be interacted with in this way, including output from **lattice** and **ggplot2** functions (see Sections 7.7 and 7.8).

It is possible to disable the **grid** display list, using the `grid.display.list()` function, in which case no grobs are stored, so these sorts of manipulations are no longer possible.

7.1.1   Standard functions and arguments

The complete set of functions that provide the ability to interact with grobs is shown in Table 7.1.

All of the functions for working with graphical output require a grob name as the first argument, to identify which grob to work with. This name will be treated as a regular expression if the `grep` argument is TRUE. If the `global` argument is TRUE then all matching grobs on the display list (not just the first) will be accessed or modified.

The following code provides a simple example. Eight concentric `circle` grobs

Table 7.1

Functions for working with grobs. Functions of the form `grid.*()` access and de-
structively modify grobs on the **grid** display list and affect graphical output. Functions of the form `*Grob()` work with user-level grobs and return grobs as their values
(they have no effect on graphical output).

Function to Work with Output	Description	Function to Work with grobs
grid.get()	Returns a copy of one or more grobs	getGrob()
grid.edit()	Modifies one or more grobs	editGrob()
grid.add()	Adds a grob to one or more grobs	addGrob()
grid.remove()	Removes one or more grobs	removeGrob()
grid.set()	Replaces one or more grobs	setGrob()

are drawn, with the first, third, fifth, and seventh circles named `"circle.odd"`
and the second, fourth, sixth, and eighth circles named `"circle.even"`. The
circles are initially drawn with decreasing shades of gray (see the left panel of
Figure 7.2).

```
> suffix <- c("even", "odd")
> for (i in 1:8)
     grid.circle(name=paste("circle.", suffix[i %% 2 + 1],
                       sep=""),
                r=(9 - i)/20,
                gp=gpar(col=NA, fill=gray(i/10)))
```

The following call to `grid.edit()` makes use of the `global` argument to
modify all grobs named `"circle.odd"` and change their fill color to a very
dark gray (see the middle panel of Figure 7.2).

```
> grid.edit("circle.odd", gp=gpar(fill="gray10"),
          global=TRUE)
```

A second call to `grid.edit()`, below, makes use of both the `grep` argument
and the `global` argument to modify all grobs with names matching the pattern
`"circle"` (all of the circles) and change their fill color to a light gray and their
border color to a darker gray (see the right panel of Figure 7.2).

```
> grid.edit("circle", gp=gpar(col="gray", fill="gray90"),
          grep=TRUE, global=TRUE)
```

Figure 7.2

Editing grobs using `grep` and `global` in `grid.edit()`. The left-hand panel shows eight separate concentric circles, with names alternating between `"circle.odd"` and `"circle.even"`, filled with progressively lighter shades of gray. The middle panel shows the use of the `global` argument to change the fill for all circles named `"circle.odd"` to black. The right-hand panel shows the use of the `grep` and `global` arguments to change all circles whose names match the pattern `"circle"` (all of the circles) to have a light gray fill and a gray border.

There are convenience functions `grid.gget()` and `grid.gedit()` that have the `grep` and `global` arguments set to `TRUE` by default.

In summary, as long as the name of a grob is known, it is possible to access that grob using `grid.get()`, modify it using `grid.edit()`, or delete it using `grid.remove()`.

The function `grid.ls()` is useful for producing a list of all grobs in the current scene, as shown by the following code.

```
> grid.ls()
```

```
circle.odd
circle.even
circle.odd
circle.even
circle.odd
circle.even
circle.odd
circle.even
```

This function is described in more detail in Section 8.4.

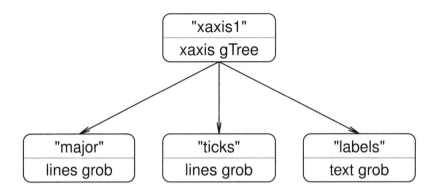

Figure 7.3
The structure of a gTree. A diagram of the structure of an xaxis gTree. There is
the xaxis gTree itself (here given the name "xaxis1") and there are its children: a
lines grob named "major", another lines grob named "ticks", and a text grob
named "labels".

7.2   Grob lists, trees, and paths

As well as basic grobs, it is possible to work with a list of grobs (a gList) or
several grobs combined together in a tree-like structure (a gTree). A gList is
just a list of several grobs (produced by the function gList()). A gTree is a
grob that can contain other grobs. Examples are the xaxis and yaxis grobs.
This section looks at how to work with gTrees.

An xaxis grob contains a high-level description of an axis, plus several child
grobs representing the lines and text that make up the axis (see Figure 7.3).

The following code draws an x-axis and creates an xaxis grob on the display
list (see the left panel of Figure 7.4). The grid.ls() function shows that the
axis1 grob has three child grobs.

```
> grid.xaxis(name="axis1", at=1:4/5)
> grid.ls()


axis1
  major
  ticks
  labels
```

The hierarchical structure of gTrees makes it possible to interact with both a high-level description, as provided by the xaxis grob, and a low-level description, as provided by the children of the gTree. The following code demonstrates an interaction with the high-level description of an xaxis grob. The xaxis gTree contains components describing where to put tick marks on the axis and whether to draw labels and so on. The code below shows the at component of an xaxis grob being modified. The xaxis grob is designed so that it modifies its children to match the new high-level description so that only three ticks are now drawn (see the middle panel of Figure 7.4).

```
> grid.edit("axis1", at=1:3/4)
```

It is also possible to access the children of a gTree. In the case of an xaxis, there are three children: a lines grob with the name "major"; another lines grob with the name "ticks"; and a text grob with the name "labels". Any of these children can be accessed by specifying the name of the xaxis grob and the name of the child in a *grob path* (gPath). A gPath is like a viewport path (see Section 6.5.3) — it is just a concatenation of several grob names. The following code shows how to access the "labels" child of the xaxis grob using the gPath() function to specify a gPath. The gPath specifies the child called "label" in the gTree called "axis1". The labels are rotated to 45 degrees (see the right panel of Figure 7.4).

```
> grid.edit(gPath("axis1", "labels"), rot=45)
```

It is also possible to specify a gPath directly as a string, for example "axis1::labels", but this is only recommended for interactive use.

7.2.1   Graphical parameter settings in gTrees

Just like any other grob, a gTree can have graphical parameter settings associated with it via a gp component. These settings affect all graphical objects that are children of the gTree, unless the children specify their own graphical parameter setting. In other words, the graphical parameter settings for a

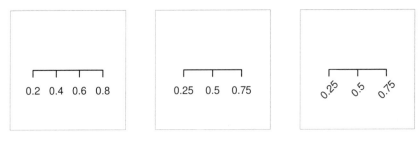

Figure 7.4
Editing a gTree. The left-hand panel shows a basic x-axis, the middle panel shows
the effect of editing the **at** component of the x-axis (all of the tick marks and
labels have changed), and the right-hand panel shows the effect of editing the **rot**
component of the **"labels"** child of the x-axis (only the angle of rotation of the
labels has changed).

gTree modify the implicit graphical context for the children of the gTree (see
page 194).

The following expression demonstrates this rule. The **gp** component of an
xaxis grob sets the drawing color to be **"gray"**. This means that all of the
children of the **xaxis** — the lines and labels — will be drawn gray.

```
> grid.xaxis(gp=gpar(col="gray"))
```

Another example of this behavior is given in Section 7.3 and the role of the
gp component in the drawing behavior of gTree objects is described in more
detail in Section 8.3.4.

7.2.2   Viewports as components of gTrees

Just like any other grob, a gTree can have a viewport (or viewport tree, or
viewport path, etc.) associated with it via a **vp** component. This viewport is
pushed before the gTree is drawn and popped afterward (see Section 6.5.4).
This means that the children of a gTree are drawn within the drawing context
defined by the viewport in the **vp** slot of the gTree (see page 199).

The following code demonstrates this rule. The **vp** component of an **xaxis**
grob specifies a viewport in the top half of the page. This means that the
children of the **xaxis** are positioned relative to that viewport.

```
> grid.xaxis(vp=viewport(y=0.75, height=0.5))
```

An example of this behavior is given in Section 7.3 and the role of the vp component in the drawing behavior of gTree objects is described in more detail in Sections 8.3.4 and 8.3.7.

7.2.3   Searching for grobs

This section provides details about how grob names and gPaths are used to find a grob.

Grobs are stored on the **grid** display list in the order that they are drawn. When searching for a matching name, the functions in Table 7.1 search the display list from the beginning. This means that if there are several grobs whose names are matched, they will be found in the order that they were drawn.

Furthermore, the functions perform a depth-first search. This means that if there is a gTree on the display list, and its name is not matched, then its children are searched for a match before any other grobs on the display list are searched.

The name to search for can be given as a gPath, which makes it possible to explicitly specify a particular child grob of a particular gTree. For example, `"axis1::labels"` specifies a grob called `"labels"` that must have a parent called `"axis1"`.

The argument `strict` controls whether a complete match must be found. By default, the `strict` argument is `FALSE`, so in the previous example, the `"labels"` child of `"axis1"` could have been accessed with the expression `grid.get("labels")`. On the other hand, if `strict` is set to `TRUE`, then simply specifying `"labels"` results in no match because there is no top-level grob with the name `"labels"`, as shown by the following code.

```
> grid.edit("labels", strict=TRUE, rot=45)


Error in
    editDLfromGPath(gPath, specs, strict, grep, global, redraw) :

  'gPath' (labels) not found
```

7.3   Working with graphical objects off-screen

Chapter 6 described **grid** functions that draw graphical output on the page or screen. All of those functions also create grobs representing the drawing and those grobs are stored on the **grid** display list.

It can also be useful to create a grob without producing any output. This section describes how to use **grid** to produce graphical objects (without drawing them). There are functions to create grobs, functions to combine them and to modify them, and the grid.draw() function to draw them.

For each **grid** function that produces graphical output, there is a counterpart that produces a graphical object and no graphical output. For example, the counterpart to grid.circle() is the function circleGrob() (see Table 6.1). Similarly, for each function that works with grobs on the **grid** display list, there is a counterpart for working with grobs off-screen. For example, the counterpart to grid.edit() is editGrob() (see Table 7.1).

The following example demonstrates the process of creating a grob and working with the grob without drawing it. The code below draws a rectangle that is as wide as a text grob, but the text is not drawn. The function textGrob() produces a **text** grob, but does not draw it.

```
> grid.rect(width=grobWidth(textGrob("Some text")))
```

It can be useful to create a grob and modify it before producing any graphical output (i.e., only draw the final result). The following code creates an axis and modifies the font face for the labels to italic before drawing the axis. The function grid.draw() is used to produce graphical output from a grob.

```
> ag <- xaxisGrob(at=1:4/5)
> ag <- editGrob(ag, "labels", gp=gpar(fontface="italic"))
> grid.draw(ag)
```

Another example of working with grobs is in the construction of gTrees. In its simplest form, a gTree is just a grouping of several grobs (more complex gTree creation is discussed later in Section 8.3).

By grouping several grobs together as a single object, the grobs can be dealt with as a single object. For example, it becomes possible to edit the graphical context for all of the grobs at once, or define the drawing context for all of the grobs at once.

When a gTree is drawn, any viewports in its **vp** component are pushed, any settings in its **gp** component are enforced, and then its children are all drawn. This means that the **vp** and **gp** components of a gTree affect where and how the children of the gTree are drawn (see Sections 7.2.1 and 7.2.2).

As an example, a boxed-text object can be created by grouping a "rect" grob and a "text" grob together as children of a gTree. This allows us to modify the color of both the rectangle and the text by modifying these features in the gTree. Similarly, it is possible to locate both the rectangle and the text by defining a viewport for the gTree.

The following code uses the `gTree()` function to create a gTree that groups a "rect" grob and a "text" grob together. There is no graphical output produced from this code. It only creates graphical objects.

```
> tg <- textGrob("sample text")
> rg <- rectGrob(width=1.1*grobWidth(tg),
                 height=1.3*grobHeight(tg))
> boxedText <- gTree(children=gList(tg, rg))
```

It is now easy to produce output including both the rectangle and the text by drawing variations on the `boxedText` grob, as demonstrated by the following code.

The first call simply draws the plain `boxedText`, which is drawn in black (see the left panel of Figure 7.5).

```
> grid.draw(boxedText)
```

The second call draws a modified `boxedText` with the drawing color set to gray (see the middle panel of Figure 7.5).

```
> grid.draw(editGrob(boxedText, gp=gpar(col="gray")))
```

The final call draws another modification of the `boxedText`, this time in a rotated viewport and with a larger font (see the right panel of Figure 7.5).

```
> grid.draw(editGrob(boxedText, vp=viewport(angle=45),
                 gp=gpar(fontsize=18)))
```

Figure 7.5
Using a gTree to group grobs. The left-hand panel shows a boxed text object (which
is a combination of a piece of text and a rectangle). The middle panel shows how
changes to the color settings in the boxed text object propagate to the components
(both the text and rectangle turn gray). The right-hand panel shows a more dramatic demonstration of the same idea as, in this case, the font size of the boxed text
is modified and it is drawn within a rotated viewport.

7.3.1   Capturing output

In the example in the previous section, several grobs are created off-screen
and then grouped together as a gTree, which allows the collection of grobs to
be dealt with as a single object.

It is also possible first to *draw* several grobs and *then* to group them. The
grid.grab() function does this by generating a gTree from all of the grobs
in the current page of output. This means that output can be captured even
from a function that produces very complex output (lots of grobs), such as
a **lattice** plot. For example, the following code draws a **lattice** plot, then
creates a gTree containing all of the grobs in the plot.

```
> bwplot(rnorm(10))
> bwplotTree <- grid.grab()
```

The grid.grab() function actually captures all of the viewports in the current
scene as well as the grobs, so drawing the gTree, as in the following code,
produces exactly the same output as the original plot.

```
> grid.newpage()
> grid.draw(bwplotTree)
```

Another function, grid.grabExpr() allows complex output to be captured
off-screen. This function takes an R expression and evaluates it. Any drawing
that occurs as a result of evaluating the expression does not produce any
output, but the grobs that would be produced are captured anyway.

The following code provides a simple demonstration. Here a **lattice** plot is captured without drawing any output.*

```
> grid.grabExpr(print(bwplot(rnorm(10))))
```

gTree[GRID.gTree.100]

Both the `grid.grab()` and `grid.grabExpr()` functions attempt to create a gTree in a sophisticated way so that it is easier to work with the resulting gTree. Unfortunately, this will not always produce a gTree that will exactly replicate the original output. These functions issue warnings if they detect a situation where output may not be reproduced correctly, and there is a `wrap` argument that can be used to force the functions to produce a gTree that is less sophisticated, but is guaranteed to replicate the original output.

7.4 Placing and packing grobs in frames

It can be useful to position the components of a plot in a way that leaves sufficient room for labels or legends. The `"grobwidth"` and `"grobheight"` coordinate systems provide a way to determine the size of a grob and can be used to achieve this sort of arrangement of components by, for example, allocating appropriate regions within a layout.

The following code demonstrates this idea. First of all, some grobs are created to use as components of a scene. The first grob, `label`, is a simple `text` grob. The second grob, `gplot`, is a gTree containing a `rect` grob, a `lines` grob, and a `points` grob that provide a simple representation of time-series data. The `gplot` has a viewport in its `vp` component and the rectangle and lines are drawn within that viewport.

---

*The expression must explicitly `print()` the **lattice** plot because otherwise nothing would be drawn (see Section 4.1).

```
> label <- textGrob("A\nPlot\nLabel ",
                    x=0, just="left")
> x <- seq(0.1, 0.9, length=50)
> y <- runif(50, 0.1, 0.9)
> gplot <-
    gTree(
      children=gList(rectGrob(gp=gpar(col="gray60",
                                      fill="white")),
                    linesGrob(x, y),
                    pointsGrob(x, y, pch=16,
                               size=unit(1.5, "mm"))),
      vp=viewport(width=unit(1, "npc") - unit(5, "mm"),
                  height=unit(1, "npc") - unit(5, "mm")))
```

The next piece of code defines a layout with two columns. The second column of the layout has its width determined by the width of the `label` grob created above. The first column will take up whatever space is left over.

```
> layout <- grid.layout(1, 2,
                        widths=unit(c(1, 1),
                          c("null", "grobwidth"),
                          list(NULL, label)))
```

Now some drawing can occur. A viewport is pushed with the layout defined above, then the `label` grob is drawn in the second column of this layout, which is exactly the right width to contain the text, and the `gplot` gTree is drawn in the first column (see Figure 7.6).

```
> pushViewport(viewport(layout=layout))
> pushViewport(viewport(layout.pos.col=2))
> grid.draw(label)
> popViewport()
> pushViewport(viewport(layout.pos.col=1))
> grid.draw(gplot)
> popViewport(2)
```

The **grid** package provides a set of functions that make it more convenient to arrange grobs like this so that they allow space for each other. The function `grid.frame()`, and its off-screen counterpart `frameGrob()`, produce a gTree with no children. Children are added to the frame using the `grid.pack()` function and the frame makes sure that enough space is allowed for the child when it is drawn. Using these functions, the previous example becomes simpler, as shown by the following code (the output is the same as Figure 7.6).

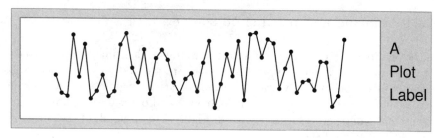

Figure 7.6
Packing grobs by hand. The scene was created using a frame object, into which the time-series plot (consisting of a rectangle, lines, and points) was packed. The text was then packed on the right-hand side, which meant that the time series plot was allocated less room in order to leave space for the text.

The big difference is that there is no need to specify a layout as an appropriate layout is calculated automatically.

The first call creates an empty frame. The second call packs `gplot` into the frame; at this stage, `gplot` takes up the entire frame. The third call packs the text label on the right-hand side of the frame; enough space is made for the text label by reducing the space allowed for the rectangle.

```
> grid.frame(name="frame1")
> grid.pack("frame1", gplot)
> grid.pack("frame1", label, side="right")
```

There are many arguments to `grid.pack()` for specifying where to pack new grobs within a frame. There is also a **dynamic** argument to specify whether the frame should reallocate space if the grobs that have been packed in the frame are modified.

Unfortunately, packing grobs into a frame like this becomes quite slow as more grobs are packed, so it is most useful for very simple arrangements of grobs or for interactively constructing a scene. An alternative approach, which is a little more work, but still more convenient than dealing directly with pushing and popping viewports (and can be made dynamic like packing), is to *place* grobs within a frame that has a predefined layout. The following code demonstrates this approach. This time, the frame is initially created with the desired layout as defined above, then the `grid.place()` function is used to position grobs within specific cells of the frame layout.

```
> grid.frame(name="frame1", layout=layout)
> grid.place("frame1", gplot, col=1)
> grid.place("frame1", label, col=2)
```

7.4.1   Placing and packing off-screen

In the previous two examples, the screen is redrawn each time a grob is packed
into the frame. It is more typical to create a frame and pack or place grobs
within it off-screen and only draw the frame once it is complete. The following
code demonstrates the use of the `frameGrob()` and `placeGrob()` functions to
achieve the same end result as shown in Figure 7.6, doing all of the construction of the frame off-screen.

```
> fg <- frameGrob(layout=layout)
> fg <- placeGrob(fg, gplot, col=1)
> fg <- placeGrob(fg, label, col=2)
> grid.draw(fg)
```

The function `packGrob()` is the off-screen counterpart of `grid.pack()`.

7.5   Other details about grobs

This section describes some important extra details about the calculation of
grob sizes and the editing of graphical contexts.

7.5.1   Calculating the sizes of grobs

As described in Section 6.3.2, the `"grobwidth"` and `"grobheight"` units, and
the `grobWidth()` and `grobHeight()` functions, provide a way to determine
the size of a grob. This section provides some more details about the correct
usage of these units.

The most important point is that the size of a grob is always calculated
relative to the current geometric and graphical context. The following code
demonstrates this point. First of all, a `text` grob and a `rect` grob are created,
and the dimensions of the `rect` grob are based on the dimensions of the text.[*]

---

[*]The `rect` grob draws two rectangles: one thick and dark gray, one white and thin.

```
> tg1 <- textGrob("Sample")
> rg1 <- rectGrob(x=rep(0.5, 2),
                  width=1.1*grobWidth(tg1),
                  height=1.3*grobHeight(tg1),
                  gp=gpar(col=c("gray60", "white"),
                          lwd=c(3, 1)))
```

Next, these two grobs are drawn in three different settings. In the first setting, the rectangle and the text are drawn in the default geometric and graphical context and the rectangle bounds the text (see the left panel of Figure 7.7).

```
> grid.draw(tg1)
> grid.draw(rg1)
```

In the second setting, the grobs are both drawn within a viewport that has cex=2. Both the text and the rectangle are drawn bigger (the calculation of the "grobwidth" and "grobheight" units takes place in the same context as the drawing of the text grob; see the middle panel of Figure 7.7).

```
> pushViewport(viewport(gp=gpar(cex=2)))
> grid.draw(tg1)
> grid.draw(rg1)
> popViewport()
```

In the third setting, the text grob is drawn in a different context than the rectangle, so the rectangle's size is "wrong" (see the right panel of Figure 7.7).

```
> pushViewport(viewport(gp=gpar(cex=2)))
> grid.draw(tg1)
> popViewport()
> grid.draw(rg1)
```

A related issue arises with the use of grob *names* when creating a "grobwidth" or "grobheight" unit (see Section 6.3.2). The following code provides a simple example.

A text grob and two rect grobs are created, with the dimensions of both rectangles based upon the dimensions of the text. One rectangle, rg1, uses a copy of the text grob in the calls to grobWidth(), and grobHeight(). The other rectangle, rg2, just uses the name of the text grob, "tg1".

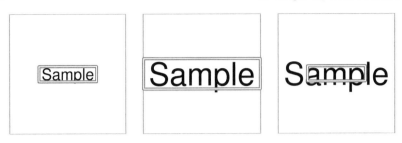

Figure 7.7

Calculating the size of a grob. In the left-hand panel, a `text` grob and a separate `rect` grob, the size of which is calculated to be the size of the `text` grob, are drawn together. In the middle panel, these objects are drawn together in a viewport with a larger font size, so they are both larger. In the right-hand panel, only the text is drawn in a viewport with a larger font size, so only the text is larger. The rectangle calculates the size of the text in a different font context.

```
> tg1 <- textGrob("Sample", name="tg1")
> rg1 <- rectGrob(width=1.1*grobWidth("tg1"),
                  height=1.3*grobHeight("tg1"),
                  gp=gpar(col="gray60", lwd=3))
> rg2 <- rectGrob(width=1.1*grobWidth(tg1),
                  height=1.3*grobHeight(tg1),
                  gp=gpar(col="white"))
```

When these rectangles and text are initially drawn, both rectangles frame the text correctly (see the left panel of Figure 7.8).

```
> grid.draw(tg1)
> grid.draw(rg1)
> grid.draw(rg2)
```

However, if the `text` grob is modified, as shown below, only the rectangle `rg1` (the dark gray rectangle) will be updated to correspond to the new dimensions of the text (see the right panel of Figure 7.8).

```
> grid.edit("tg1", grep=TRUE, global=TRUE,
            label="Different text")
```

With this approach, `"grobwidth"` and `"grobheight"` units are still evaluated in the current geometric and graphical context, but in addition, only grobs that have previously been drawn can be referred to. For example, drawing the rectangle `rg1` before drawing the text `tg1` will not work because there is no drawn grob named `"tg1"` from which a size can be calculated.

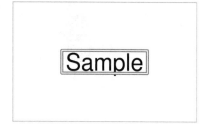 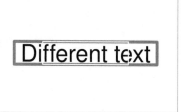

Figure 7.8
Grob dimensions by reference. In the left-hand panel there are three grobs: one
`text` grob and two `rect` grobs. The sizes of both `rect` grobs are calculated from the
`text` grob. The difference is that the white rectangle is related to the text by value
and the dark gray rectangle is related to the text by reference. The right-hand panel
shows what happens when the `text` grob is edited. Only the dark gray, by-reference,
rectangle gets resized.

```
> grid.newpage()
> grid.draw(rg1)

Error in function (name)   :
    Grob 'tg1' not found
```

7.5.2   Calculating the positions of grobs

In addition to being able to query a grob about its dimensions, it is also
possible to query a grob about its location, using `"grobx"` and `"groby"` units,
or the `grobX()` and `grobY()` functions.

Locations are calculated relative to the current geometric and graphical context, just like widths and heights, so all of the warnings from the previous
section also apply here.

The grob locations are positions on the border of a grob, given by an angle
(relative to the "center" of the grob). The following code shows a simple
example usage (see Figure 7.9). A small dot is drawn on the left and a text
label, with a surrounding box, is drawn on the right. The box grob is named
`"labelbox"`.

Figure 7.9
Calculating grob locations. The line segment is drawn from an explicit (x, y) start
location to an end location that is calculated using grobX() to give the left edge of
the box surrounding the text.

```
> grid.circle(.25, .5, r=unit(1, "mm"),
              gp=gpar(fill="black"))
> grid.text("A label", .75, .5)
> grid.rect(.75, .5,
            width=stringWidth("A label") + unit(2, "mm"),
            height=unit(1, "line"),
            name="labelbox")
```

A line segment, with an arrow, is now drawn between the dot and the left
edge of the box, using the grobX() function to determine the location of the
left edge of the box.

```
> grid.segments(.25, .5,
                grobX("labelbox", 180), .5,
                arrow=arrow(angle=15, type="closed"),
                gp=gpar(fill="black"))
```

The next example demonstrates a more complex use. This replicates an example from Figure 3.18 and demonstrates a possible use for "null" grobs.

First of all, two viewports are created, one in the top half of the page and one
in the bottom half.

```
> vptop <- viewport(width=.9, height=.4, y=.75,
                    name="vptop")
> vpbot <- viewport(width=.9, height=.4, y=.25,
                    name="vpbot")
> pushViewport(vptop)
> upViewport()
> pushViewport(vpbot)
> upViewport()
```

Now a rectangle and a line through some data are drawn in each viewport.

```
> grid.rect(vp="vptop")
> grid.lines(1:50/51, runif(50), vp="vptop")
> grid.rect(vp="vpbot")
> grid.lines(1:50/51, runif(50), vp="vpbot")
```

The next step does not draw anything, it just locates several null grobs at specific locations, two in the top viewport and two in the bottom viewport.

```
> grid.null(x=.2, y=.95, vp="vptop", name="tl")
> grid.null(x=.4, y=.95, vp="vptop", name="tr")
> grid.null(x=.2, y=.05, vp="vpbot", name="bl")
> grid.null(x=.4, y=.05, vp="vpbot", name="br")
```

Finally, a polygon is drawn that spans *both* viewports. The first two vertices of the polygon are calculated from the positions of the two null grobs in the top viewport and the second two vertices of the polygon are calculated from the positions of the two null grobs in the bottom viewport.

```
> grid.polygon(unit.c(grobX("tl", 0),
                      grobX("tr", 0),
                      grobX("br", 0),
                      grobX("bl", 0)),
               unit.c(grobY("tl", 0),
                      grobY("tr", 0),
                      grobY("br", 0),
                      grobY("bl", 0)),
               gp=gpar(col="gray", lwd=3))
```

The final result is shown in Figure 7.10.

7.5.3   Editing graphical context

When a grob is edited using `grid.edit()` or `editGrob()`, the modification of a `gp` component is treated as a special case. Only the graphical parameters that are explicitly given new settings are modified. All other settings remain untouched. The following code provides a simple example.

A circle is drawn with a gray fill color (see the left panel of Figure 7.11), then the border of the circle is made thick (see the middle panel of Figure 7.11) and the fill color remains the same. Finally, the border is changed to a dashed

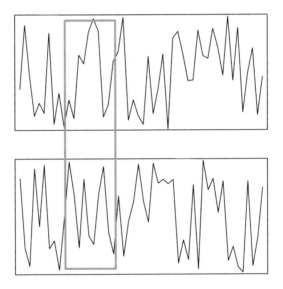

Figure 7.10
Calculating null grob locations. The two line plots are drawn in separate viewports.
The thick gray rectangle is drawn relative to the locations of four null grobs, two of
which are located in the top viewport and two of which are located in the bottom
viewport.

 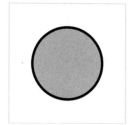 

Figure 7.11
Editing the graphical context. The left-hand panel shows a circle with a solid, thin black border and a gray fill. The middle panel shows the effect of making the border thicker. The important point is that the other features of the circle are not affected (the border is still solid and the fill is still gray). The right-hand panel shows another demonstration of the same idea, with the border now drawn dashed (but the border is still thick and the fill is still gray).

line type, but it stays thick (and the fill remains gray — see the right panel of Figure 7.11).

```
> grid.circle(r=0.3, gp=gpar(fill="gray80"),
             name="mycircle")
> grid.edit("mycircle", gp=gpar(lwd=5))
> grid.edit("mycircle", gp=gpar(lty="dashed"))
```

7.6 Saving and loading grid graphics

The best way to create a persistent record of a **grid** plot is to record in a text file the R code that was used to create the plot. The code can then be run again, e.g., using `source()`, to reproduce the output.

It is also possible to save grobs in R's binary format using the `save()` function. The grobs can then be loaded, using `load()`, and redrawn using `grid.draw()`. For the purpose of saving an entire scene, it may be more useful to save and load a gTree created by the `grid.grab()` function (see Section 7.3.1).

A possible danger with saving a **grid** grob is that methods specific to that grob are not automatically recorded, so the grob may not behave correctly when loaded into a different session. This will only be an issue for grobs that are not predefined by **grid** (see Chapter 8, particularly Section 8.3).

7.7   Working with lattice grobs

The output from a **lattice** function is fundamentally just a collection of **grid** viewports and grobs. Section 6.8 described some examples of making use of the **grid** viewports that are set up by a **lattice** plot to add extra output. This section looks at some examples of working with the grobs that are created by a **lattice** plot.

The following code creates a **lattice** scatterplot to work with.

```
> angle <- seq(0, 2*pi, length=21)[-21]
> x <- cos(angle)
> y <- sin(angle)

> xyplot(y ~ x, aspect=1,
         xlab="displacement",
         ylab="velocity")
```

The `grid.ls()` function shows the set of graphical primitives that have been created for this plot.

```
> grid.ls()
```

```
GRID.rect.156
plot_01.xlab
plot_01.ylab
GRID.segments.157
GRID.segments.158
GRID.text.159
GRID.segments.160
GRID.text.161
GRID.segments.162
GRID.points.163
GRID.rect.164
```

The grobs created by other people's functions will not necessarily provide useful names for all components that are drawn, but in this case, it is easy to spot which components provide the x-axis label and y-axis label for the plot.

The following code edits the axis labels to change the font to a "mono" family and to position the labels at the ends of the axes (see Figure 7.12).

```
> grid.edit("[.]xlab$", grep=TRUE,
            x=unit(1, "npc"), just="right",
            gp=gpar(fontfamily="mono"))
> grid.edit("[.]ylab$", grep=TRUE,
            y=unit(1, "npc"), just="right",
            gp=gpar(fontfamily="mono"))
```

Other grob operations are also possible. For example, the following code removes the labels from the plot.

```
> grid.remove(".lab$", grep=TRUE, global=TRUE)
```

Finally, it is possible to group all of the grobs from a **lattice** plot together using `grid.grab()`. This creates a gTree that can then be used as a component in creating another picture.

7.8 Working with ggplot2 grobs

Like **lattice**, **ggplot2** creates lots of **grid** grobs when it draws a plot and these grobs can be manipulated using **grid** functions.

The following code uses **ggplot2** to create a scatterplot with a linear model line of best fit.

```
> ggplot(mtcars2, aes(x=disp, y=mpg)) +
      geom_point() +
      geom_smooth(method=lm)
```

The next code navigates down to the plot region and queries the grob that represents the line of best fit, using `grobX()` and `grobY()`, to determine a location on the line. This location is used to draw an arrow that points from a text label to the line of best fit (see Figure 7.13).

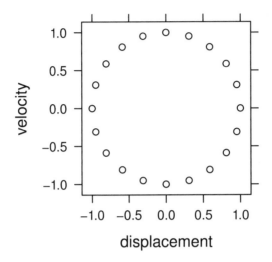

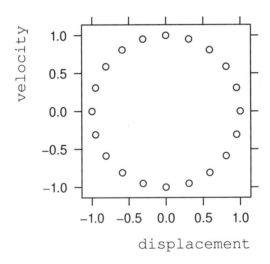

Figure 7.12
Editing the grobs in a **lattice** plot. The top plot is an initial scatterplot produced
using the **lattice** function xyplot(). The bottom plot shows the effect of editing
the **grid text** grobs that represent the labels on the plot (the labels are relocated
at the ends of the axes and are drawn in a monospace font).

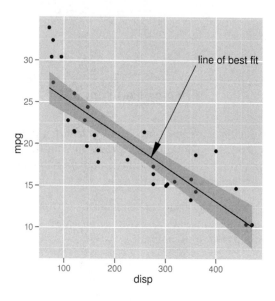

Figure 7.13
Working with **ggplot2** grobs. A **ggplot2** scatterplot is drawn and then a line is
added with an end point that is calculated from the grob that represents the smooth
line on the plot.

```
> downViewport("panel-3-3")
> sline <- grid.get(gPath("smooth", "polyline"),
                    grep=TRUE)
> grid.segments(.7, .8,
                grobX(sline, 45), grobY(sline, 45),
                arrow=arrow(angle=10, type="closed"),
                gp=gpar(fill="black"))
> grid.text("line of best fit", .71, .81,
            just=c("left", "bottom"))
```

Determining the name of the correct grob in this example required an inspec-
tion of the output from `grid.ls()`. Section 8.4 provides more examples of
how to explore **grid** grobs and viewports.

Chapter summary

As well as producing graphical output, all **grid** functions create grobs (graphical objects) that contain descriptions of what has been drawn. These grobs may be accessed, modified, and even removed, and the graphical output will be updated to reflect the changes.

There are also **grid** functions for creating grobs without producing any graphical output. A complete description of a plot can be produced by creating, modifying, and combining grobs off-screen.

A gTree is a grob that can have other grobs as its children. A gTree can be useful for grouping grobs and for providing a high-level interface to a group of grobs.

The **lattice** and **ggplot2** plotting functions generate large numbers of **grid** grobs. These grobs may be manipulated just like any other grobs to access, edit, and delete parts of a **ggplot2** or **lattice** plot.

8

Developing New Graphics Functions and Objects

Chapter preview

This chapter looks in depth at the task of writing graphical functions for others to use.

There are important guidelines for writing simple functions whose main purpose is to produce graphical output. There is an emphasis on making sure that other users can annotate the output produced by a function and that other users can make use of the function as a component in larger or more complex plots.

There is also a discussion on how to create a new class of graphical object. This is important for allowing users to interactively edit output, to ask questions such as how much space a graphical object requires, and to be able to combine graphical objects together in a gTree.

This chapter addresses the issue of developing graphics functions for others to use. This will involve a discussion of some of the lower-level details of how **grid** works as well as some more abstract ideas of software design. A basic understanding of programming concepts is recommended, and the later sections assume an understanding of object-oriented concepts such as classes and methods.

Important low-level details of the **grid** graphics system and important design considerations are introduced in increasing levels of complexity to allow developers to construct simple graphics functions at first. Readers aiming to design a new fully featured **grid** graphical object should read the entire chapter.

8.1   An example

In order to provide concrete examples of the concepts described in this chapter, a set of graphical functions and objects will be developed for the purpose of producing plots of oceanographic data.

An example of the final output that is desired is shown in Figure 8.1. Sections 8.2 to 8.3.9 go through the process of creating functions and objects to produce this output.

The data are measurements of fluorescence calculated at the thermocline (point of maximum temperature gradient) for 87 measuring stations off the coast of South Australia. The values plotted in the image are from a prediction surface based on an analysis using the `Krig()` function in the `fields` package.[*]

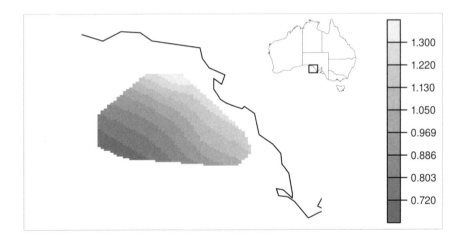

Figure 8.1
A plot of oceanographic data. The plot consists of a section of South Australian coastline, an image representing fluorescence at the thermocline, a small map to indicate where the main plot region is in Australia, and a legend to map the gray scale to fluorescence values.

---

[*]The data for the prediction surface are available as the data set `fluoro.predict`. Sam McClatchie provided the data and the original motivation to look at oceanographic plots in R.

8.1.1 Modularity

One decision can be made before writing a single line of code: the code should be modular. This means that the code should consist of several small functions, each of which produces a well-defined, self-contained piece of graphical output. It would be a bad idea to create Figure 8.1 in one big function. Such a function would be unlikely to be very flexible, would be very hard to maintain (it is easier to see what is going on in smaller functions), and would be very hard to debug (it is much easier to test small functions with a clear, simple purpose).

The following sections look at writing several simple functions, each of which produces a conceptually separate part of the final plot. One possible breakdown of Figure 8.1 involves the following elements: two maps of Australia (one just a piece of the coastline), an array of colored rectangles (an *image*), and a legend. Immediately, the focus is on producing much more basic graphical output. If some useful functions are created for these, the functions will provide much more reusable graphical elements that could be combined in other ways to create all sorts of other plots (for example, see Section 8.3.10 and Figure 8.18).

8.2 Simple graphics functions

The simplest approach is to write a graphics function just for its side effect of producing graphical output (i.e., using **grid** graphics functions as described in Chapter 6). The first example will be a simple graphics function to produce an image. The code in Figure 8.2 provides code defining a function `grid.imageFun()` for this purpose.

This function takes arguments to describe the number of rows and columns in the image (`nrow` and `ncol`), the colors to use for each cell in the image (`cols`), and the order in which those colors should be applied to the cells (`byrow`). Output is produced by a call to the `grid.rect()` function (line 12), which draws a rectangle for each cell in the image.

This function can be used to draw an array of rectangles just like any other plotting function. An example usage is given in the following code. First, a set of gray scale colors are defined (these will be used throughout the chapter).

```
> grays <- gray(0.5 + (rep(1:4, 4) - rep(0:3, each=4))/10)
```

```
1 grid.imageFun <- function(nrow, ncol, cols,
2                                 byrow=TRUE) {
3   x <- (1:ncol)/ncol
4   y <- (1:nrow)/nrow
5   if (byrow) {
6     right <- rep(x, nrow)
7     top <- rep(y, each=ncol)
8   } else {
9     right <- rep(x, each=nrow)
10    top <- rep(y, ncol)
11  }
12  grid.rect(x=right, y=top,
13    width=1/ncol, height=1/nrow,
14    just=c("right", "top"),
15    gp=gpar(col=NA, fill=cols),
16    name="image")
17 }
```

Figure 8.2
A grid.imageFun() function. This function draws an array of nrow by ncol rectangles filled with the specified colors.

Now two images are drawn with the same colors, but different byrow settings (see Figures 8.3a and 8.3b).

```
> grid.imageFun(4, 4, grays)
```

```
> grid.imageFun(4, 4, grays, byrow=FALSE)
```

There is an obvious deficiency in this function because it does not perform any checking of its arguments to ensure that the correct information is being passed to it. For example, there is no check that nrow and ncol are numeric values of length 1. In general, in order to reduce the size and complexity of the code chunks, the examples will leave out input-checking code. This issue is addressed more seriously in the context of developing new graphical objects in Section 8.3.3.

The grid.imageFun() example shows that it is quite straightforward to create a new graphics function that just produces output. However, there are three important things to keep in mind when writing such a function: other people might want to embed the output from your function as an element within a more complex scene; other people might want to embed their output within the output from your function; and other people might want to manipulate

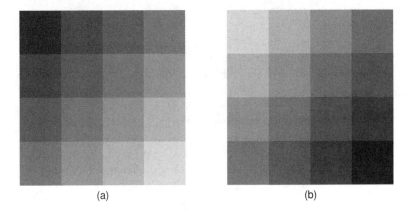

(a)   (b)

Figure 8.3
Output from the `grid.imageFun()` function. The two images use the same set of colors, but have different orientations. Image (a) has `byrow=TRUE` and image (b) has `byrow=FALSE`.

the output from your function. The following sections look at how you should design your function so that these tasks are straightforward for other people.

8.2.1   Embedding graphical output

The **grid** system is designed to allow graphical output to be embedded within other graphical output. All drawing occurs within the current viewport and no assumptions are made about the position or size of that viewport. New **grid** functions should be written with this in mind and it should not be assumed that output is being drawn into the entire device.

The `grid.imageFun()` function demonstrates this idea; this function just draws rectangles within the current viewport, wherever that may be and however large it may be.

On the other hand, it is sometimes important to enforce certain constraints on how graphical output is drawn. A good example is in the drawing of maps. Usually, a map is drawn with a specific aspect ratio so that, for example, 1 unit in the x-dimension has the same physical size as 1 unit in the y-dimension. In such cases, it may be necessary for a function to push its own viewports to enforce an aspect ratio before performing any drawing. A function to draw a map of Australia will be developed in order to demonstrate this idea.

The package **oz** provides data for drawing maps of Australia. The `ozRegion()`

```
 1 grid.ozFun <- function(ozRegion) {
 2   pushViewport(
 3     viewport(name="ozlay",
 4              layout=grid.layout(1,1,
 5                          widths=diff(ozRegion$rangex),
 6                          heights=diff(ozRegion$rangey),
 7                          respect=TRUE)))
 8   pushViewport(viewport(name="ozvp",
 9                          layout.pos.row=1,
10                          layout.pos.col=1,
11                          xscale=ozRegion$rangex,
12                          yscale=ozRegion$rangey,
13                          clip=TRUE))
14   index <- 1
15   for(i in ozRegion$lines) {
16     grid.lines(i$x, i$y, default.units="native",
17                name=paste("ozlines", index, sep=""))
18     index <- index + 1
19   }
20   upViewport(2)
21 }
```

Figure 8.4
A grid.ozFun() function. This function draws a map of Australia or some part
thereof.

function in the oz package returns an object of class "ozRegion" containing
x-axis and y-axis ranges, and a list of x-locations and y-locations to draw map
lines. The grid.ozFun() shown in Figure 8.4 makes use of ozRegion() to
draw a map of Australia using **grid**.

The most important part of this function is the pushing of viewports that
establish the correct aspect ratio for drawing the map (lines 2 to 13). The
first viewport contains a layout with a single cell set to the correct aspect ratio
and the second viewport occupies that cell and sets the appropriate "native"
coordinate system for the map. This allows the map to be drawn within any
viewport, but retain the appropriate shape.

The rest of the grid.ozFun() function just draws the lines representing the
Australian coastline (and state boundaries) using grid.lines().

The following code shows an example of the grid.ozFun() function being
used to draw all of Australia (see Figure 8.5). The map is not distorted even
though the region it is drawn in (indicated by the gray rectangle) is very wide.

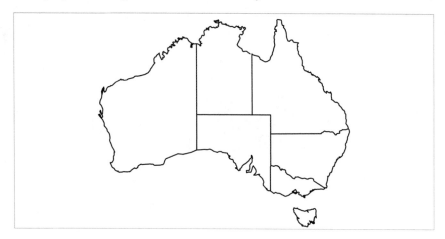

Figure 8.5

Example output from `grid.ozFun()`. By default it draws all of Australia.

```
> grid.ozFun(ozRegion())
```

8.2.2   Facilitating annotation

In addition to being able to produce graphical output within any context, it is vital that further graphical output can be added to the output of a graphical function. Again, the **grid** system is designed to facilitate this, by allowing navigation between viewports.

In this context, there are two important features of the `grid.ozFun()` function defined in Figure 8.4: the viewports that are pushed have names, `"ozlay"` and `"ozvp"` (lines 3 and 8); and the function calls `upViewport()` (not `popViewport()`) when it has finished drawing (line 20). These features mean that the viewports are available and accessible for other code to use after the `grid.ozFun()` function has done its drawing.

The following code provides an example of annotation using the `grid.imageFun()` function to add an image to output from the `grid.ozFun()` function (see Figure 8.6). In this example, only a small part of the South Australian coastline is used (the coastline close to the area where fluorescence data were gathered).

First of all, the latitude and longitude ranges are set up for the map (`mapLong` and `mapLat`) and for the image (`imageLong` and `imageLat`). Also, the set of colors for the image are calculated (`imageCols`). The prediction surface

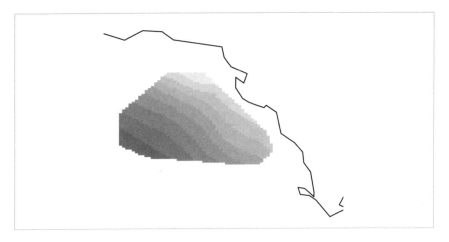

Figure 8.6
Annotating `grid.ozFun()` output.   An image has been added using the
`grid.imageFun()` function.

to be plotted is in a variable called `fluoro.predict`, which has components
`x`, `y`, and `z` for the longitude, latitude, and predicted fluorescence value, respectively. These ranges and colors will be used throughout the rest of the
chapter.

```
> mapLong <- c(132, 136)
> mapLat <- c(-35, -31.5)
> imageLong <- range(fluoro.predict$x)
> imageLat <- range(fluoro.predict$y)
> zbreaks <- seq(min(fluoro.predict$z, na.rm=TRUE),
                 max(fluoro.predict$z, na.rm=TRUE),
                 length=10)
> zcol <- cut(fluoro.predict$z, zbreaks,
              include.lowest=TRUE, labels=FALSE)
> ozgrays <- gray(0.5 + 1:9/20)
> imageCols <- ozgrays[zcol]
```

Now, the map and image can be drawn. The map is drawn first which produces
the coast line of South Australia and sets up the viewports `"ozlay"` and
`"ozvp"`.

```
> grid.ozFun(ozRegion(xlim=mapLong, ylim=mapLat))
```

The function `downViewport()` is used to navigate down to the viewport
`"ozvp"`, which has scales set up representing the latitude and longitude of

the map. This is only possible because the `grid.ozFun()` function specified useful names for the viewports it set up.

```
> downViewport("ozvp")
```

A further viewport is pushed to occupy the region where the image should be drawn and the image is drawn within that viewport.

```
> pushViewport(viewport(y=min(imageLat),
                        height=abs(diff(imageLat)),
                        x=max(imageLong),
                        width=abs(diff(imageLong)),
                        default.units="native",
                        just=c("right", "bottom")))
> grid.imageFun(50, 50, col=imageCols)
> upViewport(0)
```

8.2.3 Editing output

In addition to being able to add further output to a plot, it is useful to make it easy for others to modify the existing elements of a plot. The important step in this case is to provide a name for each piece of graphical output that your function produces.

The `grid.imageFun()` function uses the name `"image"` for the set of rectangles that it draws (line 16 in Figure 8.2) and the `grid.ozFun()` function names each map border that it draws `"ozlinesi"`, where i varies from 1 to the number of borders drawn (line 17 in Figure 8.4).

These names are useful for interacting with the output from these functions, particularly for the purpose of editing the output. The following code presents a couple of examples of modifying the plot produced in Figure 8.6. The first edit reverses the set of colors used in the image. The second edit changes the color of all map borders to gray and makes the borders thicker (see Figure 8.7).

```
> grid.edit("image", gp=gpar(fill=rev(ozgrays)[zcol]))
> grid.gedit("^ozlines[0-9]+$", gp=gpar(col="gray", lwd=2))
```

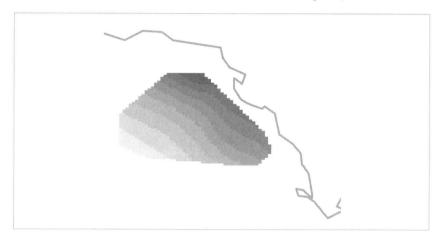

Figure 8.7
Editing `grid.ozFun()` output. Compared to Figure 8.6, the colors of the image have
been reversed and the Australian coastline is thicker and colored gray.

8.2.4   Absolute versus relative sizes

Another thing to consider when designing a graphics function is whether to
use absolute or relative coordinate systems and graphical parameters for sizing
graphical output. If absolute coordinates systems such as `"in"`, `"cm"`, or `"mm"`
are used to size output, then the output will remain that size no matter how
large or small the surrounding viewport is made. This is also true of graphical
parameters such as `fontsize` (which specifies the size of text in points), and
`lwd`. If, on the other hand, relative coordinates such as `"npc"` or `"native"`
are used for sizing output, the output will resize with its container. Graphical
parameters that are relative like this are `cex` for sizing text and `lex` for line
width.

In general, absolute sizes are more appropriate for producing or fine-tuning a
piece of output for a specific use (e.g., a figure in an article). Relative sizes
are more appropriate when designing general graphics functions for others to
use, where it is unknown how large the final output will be. One possible
exception to this rule is the sizing of text. It is reasonable to set text size in
absolute terms (i.e., a particular point size) in order to ensure that the text
is legible.

The coordinate systems `"char"`, `"line"`, `"strwidth"`, `"strheight"`, `"grobx"`,
`"groby"`, `"grobwidth"`, and `"grobheight"` are considered to be relative because they depend on the size of other output.

8.3 Graphical objects

A properly written graphics function can be very useful if it can be reused in other plots and arbitrarily added to or modified as described in previous sections. There are, however, a number of benefits to be gained from also creating a graphical object, or grob, to represent the output that your function produces.

The following sections consider again the development of functions to produce maps and images, but this time with an emphasis on creating *objects* that represent the output, rather than just producing output.

Defining new grobs involves working with classes and generic functions. This section assumes a familiarity with the basic ideas of object-oriented programming and its implementation in S3 classes and methods.

The design of classes and methods is a reasonably sophisticated process, there are often a number of possible designs to choose from, and it can be difficult to determine a "best solution." This means that it is impossible to provide a single definitive statement about how a new graphical object should be developed. Instead, this section presents a number of examples with several different implementations and there is a discussion of the advantages and disadvantages of different approaches.

8.3.1 Overview of creating a new graphical class

There are two main steps involved in defining a new graphical class. First of all, the structure of the class must be described. This consists of specifying the components of the new class: what information will be stored in objects of that class. For example, the "rect" class has components x, y, width, height, and just that describe the location and size of the rectangle.

The functions grob() and gTree() are used to define the structure for a new graphical class (as described in more detail in the next section). These functions ensure that all grobs have a number of standard components. For example, all grobs must have gp, vp, and name components. In addition, all classes derived from "gTree" (via the gTree() function) also have components children and childrenvp that describe the children of the gTree and how those children are drawn (see Section 8.3.4).

The second step in defining a new graphical class is to define the behavior of the class. This consists of writing methods for several important generic

functions. Methods can be written to control the validation of a grob, how a
grob is drawn, and what happens when a grob is modified. It is also possible
to write methods that allow a grob to be queried for its location and size.
These generic functions are described in Sections 8.3.4 to 8.3.7.

8.3.2   Defining a new graphical class

The code in Figure 8.8 gives an example of defining a new graphical class.
An "imageGrob" class is defined, which contains a description of the image
output generated by the grid.imageFun() function that was defined earlier.

The imageGrob() function calls the function gTree() to create an object
of a new class, "imageGrob". An imageGrob is a gTree with several components that provide a high-level description of an image (ncol, nrow, cols, and
byrow). There is also a single child, which is a rect grob, representing the
rectangles that will be drawn to produce the image. The imageGrob() function also provides the standard gp, vp, and name components, which should
be available for all grobs.

The makeImageRect() function generates a rect grob from a high-level image
description. This is very similar to the function grid.imageFun(), but it
produces an object containing a description of some rectangles, rather than
drawing the rectangles, and it calls rectGrob() rather than grid.rect() (line
11). This function is not intended to be used directly — it is just a "helper
function" for the main imageGrob() function. This is an example of modular
code that makes it easier to read the main function and it will be used later
when some other details of creating graphical objects are considered.

The grid.imageGrob() function is just a convenience for producing graphical
output from an imageGrob grob; it just creates an appropriate grob and draws
it. The following code produces the same result as Figure 8.3a.

```
> grid.imageGrob(4, 4, grays)
```

There are now functions that define a new class and create an object of that
class. Sections 8.3.3 to 8.3.7 describe how to define appropriate behavior for
the new class so that it draws correctly and responds appropriately to being
modified.

Summary of creating a new graphical class

A new class is derived from the "grob" class using the grob() function, or
from the "gTree" class using the gTree() function (e.g., line 20 in Figure
8.8). This will ensure standard behavior when drawing and editing grobs,

```
1 makeImageRect <- function(nrow, ncol, cols, byrow) {
2   xx <- (1:ncol)/ncol
3   yy <- (1:nrow)/nrow
4   if (byrow) {
5     right <- rep(xx, nrow)
6     top <- rep(yy, each=ncol)
7   } else {
8     right <- rep(xx, each=nrow)
9     top <- rep(yy, ncol)
10   }
11   rectGrob(x=right, y=top,
12           width=1/ncol, height=1/nrow,
13           just=c("right", "top"),
14           gp=gpar(col=NA, fill=cols),
15           name="image")
16 }

18 imageGrob <- function(nrow, ncol, cols, byrow=TRUE,
19                       name=NULL, gp=NULL, vp=NULL) {
20   igt <- gTree(nrow=nrow, ncol=ncol,
21               cols=cols, byrow=byrow,
22               children=gList(makeImageRect(nrow, ncol,
23                                            cols, byrow)),
24               gp=gp, name=name, vp=vp,
25               cl="imageGrob")
26   igt
27 }

29 grid.imageGrob <- function(...) {
30   igt <- imageGrob(...)
31   grid.draw(igt)
32 }
```

Figure 8.8

An "imageGrob" class. This is a grob-based equivalent of grid.imageFun().

calculating the size of grobs, and so on (Sections 8.3.3 to 8.3.7 provide detailed information about the default behavior of grobs). Apart from the `cl` argument that specifies the name of the new class (line 25 in Figure 8.8), the arguments to these functions provide a list of components for the new class.

There are some standard components common to all grobs: `gp`, `vp`, and `name`. It is sensible to make these available via the constructor function for your new class (e.g., line 19 in Figure 8.8).

The `gp` component is designed to contain a `gpar` object, which is a set of graphical parameter settings; the `vp` component is designed to hold a viewport or a viewport path; and the `name` component provides the name for the grob. All of these are validated automatically and are used in the drawing and editing of the grob (see Section 8.3.3).

The `"gTree"` class defines two more standard components: the `children` component contains the children of the gTree (as a gList) and the `childrenvp` component contains viewports for the children to be drawn within (used in the drawing of the children of the gTree). An example of the use of the `children` component is shown in Figure 8.8 on line 22.

All other components are at the discretion of the class designer.

Having defined a new graphical class, it is then necessary to write one or more methods for some important generic functions as described in the following sections.

8.3.3   Validating grobs

This section describes the `validDetails()` function, which is important for ensuring that the components of a grob contain valid values.

The code examples used in the examples of simple graphics functions ignored the issue of checking user input to ensure that valid values are supplied for arguments or components. This issue becomes particularly important when dealing with grobs because it is not only possible to supply invalid values when a grob is first created, but also whenever a grob is modified via `grid.edit()` or `editGrob()`.

Default validating behavior

When a grob is created or modified, it is automatically validated. The validation checks that the `gp`, `vp`, and `name` components of a grob are sensible (and for a gTree, the `children` and `childrenvp` components are also checked) and then the `validDetails()` generic function is called. By default this function

```
 1 validDetails.imageGrob <- function(x) {
 2   if (!is.numeric(x$nrow) || length(x$nrow) > 1 ||
 3       !is.numeric(x$ncol) || length(x$ncol) > 1)
 4     stop("nrow and ncol must be numeric and length 1")
 5   if (!is.logical(x$byrow))
 6     stop("byrow must be logical")
 7   x
 8 }

10 validDetails.ozGrob <- function(x) {
11   if (!inherits(x$ozRegion, "ozRegion"))
12     stop("Invalid ozRegion")
13   x
14 }
```

Figure 8.9
Some `validDetails()` methods. These are called when an `imageGrob` or an `ozGrob` is first created, or when such an object is modified using `grid.edit()`.

does nothing. A new class should define a method to check the components that are specific to that class.

The `imageGrob` example

Figure 8.9 shows a `validDetails()` methods for the `"imageGrob"` class (Figure 8.9 also shows a method for the `"ozGrob"` class, which is introduced in the next section). The `validDetails()` method for the `"imageGrob"` class (lines 1 to 8) checks that the **nrow** and **ncol** components are numeric and of length 1 and that the **byrow** component is a logical vector. The return value of the method is the validated `imageGrob` (line 7). All `validDetails()` methods must do this whether they modify the grob or not.

With these validation methods defined, both the creation and the modification of an `imageGrob` will perform checks to ensure that the components of the `imageGrob` contain valid values. The following code demonstrates the validation at work. First of all, the creation of an `imageGrob` fails because **byrow** is not a logical value.

```
> grid.imageGrob(4, 4, grays, byrow="what?")
```

```
Error in validDetails.imageGrob(x) :
    byrow must be logical
```

```
 1 makeOzViewports <- function(ozRegion) {
 2   vpStack(viewport(name="ozlay", layout=grid.layout(1, 1,
 3                        widths=diff(ozRegion$rangex),
 4                        heights=diff(ozRegion$rangey),
 5                        respect=TRUE)),
 6           viewport(name="ozvp", layout.pos.row=1,
 7                    layout.pos.col=1,
 8                    xscale=ozRegion$rangex,
 9                    yscale=ozRegion$rangey,
10                    clip=TRUE))
11 }

13 makeOzLines <- function(ozRegion) {
14   numLines <- length(ozRegion$lines)
15   lines <- vector("list", numLines)
16   index <- 1
17   for(i in ozRegion$lines) {
18     lines[[index]] <- linesGrob(i$x, i$y,
19                        default.units="native",
20                        vp=vpPath("ozlay", "ozvp"),
21                        name=paste("ozlines", index, sep=""))
22     index <- index + 1
23   }
24   do.call("gList", lines)
25 }

27 ozGrob <- function(ozRegion, name=NULL, gp=NULL, vp=NULL) {
28   gTree(ozRegion=ozRegion, name=name, gp=gp, vp=vp,
29     childrenvp=makeOzViewports(ozRegion),
30     children=makeOzLines(ozRegion),
31     cl="ozGrob")
32 }

34 grid.ozGrob <- function(...) {
35   grid.draw(ozGrob(...))
36 }
```

Figure 8.10

An "ozGrob" class. This is a grob-based equivalent of grid.ozFun().

An ozGrob is a gTree with a single component, ozRegion, which contains a description of the region of Australia to map (line 28 in Figure 8.10). An ozGrob also has a number of children, all of which are lines grobs representing the coastline and state boundaries to draw (line 30), and an ozGrob has a viewport stack in its childrenvp component (line 29). These viewports create a region with the right aspect ratio for drawing a map and the children of the ozGrob are all created with viewport paths to specify that they should be drawn within this region (line 20).

When an ozGrob is drawn, the viewports in its childrenvp component are pushed as part of the default drawing behavior for gTrees, then the grobs in its children component are drawn. Each child has a vp component indicating which viewport to navigate to before drawing.

The grid.ozGrob() function is just a convenient front-end for drawing an ozGrob. This can be used just like the function grid.ozFun() to draw some or all of a map of Australia. The following code produces exactly the same output as shown in Figure 8.5. There are more examples using ozGrob objects in later sections.

```
> grid.ozGrob(ozRegion())
```

An ozImage example

Both the "imageGrob" class and the "ozGrob" class are derived from the "gTree" class. This means that they have other grobs as children and the default drawing behavior for gTrees draws those children correctly when the imageGrob or ozGrob is drawn. This section looks at an example where a drawDetails() method has to be written in order to produce any output.

A typical reason for needing to write a drawDetails() method is that your new class does not have a fixed set of grobs as children. Axes that must calculate tick marks on the fly are a good example (it is only possible to figure out how many tick marks to draw and where to locate them when the axis is actually drawn).

In order to demonstrate the definition of a drawDetails() method, an "ozImage" class will be defined. This class combines an ozGrob and an imageGrob and has to do the drawing itself to get them combining correctly (for producing output like that in Figure 8.6).

The code in Figure 8.11 shows the definition of an "ozImage" class. An ozImage is just a grob (it has no children; lines 3 to 5) so without a drawDetails() method, it would produce no output. In order to produce output when an ozImage is drawn, a drawDetails() method is defined for the "ozImage" class (lines 8 to 21). This method creates an ozGrob and

```
1 ozImage <- function(mapLong, mapLat,
2                      imageLong, imageLat, cols) {
3   grob(mapLong=mapLong, mapLat=mapLat,
4        imageLong=imageLong, imageLat=imageLat, cols=cols,
5        cl="ozImage")
6 }

8 drawDetails.ozImage <- function(x, recording) {
9   grid.draw(ozGrob(ozRegion(xlim=x$mapLong,
10                             ylim=x$mapLat)))
11  depth <- downViewport(vpPath("ozlay", "ozvp"))
12  pushViewport(viewport(y=min(x$imageLat),
13                        height=diff(range(x$imageLat)),
14                        x=max(x$imageLong),
15                        width=diff(range(x$imageLong)),
16                        default="native",
17                        just=c("right", "bottom")))
18  grid.draw(imageGrob(50, 50, col=x$col))
19  popViewport()
20  upViewport(depth)
21 }
```

Figure 8.11
An "ozImage" class. This combines an imageGrob with an ozGrob to make a larger,
more complex grob.

draws it (lines 9 to 10), then navigates down to the "ozvp" viewport, pushes
a viewport within which to draw the image, creates and draws an imageGrob
(line 18), and finally navigates back up to the viewport that it started in.

With this class defined, Figure 8.6 can be produced as follows.

```
> grid.draw(ozImage(mapLong, mapLat,
                    imageLong, imageLat, imageCols))
```

An important point about drawDetails() methods is that none of the drawing and viewport operations within a drawDetails() method are recorded on
the display list. For example, as a result of the above code, there is an ozImage
grob on the display list, but there is neither an ozGrob nor an imageGrob on
the display list. This has implications for editing output which are discussed
in the next section.

8.3.5 Editing grobs

This section describes the editDetails() generic function, which is important for ensuring that a grob responds appropriately when it is edited. It is particularly important for classes derived from "gTree" to ensure that the children of the gTree are updated when the high-level components of the gTree are modified.

One advantage of defining a grob to represent graphical output is that the grob provides a high-level interface to the graphical output. For example, an imageGrob contains components that describe an image in terms of how many rows and columns it has. The low-level description of the precise location of individual rectangles within the image is left to the lower-level rect grob. This means that it is possible to modify the high-level description in order to change the graphical output. For example, the number of rows in an imageGrob could be modified simply by changing the high-level nrow component rather than by having to modify the location and size of all of the low-level rectangles. Unfortunately, modifying the high-level description of an imageGrob as it has been defined so far will have no effect on the output because it will have no effect on the children of the imageGrob. For gTrees with children, it is necessary to provide instructions for how to change the children when the high-level description is modified.

Default editing behavior

The grid.edit() function and the editGrob() function are used to modify a grob.

When a grob is modified, the components of the grob are set to the new values and the editDetails() generic function is called. The default behavior is to do nothing, but a class can define a method which, for example, propagates a change to its children.

The imageGrob example

In the case of an imageGrob, an editDetails() method is required to ensure that the child rect corresponds to the high-level description in the imageGrob. Figure 8.12 shows code defining an editDetails() method for the "imageGrob" class (lines 1 to 10). This method totally re-creates the child rect grob if any of the ncol, nrow, or byrow arguments are modified (lines 2 to 5), and edits the child rect grob if the cols argument is modified (lines 6 to 8). A very important feature of the function is that it returns the modified grob (line 9). All editDetails() methods must do this.

```
1 editDetails.imageGrob <- function(x, specs) {
2   if (any(c("ncol", "nrow", "byrow") %in% names(specs))) {
3     x <- addGrob(x, makeImageRect(x$nrow, x$ncol,
4                                   x$cols, x$byrow))
5   }
6   if (any(c("cols") %in% names(specs))) {
7     x <- editGrob(x, "image", gp=gpar(fill=x$cols))
8   }
9   x
10 }

12 editDetails.ozGrob <- function(x, specs) {
13   if ("ozRegion" %in% names(specs)) {
14     x$childrenvp <- makeOzViewports(x$ozRegion)
15     x <- setChildren(x, makeOzLines(x$ozRegion))
16   }
17   x
18 }
```

Figure 8.12
Some editDetails() methods for imageGrob and ozGrob objects. These will be run
when such objects are modified using grid.edit().

The ozGrob example

Figure 8.12 also shows an editDetails() method for the "ozGrob" class.
This method ensures that changes to the ozRegion component of an ozGrob
will be reflected in the children of the ozGrob by completely re-creating the
childrenvp and children components of the ozGrob.

The ozImage example

There is no method for the "ozImage" class because it is only a grob and
there are no children to propagate changes to. The output of an ozImage is
re-created by its drawDetails() method whenever an ozImage is edited.

The imageGrob example again

With the editDetails() method defined for the "imageGrob" class, it is
possible to edit an imageGrob. The following code creates an image (see
Figure 8.13a) and then modifies the orientation via the high-level description in the imageGrob. The changes are passed on to the rect child by the

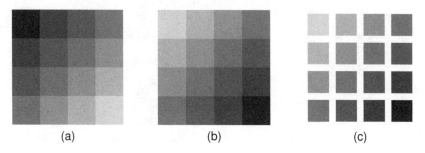

(a)   (b)   (c)

Figure 8.13

Editing an `imageGrob`. Panel (a) shows a simple `imageGrob`. In panel (b), the `imageGrob` has been modified by changing the `byrow` argument. In panel (c), the `rect` grob within the `imageGrob` has been modified to change the borders of each rectangle in the image to be thick and white.

`editDetails()` method (see Figure 8.13b).

```
> grid.imageGrob(4, 4, grays, name="imageGrob")
```

```
> grid.edit("imageGrob", byrow=FALSE)
```

A gTree with child grobs not only allows interaction with a high-level description of graphical output, but it also makes it possible to access the low-level description as well through the child grobs. For example, it is possible to edit the low-level `rect` child of an `imageGrob`. The following code modifies the image drawn in the previous example to change the borders of the child rectangles to be white and very wide (see Figure 8.13c). The only important design aspect here is that the `"imageGrob"` class provides a name, `"image"`, for the `rect` child. This makes it possible to specify a gPath to the `rect`.

```
> grid.edit("imageGrob::image", gp=gpar(col="white", lwd=6))
```

The difference between editing the low-level `rect` object and re-creating it is that when the `rect` object is edited it will retain any other customizations that have been made to it. The following code gives a small example. First of all an `imageGrob` object is drawn (the output is identical to Figure 8.13a).

```
> grid.imageGrob(4, 4, grays, name="imageGrob")
```

In the first edit, the low-level `rect` object is edited so that the borders of the individual rectangles are drawn white (see Figure 8.14a).

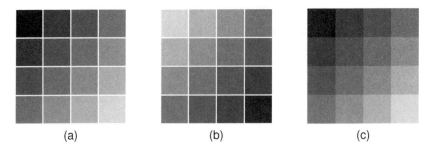

(a)   (b)   (c)

Figure 8.14
Low-level editing of an `imageGrob`. Panel (a) shows an `imageGrob` with the `rect` grob modified so that all of the rectangles in the image have a white border. In panel (b), the `cols` argument of the `imageGrob` has been modified, which changes the fill color of each rectangle, but does not alter the borders (which are still white). In panel (c), the `byrow` argument of the `imageGrob` has been modified and this causes the `rect` grob to be remade so the white borders are lost.

```
> grid.edit("imageGrob::image", gp=gpar(col="white"))
```

The second edit modifies the high-level `cols` component in the `imageGrob` object, but because this only edits the low-level `rect` object the individual rectangle borders are retained (see Figure 8.14b).

```
> grid.edit("imageGrob", cols=rev(grays))
```

In the final edit, the high-level `byrow` argument is edited, which causes the low-level `rect` object to be re-created and the individual borders are lost (see Figure 8.14c).

```
> grid.edit("imageGrob", byrow=FALSE)
```

The `ozImage` example again

It is worth noting that for grobs that do not have any children, but produce output by creating and drawing grobs in a `drawDetails()` method, it is not actually possible to perform this sort of low-level editing. For example, it is not possible to edit the low-level `imageGrob` or `ozGrob` of an `ozImage` because the `imageGrob` and `ozGrob` are never stored anywhere. Some solutions to this problem are discussed in Section 8.3.11.

8.3.6 Querying grobs

This section describes the functions xDetails() and yDetails(), which are useful for determining a location on the boundary of a grob, and the functions widthDetails() and heightDetails(), which are useful for calculating the amount of space that a grob requires for drawing.

Every new graphical class should have appropriate behavior defined for validating, drawing, and editing objects of the class. This section looks at defining behavior for querying a grob for its location and size, which only makes sense in some cases.

This is most often used for providing column widths and row heights for **grid** layouts, but can also be used to position one grob relative to another, for example, to draw a rectangular border around a piece of text or to draw a line from one grob to another.

Default sizing behavior

The calculations of "grobx", "groby", "grobwidth", and "grobheight" units call the generic functions, xDetails(), yDetails(), widthDetails(), and heightDetails(), respectively, which should return a unit describing a location on the boundary of the grob or the width and height of the grob.

The default xDetails() and yDetails() methods return unit(0.5, "npc"), which means that locations default to the center of the current viewport.

The default methods for widthDetails() and heightDetails() return the value unit(1, "null"), which means that the corresponding widths and heights get an equal share of the available space within layouts (see Section 6.5.6). Outside of a layout, the default width and height both evaluate to zero.

There are predefined methods for most primitives and for the "frame" class (see Section 7.4).

A ribbonLegend example

In order to demonstrate the use of these methods, a "ribbonLegend" class will be defined (see Figures 8.15 and 8.16).

A ribbonLegend consists of several components describing the number of levels to represent in the legend and the colors to use for each level (line 6 in Figure 8.16). It is also possible to control the amount of empty space to leave around the legend (line 3). The children of a ribbonLegend are a rect grob to draw rectangles for the levels and a `lines` grob and a `text` grob to show the legend scale (see the `ribbonKids()` function in Figure 8.15). There is a `childrenvp` component that contains a `vpTree`: the parent viewport defines a layout and then two child viewports occupy column 2 and column 3 of row 2 of that layout (see the `ribbonVps()` function in Figure 8.15). The layout is created with column widths based on the space needed for the scale labels (in column 3), one line of text for the "ribbon" of rectangles (in column 2), and the required empty space around the outside (columns 1 and 4).

A `widthDetails()` method is defined for the `"ribbonLegend"` class in Figure 8.16 (lines 12 to 14). The width of a `ribbonLegend` is calculated as the sum of the widths of the layout in the top viewport of the `childrenvp` component (line 13), which reflects the space required to draw the legend.

This class can be used to produce a ribbon legend as shown at the right-hand side of Figure 8.1. If a ribbon legend is used as the data for a `"grobwidth"` unit, it will request enough space to draw itself. Examples are given in Sections 8.3.9 and 8.3.10.

8.3.7   Pre-drawing and post-drawing

There are two more generic functions that have not yet been mentioned: `preDrawDetails()` and `postDrawDetails()`. These functions are called as part of the default drawing behavior (see Section 8.3.4).

Default pre/post-drawing behavior

The pushing and popping of viewports in `vp` components described in the default drawing behavior (Section 8.3.4) includes a call to the `preDrawDetails()` generic function (after the `vp` component has been pushed, but before the `drawDetails()` method is called) and a call to the `postDrawDetails()` generic function (after the `drawDetails()` method has been called, but before the `vp` component is popped).

By default, the generic functions do nothing, but a new graphical class can define a method to perform additional pushing, popping, and navigation of viewports if required.

The pre-drawing and post-drawing are separate actions from the actual drawing because they are also performed during the evaluation of `"grobwidth"` and `"grobheight"` units. This is done so that the size of a grob is calculated based on the context in which it is drawn (i.e., so that the size corresponds to the actual size of the graphical output). This means that any pushing (and popping) of viewports for a new grob class must be performed in a

```
 1 calcBreaks <- function(nlevels, breaks, scale) {
 2   if (is.null(breaks)) {
 3     seq(min(scale), max(scale), diff(scale)/nlevels)
 4   } else {
 5     breaks
 6   }
 7 }

 9 ribbonVps <- function(nlevels, breaks, margin, scale) {
10   breaks <- format(signif(calcBreaks(nlevels, breaks, scale),
11                           3))
12   vpTree(
13     viewport(name="layout", layout=
14       grid.layout(3, 4,
15         widths=unit.c(margin, unit(1, "line"),
16                       max(unit(0.8, "line") +
17                           stringWidth(breaks)), margin),
18         heights=unit.c(margin, unit(1, "null"), margin))),
19     vpList(viewport(layout.pos.col=2, layout.pos.row=2,
20                     yscale=scale, name="ribbon"),
21            viewport(layout.pos.col=3, layout.pos.row=2,
22                     yscale=scale, name="labels")))
23 }

25 ribbonKids <- function(nlevels, breaks, cols, scale) {
26   breaks <- calcBreaks(nlevels, breaks, scale)
27   nb <- length(breaks)
28   tickloc <- breaks[-c(1, nb)]
29   gList(rectGrob(y=unit(breaks[-1], "native"),
30                  height=unit(diff(breaks), "native"),
31                  just="top", gp=gpar(fill=cols),
32                  vp=vpPath("layout", "ribbon")),
33         segmentsGrob(x1=unit(0.5, "line"),
34                      y0=unit(tickloc, "native"),
35                      y1=unit(tickloc, "native"),
36                      vp=vpPath("layout", "labels")),
37         textGrob(x=unit(0.8, "line"),
38                  y=unit(tickloc, "native"),
39                  just="left",
40                  label=format(signif(tickloc, 3)),
41                  vp=vpPath("layout", "labels")))
42 }
```

Figure 8.15
Helper functions for a "ribbonLegend" class. The class itself is defined in Figure 8.16.

```
 1 ribbonLegend <- function(nlevels=NULL, breaks=NULL, cols,
 2                           scale=range(breaks),
 3                           margin=unit(0.5, "line"),
 4                           gp=NULL, vp=NULL, name=NULL) {
 5   gTree(
 6     nlevels=nlevels, breaks=breaks, cols=cols, scale=scale,
 7     children=ribbonKids(nlevels, breaks, cols, scale),
 8     childrenvp=ribbonVps(nlevels, breaks, margin, scale),
 9     gp=gp, vp=vp, name=name, cl="ribbonLegend")
10 }

12 widthDetails.ribbonLegend <- function(x) {
13    sum(layout.widths(viewport.layout(x$childrenvp[[1]])))
14 }
```

Figure 8.16
A "ribbonLegend" class. This consists of a "ribbon" of rectangles filled with the specified colors, plus an axis showing the scale.

preDrawDetails() (and postDrawDetails()) method if they will have any effect on the calculations on the size of the grob. An example of a class that has its own preDrawDetails() and postDrawDetails() methods is the "frame" grob class (see Section 7.4).

8.3.8   Summary of graphical object methods

Defining the behavior for a new graphical class requires writing one or more methods for the standard **grid** generic functions:

- **Always** write a constructor function for the class to generate a grob or a gTree containing the description of what to draw.
- **Always** write a validDetails() method for checking the validity of values in the non-standard components of the class.
- **Sometimes** write a drawDetails() method to specify how to draw the class.
- **Rarely** write preDrawDetails() and postDrawDetails() methods if the drawing involves pushing viewports that affect the determination of the size of the graphical output.
- **Always (for gTrees)** write an editDetails() method so that changes to the high-level description are propagated to child grobs.
- **Sometimes** write xDetails() and yDetails() methods if the boundary of the graphical output can be sensibly determined.

- **Sometimes** write `widthDetails()` and `heightDetails()` methods if the size of the graphical output can be sensibly determined.

8.3.9 Completing the example

Almost all of the components required to produce Figure 8.1 have now been defined. The following code produces an `ozImage` that contains a portion of the South Australian coastline and an image representing fluorescence just off the coast.

```
> ozimage <- ozImage(mapLong, mapLat,
                     imageLong, imageLat, imageCols)
```

This next piece of code creates a `ribbonLegend` for the image plot.

```
> ribbonlegend <- ribbonLegend(breaks=zbreaks,
                              cols=ozgrays,
                              scale=range(zbreaks),
                              gp=gpar(cex=0.7))
```

The final piece required is the ability to draw a "key" consisting of a map of Australia with a rectangle to indicate the region drawn in the main image. The code in Figure 8.17 defines an `"ozKey"` class for this purpose. An `ozKey` is a gTree with an `ozGrob` and a `rect` grob as children. An `ozKey` draws its children within a viewport, the location and size of which are specified when the `ozKey` is constructed. A `drawDetails()` method is not required for this class because the default drawing behavior for gTrees is sufficient. A `validDetails()` method should be written and an `editDetails()` method would be required for an `ozKey` to respond properly to editing; however, these are left as an exercise for the reader.

The following code constructs an `ozKey` grob for the image plot.

```
> ozkey <- ozKey(x=unit(1, "npc") - unit(1, "mm"),
                y=unit(1, "npc") - unit(1, "mm"),
                width=unit(3.5, "cm"),
                height=unit(2, "cm"),
                just=c("right", "top"),
                mapLong, mapLat)
```

Finally, the `ozImage`, `ozKey`, and `ribbonLegend` grobs are used to construct Figure 8.1 as follows.

```
 1 ozKey <- function(x, y, width, height, just,
 2                    mapLong, mapLat) {
 3   gTree(childrenvp=viewport(name="ozkeyframe",
 4                             x=x, y=y, just=just,
 5                             width=width, height=height),
 6         children=gList(ozGrob(ozRegion(), vp="ozkeyframe",
 7                               gp=gpar(lwd=0.1)),
 8                        rectGrob(x=mean(mapLong),
 9                                 y=mean(mapLat),
10                                 width=abs(diff(mapLong)),
11                                 height=abs(diff(mapLat)),
12                                 default.units="native",
13                                 gp=gpar(lwd=1),
14                                 vp=vpPath("ozkeyframe",
15                                           "ozlay", "ozvp"))))
16 }
```

Figure 8.17

An "ozKey" class. This consists of a map of Australia with a rectangle superimposed.

```
> fg <- frameGrob()
> fg <- packGrob(fg, ozimage)
> fg <- placeGrob(fg, ozkey)
> fg <- packGrob(fg, ribbonlegend, "right")
> grid.draw(fg)
```

This makes use of the fact that the ribbonLegend will be allocated the correct width because it has a widthDetails() method defined.

8.3.10   Reusing graphical elements

Having created all of these graphical objects, there exists a set of graphical elements that produce useful graphical output, that can have other output added to them, that can be modified at different levels, *and* that can be embedded in other settings. This means that it is possible to use them to produce a quite different sort of plot. Figure 8.18 shows an example that combines an ozGrob and several ribbonLegends in a quite different way.

First of all, a map of Australia is drawn, then a ribbonLegend grob is drawn for each city to show the range of average monthly temperatures. The data are minimum and maximum monthly temperatures at six major cities spread around Australia available as the data set ozTemp.

```
> grid.ozGrob(ozRegion())
> downViewport("ozvp")
> for (i in 1:(dim(ozTemp)[1])) {
    grid.points(ozTemp$long[i], ozTemp$lat[i], pch=16)
    rl <- ribbonLegend(breaks=c(min(ozTemp$min),
                               ozTemp$min[i],
                               ozTemp$max[i],
                               max(ozTemp$max)),
                       cols=c("white", "gray", "white"),
                       gp=gpar(cex=.7))
    pushViewport(viewport(x=unit(ozTemp$long[i], "native"),
                          y=unit(ozTemp$lat[i], "native"),
                          height=unit(1, "in"),
                          width=grobWidth(rl),
                          clip="off"))
    grid.circle(r=1,
                gp=gpar(col="gray", fill="white", alpha=0.8))
    grid.draw(rl)
    popViewport()
  }
> upViewport(0)
```

8.3.11   Other details

This section describes some more detailed and abstract issues that can arise in the design of a new graphical object.

Extending other grobs

From a design point of view, the correct implementation of the "imageGrob" class (Figure 8.8) probably would have been to derive it from the predefined "rect" class (rather than deriving it from the "gTree" class and having it *contain* a rect object as its child). In other words, from a design perspective, an image *is* a rect, simply with a different parameterization.

Unfortunately, this sort of implementation is awkward because **grid** uses the S3 class system, which does not support inheritance of structure. Deriving an "image" class from the "rect" grob class would mean that generic functions are automatically inherited, but the inheritance of structure (components) has to be done by hand. This requires knowledge of the internal structure of the "rect" class and will fail if there are any changes to that structure, so this approach is not recommended.

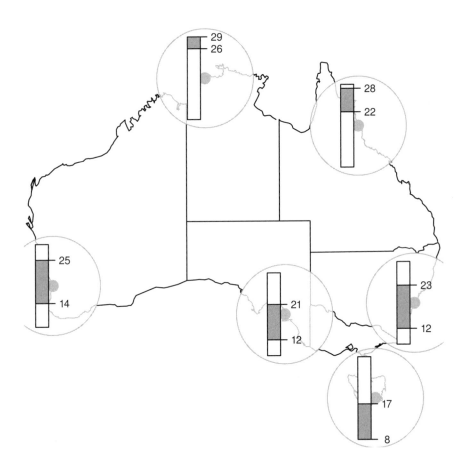

Figure 8.18

A plot of temperature data for several cities in Australia. This plot was composed using the same `ozGrob` and `ribbonLegend` grobs that were used to construct the oceanographic plot in Figure 8.1.

This means that the standard approach has to be to derive new classes directly from the "grob" class or the "gTree" class, not from other graphical classes.

Display lists

As mentioned in Section 9.6, R's graphics engine maintains a display list, which is a record of all graphical output on a page, and this is used to redraw a scene if a page is resized (among other things). The output from both traditional and **grid** graphics functions is recorded on this display list.

The **grid** package also maintains its own separate display list, which is used for accessing grobs in the current scene and for redrawing the current scene after editing (i.e., after a call to grid.edit()). The **grid** display list can be replayed explicitly using the grid.refresh() function.

The **grid** display list can be disabled using grid.display.list(), which saves on **grid**'s memory usage, but disables **grid**'s ability to modify and redraw a scene. If the **grid** display list is disabled, the functions grid.edit(), grid.get(), grid.add(), and grid.remove() will no longer work.

It is possible to record **grid** output only on the **grid** display list with the engine.display.list() function, as shown by the following code. Redrawing will be slightly slower, but this avoids the memory cost of having output recorded on both the **grid** display list and the graphics engine display list.

```
> engine.display.list(FALSE)
```

This action only affects the recording of **grid** operations on the graphics engine display list; traditional graphics output is still recorded on the graphics engine display list.

Calculations during drawing

With **grid** units and layouts, it is possible to specify quite complex arrangements of output in a "declarative" manner. For example, the idea that a particular region should be square (have an aspect ratio of 1) can be expressed at a high level, by specifying both width and height as unit(1, "snpc"), and the system will ensure that this occurs. There is no need to calculate the physical dimensions of the current viewport and from those determine how to make a square region.

It is, however, sometimes necessary to perform calculations by hand. For example, consider the problem of splitting text into several lines based on the width of the available space. The code in Figure 8.19 defines a function,

```
 1 splitString <- function(text) {
 2   strings <- strsplit(text, " ")[[1]]
 3   newstring <- strings[1]
 4   linewidth <- stringWidth(newstring)
 5   gapwidth <- stringWidth(" ")
 6   availwidth <-
 7     convertWidth(unit(1, "npc"),
 8                       "in", valueOnly=TRUE)
 9   for (i in 2:length(strings)) {
10     width <- stringWidth(strings[i])
11     if (convertWidth(linewidth + gapwidth + width,
12                       "in", valueOnly=TRUE) <
13         availwidth) {
14       sep <- " "
15       linewidth <- linewidth + gapwidth + width
16     } else {
17       sep <- "\n"
18       linewidth <- width
19     }
20     newstring <- paste(newstring, strings[i], sep=sep)
21   }
22   newstring
23 }
```

Figure 8.19
A splitString() function. This function takes a piece of text and splits it into multiple lines so that the text will fit (horizontally) within the current viewport. Validation checks (e.g., whether strings is a character vector of length at least 2) have not been included.

splitString(), to perform this operation (in a very simple-minded way). The important part of this function is the use of the convertWidth() function to obtain the size of the current line of text in inches (line 11) for comparison with the size of the current region in inches (lines 6 to 8).

The following code uses the splitString() function to draw some text within the current viewport (see the left-hand panel in Figure 8.20).

```
> text <- "The quick brown fox jumps over the lazy dog."
> grid.text(splitString(text),
            x=0, y=1, just=c("left", "top"))
```

There is a problem with the above code. If it is used to draw into a window and then the window is resized, the calculations are not rerun and the line

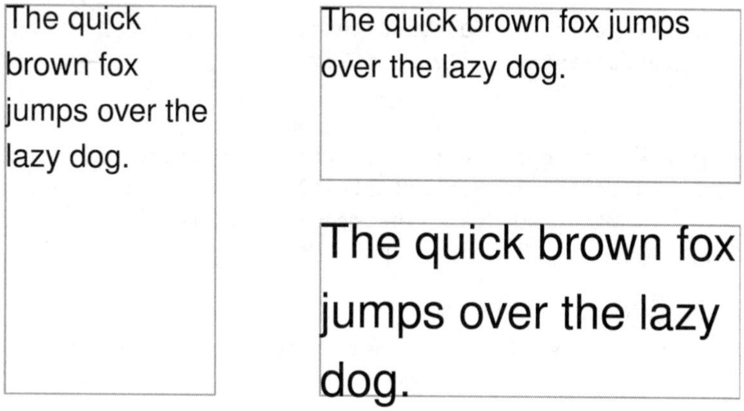

Figure 8.20
Performing calculations before drawing. If the drawing of a grob depends on calculations (in this case, calculations to split text into multiple lines to fit horizontally within the current viewport), the calculations should be included within a drawDetails() method. This means that the calculations will be rerun if the device is resized (left panel versus top-right panel) or if the grob is edited to make the font size larger (top-right panel versus bottom-right panel).

splitting becomes incorrect.

The issue is that only drawing actions are recorded on the display list, not any calculations leading up to the drawing. Anything that works off the display list (like redrawing after a resize) only reruns drawing actions.

There are two solutions to this problem. One solution rests on the fact that all code within a drawDetails() method (or a preDrawDetails() or postDrawDetails() method) is captured on the graphics engine display list. The code in Figure 8.21 uses this fact to create a "splitText" class with a drawDetails() method that performs the calculations.

A splitText grob will recalculate the line breaks when a window is resized (see the top-right panel of Figure 8.20).

```
> splitText <- splitTextGrob(text, name="splitText")
> grid.draw(splitText)
```

Another advantage of creating a grob with a drawDetails() method is that it is possible to edit the grob and have the calculations updated (see the bottom-right panel of Figure 8.20).

```
> grid.edit("splitText", gp=gpar(cex=1.5))
```

```
1 splitTextGrob <- function(text, ...) {
2   grob(text=text, cl="splitText", ...)
3 }

5 drawDetails.splitText <- function(x, recording) {
6   grid.text(splitString(x$text),
7             x=0, y=1, just=c("left", "top"))
8 }
```

Figure 8.21
A "splitText" class. The drawDetails method for the class recalculates where to place line breaks in the text, based on the current viewport size.

The other way to encapsulate calculations with drawing operations is to use the grid.record() function, as shown by the following code.

```
> grid.record({
                grid.text(splitString(text),
                          x=0, y=1, just=c("left", "top"))
                },
                list(text=text))
```

This is convenient for writing code purely for its side effect (i.e., without having to deal explicitly with grobs), but it provides less control over the design of the object that is created. There is also a recordGrob() function that simply creates a grob encapsulating the calculations and drawing operations without drawing anything.

Avoiding argument explosion

Very complex or high-level graphics functions and objects are usually composed of several lower-level elements, which in turn may be composed of several even-lower-level elements. For example, a scatterplot matrix is composed of several scatterplots and each scatterplot contains axes, labels, and data symbols.

Ideally, it should be possible to control any aspect of a graphical scene. In terms of writing code, this means that an argument or component should be supplied to allow the user to specify a customized value for any parameter of the scene.

At the level of graphical primitives, parameters consist of such things as the locations of lines, the color of lines, and the line thickness. At a higher level, for example for axes, there are higher-level parameters, such as where to place tick marks, but it is also desirable to still be able to control the individual elements of the axis.

It is tempting to simply provide arguments for the elements of an axis as arguments of the axis itself. An example is where an axis could have a `rot` argument to specify the angle of rotation of the tick mark labels, but this approach quickly runs into difficulties. For one thing, ambiguities can easily arise. If an axis had an overall label it is unclear whether the `rot` argument would apply to the tick mark labels or to the overall label. Another problem is that as elements become more complex, the number of parameters required for all subelements grows alarmingly. Consider the number of separate arguments required to individually specify the angle of rotation for tick mark labels on all scatterplots within a scatterplot matrix!

The **grid** package provides several features that can help to solve this problem. The functions `grid.edit()` and `editGrob()` (see Section 7.1) make it possible to access the lower-level elements of an object using a gPath. For example, in the following code, an x-axis is created and then the labels on the tick marks are rotated by editing the `rot` component of the `text` grob called `"labels"` that is a child of the `xaxis` grob.

```
> grid.xaxis(at=1:3/4, name="xaxis1")
> grid.edit("xaxis1::labels", rot=45)
```

More complex is the case where a grob calculates its children on the fly. This typically occurs when a grob has no permanent children to access via a gPath and this will often correspond to a grob that has a `drawDetails()` method.

The functions `gEdit()` and `gEditList()` allow the user to specify one or more edit operations and the functions `applyEdit()` and `applyEdits()` apply those operations to a grob. These can be used to specify editing actions that should be applied on the fly. The following code demonstrates their use. In this case, an x-axis is created without specifying the `at` argument, which means that tick marks are calculated on the fly. The `edits` argument is used to specify modifications to the labels that will be generated on the fly.

```
> grid.xaxis(name="xaxis1")
> grid.edit("xaxis1", edits=gEdit("labels", rot=45))
```

This approach is similar to the concept of panel functions (see Sections 3.4.6 and 4.7).

Mixing graphical functions and graphical objects

This chapter has addressed two main ways in which to develop new graphical functionality: as a graphics function, purely for the side effect of producing output (see Section 8.2); and as a graphical object (Section 8.3). There has also been an emphasis on producing reusable graphical elements, a corollary of which is that existing graphical elements should be used where possible in the construction of new graphical elements.

There is no way to force other developers to create graphical objects rather than graphical functions, so it is necessary to be able to make use of both existing functions and existing objects whether constructing a new function or a new object.

In order to discuss each of the four possible situations (new functions from existing functions, new functions from existing grobs, new grobs from existing functions, and new grobs from existing grobs) the following paragraphs consider the simple case of drawing a "face," which consists of a rectangle for the border, two circles for eyes, and a line for the mouth (see Figure 8.22 for examples).

Defining a new graphics function is straightforward whether using existing graphics functions or existing graphical objects. Figure 8.23 defines two new graphical functions to draw a face. The function faceA() demonstrates the most straightforward case of a graphics function that includes calls to other graphics functions to produce output. The function faceB() shows a graphics function making use of existing graphical objects, which is done by just passing the result of the object constructor functions to the function grid.draw().

Developing a new graphical object can be a bit trickier, but there are several tools to help out. Figure 8.24 defines three functions for creating a new graphical object to represent a face. The function faceC() shows the simplest case, where a gTree is built from existing graphical objects, by just creating the appropriate objects as children of the gTree.

The functions faceD() and faceE() demonstrate the harder problem of creating a new graphical object using only existing graphics functions. In the case of faceD(), the output of the graphics functions is captured as a gTree using the grid.grabExpr() function. The faceE() function shows a different approach: it creates a grob with a special class, "face", and wraps the existing graphics functions in a drawDetails() method for that new class.

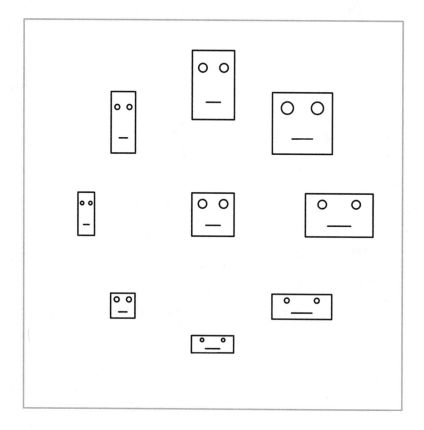

Figure 8.22
Drawing faces. Examples of the output that could be produced using the graphics functions and graphical objects defined in Figures 8.23 and 8.24.

```
 1 faceA <- function(x, y, width, height) {
 2   pushViewport(viewport(x=x, y=y,
 3                              width=width, height=height))
 4   grid.rect()
 5   grid.circle(x=c(0.25, 0.75), y=0.75, r=0.1)
 6   grid.lines(x=c(0.33, 0.67), y=0.25)
 7   popViewport()
 8 }

10 faceB <- function(x, y, width, height) {
11   pushViewport(viewport(x=x, y=y,
12                              width=width, height=height))
13   grid.draw(rectGrob())
14   grid.draw(circleGrob(x=c(0.25, 0.75), y=0.75, r=0.1))
15   grid.draw(linesGrob(x=c(0.33, 0.67), y=0.25))
16   popViewport()
17 }
```

Figure 8.23
Some face functions. Some different ways to implement a new graphics function to draw a "face." The function faceA() makes use of existing graphics functions. The function faceB() makes use of existing graphical objects.

```
 1 faceC <- function(x, y, width, height) {
 2   gTree(childrenvp=viewport(x=x, y=y,
 3                             width=width, height=height,
 4                             name="face"),
 5         children=gList(rectGrob(vp="face"),
 6                        circleGrob(x=c(0.25, 0.75),
 7                                   y=0.75, r=0.1, vp="face"),
 8                        linesGrob(x=c(0.33, 0.67), y=0.25,
 9                                  vp="face")))
10 }

12 faceD <- function(x, y, width, height) {
13   grid.grabExpr({
14                 pushViewport(viewport(x=x, y=y,
15                                       width=size,
16                                       height=size))
17                 grid.rect()
18                 grid.circle(x=c(0.25, 0.75),
19                             y=0.75, r=0.1)
20                 grid.lines(x=c(0.33, 0.67), y=0.25)
21                 popViewport()
22                 })
23 }

25 drawDetails.face <- function(x, recording) {
26   pushViewport(viewport(x=x$x, y=x$y,
27                         width=x$width, height=x$height))
28   grid.rect()
29   grid.circle(x=c(0.25, 0.75), y=0.75, r=0.1)
30   grid.lines(x=c(0.33, 0.67), y=0.25)
31   popViewport()
32 }

34 faceE <- function(x, y, width, height) {
35   grob(x=x, y=y, width=width, height=height, cl="face")
36 }
```

Figure 8.24
Some face objects. Some different ways to implement a new graphical object to represent a "face." The function faceC() makes use of existing graphical objects. The function faceD() makes use of existing graphics functions by capturing their output as a gTree. The function faceE() makes use of existing graphics functions by creating a new class of grob with a special drawDetails() method.

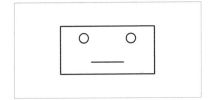

Figure 8.25

A simple drawing of a face on an otherwise blank page.

8.4   Debugging grid

When developing graphics functions or graphics objects with **grid**, it is very easy to get a bit lost. The only way to tell if code is doing the right thing is to view the graphical output that it produces. If the code is wrong, a common outcome to be faced with is a completely blank page.

This section describes some functions that are useful for querying the current state of the **grid** graphics system, which is useful for investigating the structure of the output that is produced by a function and for understanding what has gone wrong when a function has produced unexpected output.

These functions are also useful for exploring the **grid** objects that have been created by someone else's code, for example, a **ggplot2** or **lattice** plot.

As a simple example to demonstrate the use of these functions, the following code draws a simple face on an otherwise blank page (see Figure 8.25).

```
> grid.newpage()
> grid.draw(faceC(.5, .5, .5, .5))
```

The grid.ls() function can be used to list all graphical objects that have been drawn on a page. In this case, there is a single gTree, with a rect grob and a circle grob and a lines grob as its children.

```
> grid.ls()
```

```
GRID.gTree.232
   GRID.rect.233
   GRID.circle.234
   GRID.lines.235
```

The `grid.ls()` function can also be used to list the viewports that have been created on the page. The function works off the **grid** display list, so it shows every time a viewport is pushed and every time that a navigation down or back up the viewport tree occurs. In this case, a viewport called `face` has been pushed once (before the gTree drew any of its children) and then navigated down to three times (once for each time a child was drawn), with navigation back up occurring between each child.

```
> grid.ls(viewports=TRUE, grobs=FALSE)
```

```
ROOT
  face
   1
  face
   1
  face
   1
  face
   1
```

There is some flexibility in how the information is displayed. For example, the following code shows both viewports and grobs and prints full paths to both.

```
> grid.ls(viewports=TRUE, print=pathListing)
```

```
ROOT
ROOT            | GRID.gTree.240
ROOT::face
ROOT::face::1
ROOT::face
ROOT::face      | GRID.gTree.240::GRID.rect.241
ROOT::face::1
ROOT::face
ROOT::face      | GRID.gTree.240::GRID.circle.242
ROOT::face::1
ROOT::face
ROOT::face      | GRID.gTree.240::GRID.lines.243
ROOT::face::1
```

Furthermore, the result of the `grid.ls()` function is a list of information about the grobs and viewports on the current page, which can be used in

further processing, for example, to search for objects of a certain name or
objects of a certain type.

The `current.vpTree()` function provides an alternative way to list the viewports on the current page. One difference is that the result of this function is
an actual viewport tree object, not just the names of the viewports.

```
> current.vpTree()
```

viewport [ROOT] -> (viewport [face])

Finally, there is a `showViewport()` function that *draws* a representation of
the viewports on the current page. For example, the following code draws the
viewports that are produced by **lattice** when drawing a simple scatterplot
(see Figure 8.26; the actual lattice plot itself is the one in Figure 4.1).

```
> xyplot(pressure ~ temperature, pressure)
```

```
> showViewport(newpage=TRUE)
```

There are various options for controlling how viewports are shown and which
viewports are shown. For example, the following code only shows the "leaf"
viewports and shows them side-by-side rather than all overlapped (see Figure
8.27).

```
> showViewport(newpage=TRUE, leaves=TRUE)
```

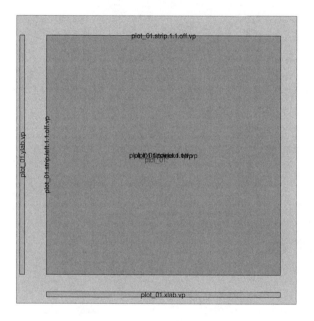

Figure 8.26

A diagram of the viewports that **lattice** creates when drawing a simple scatterplot.
Each viewport is represented by a rectangle and labeled with its name. Because
several of the viewports overlap, some of the labels are not legible.

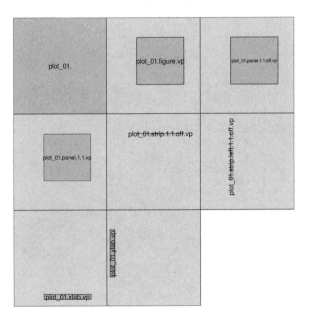

Figure 8.27
A diagram of the "leaf" viewports that **lattice** creates when drawing a simple scatterplot. Each viewport is drawn as a rectangle within a separate panel, labeled with its name. This is an alternative view to Figure 8.26.

Chapter summary

It is possible to write simple **grid** graphics functions for the purpose of producing graphical output. Such functions should not assume that they have the entire device to draw into. They should only assume that they are drawing within a **grid** viewport. Naming any viewports created in the function and using `upViewport()` rather than `popViewport()` makes it possible for others to annotate the graphical output produced by the function. Naming all grobs produced by the function makes it possible for others to edit the output from the function (or remove grobs or add grobs or extract grobs).

Creating a graphical object, either a grob or a gTree, to represent the output generated by the function requires extra effort to set up methods for the new graphical object class, but provides additional benefits. Most graphical objects will be gTrees consisting of a high-level description plus several child grobs representing the output produced. A gTree makes it possible for others to interact with the high-level description, while still being able to access the low-level element grobs. A grob can also be useful to provide information about the amount of space required to produce graphical output. Finally, a grob makes it possible for others to create higher-level gTrees with the grob as a child element.

Part III

THE GRAPHICS ENGINE

9

Graphics Formats

Chapter preview

This chapter describes how to produce graphical output in different formats. The output of graphics functions is typically drawn on screen initially, but this chapter describes how to save plots to files on disk. There is a discussion of the advantages and disadvantages of the various formats for different purposes. The same R code will sometimes produce slightly different output on different formats, so these differences are also described.

This part of the book is devoted to the core graphics engine in R, which is provided by the **grDevices** package. The information in this chapter and the next applies to almost all graphics functions and packages mentioned in this book.

The **grDevices** package is part of the standard R installation and is normally loaded by default in every R session. In a non-standard installation, it may be necessary to make the following call in order to access core graphics functions (if the **grDevices** package is already loaded, this will not do any harm).

```
> library(grDevices)
```

The graphics engine provides two main facilities for almost all graphics functions in R: support for producing output in different graphics formats, which is described in this chapter, and support for specifying values for graphical parameters, such as colors and fonts, which is described in Chapter 10.

9.1   Graphics devices

Throughout Parts I and II of this book, there have been vague statements about graphical output being drawn on a "page" or a "screen." This chapter addresses the issue of where graphical output appears and how it gets recorded.

In a typical interactive R session, a graphics window is automatically opened the first time that a graphics function is called and a plot is drawn on screen in this window. So for simple usage, there is no need for the user to decide where graphics output should go because there is a sensible default. However, for the purposes of producing a report, for example, in a PDF document, drawing a plot on screen is not very helpful. Instead, the plot needs to be saved in a PDF format, in a file on a hard disk. This section describes how to direct graphical output to a file rather than to the screen and how to specify the format of that file.

In R's terminology, graphical output is directed to a particular *graphics device*. In general, a graphics device must first be *opened*, then any subsequent calls to graphics functions produce output on that device. The dev.new() function opens the default device, as given by options("device"), but each device also has its own specific function. For example, the pdf() function opens a file and stores graphics output in a PDF format. A full list is given in the next section. For file-based devices, it is also important to *close* the device using the dev.off() function once all graphical output is complete.

The following code shows how to produce a simple scatterplot in PDF format. The output is stored in a file called myplot.pdf.

```
> pdf(file="myplot.pdf")
> plot(pressure)
> dev.off()
```

A simple modification of this pattern produces the same output in PNG format (in a file called myplot.png), as shown below.

```
> png(file="myplot.png")
> plot(pressure)
> dev.off()
```

It is possible to have more than one device open at the same time, but only one device is currently *active* and all graphics output is sent to that device.

If multiple devices are open, there are functions to control which device is active. The list of open devices can be obtained using `dev.list()`. This gives the name (the device format) and number for each open device. The function `dev.cur()` returns this information only for the currently active device. The `dev.set()` function can be used to make a device active, by specifying the appropriate device number and the functions `dev.next()` and `dev.prev()` can be used to make the next/previous device on the device list the active device.

The `dev.size()` function can be used to obtain the size of the current device, in either inches, centimeters, or pixels.

All open devices can be closed at once using the function `graphics.off()`. When an R session ends, all open devices are closed automatically.

9.2 Graphical output formats

Table 9.1 gives a full list of functions that open devices and the output formats that they correspond to.

All of these functions provide several arguments to allow the user to specify things such as the physical size of the window or document that is created.

Due to differences between graphics formats, it is very unlikely that the same R code will produce *identical* results on different devices. For example, a PDF version of a plot is unlikely to appear identical to a PNG version of the same plot.

Some of the distinct features of the various graphics formats are discussed further in the following sections.

9.2.1 Vector formats

Graphics devices can be divided into two main groups: *vector* formats and *raster* formats. In a vector format, an image is described by a set of mathematical shapes, for example, a line segment from one (x, y) location to another. In a raster format, an image consists of an array of *pixels*, with information such as color recorded for each pixel. The vector-format version for drawing a line segment might look something like the following, which involves just the end points of the line that should be drawn.

Table 9.1

Graphics formats that R supports and the functions that open an appropriate graphics device.

Function	Graphical Format
Screen Devices	
x11() or X11()	X Window window (Cairo graphics)
windows()	Microsoft Windows window
quartz()	MacOS X Quartz window
File Devices	
postscript()	Adobe PostScript file
pdf()	Adobe PDF file
svg()	SVG file (Linux and MacOS X only)
win.metafile()	Windows Metafile file (Windows only)
png()	PNG file
jpeg()	JPEG file
tiff()	TIFF file
bmp()	BMP file
pictex()	LaTeX PicTeX file
xfig()	xfig FIG file
bitmap()	Multiple formats via Ghostscript

```
2 2 moveto
8 6 lineto
```

By contrast, a raster format version of the same line might look like the following, which involves specifing which pixels should be drawn to show the line.

```
0 0 0 0 0 0 0 0 0 0
0 0 0 0 0 0 0 0 1 0
0 0 0 0 0 0 1 1 0 0
0 0 0 0 1 1 0 0 0 0
0 0 1 1 0 0 0 0 0 0
0 1 0 0 0 0 0 0 0 0
0 0 0 0 0 0 0 0 0 0
```

Vector formats include PDF, PostScript, and SVG. Examples of raster formats are PNG, JPEG, TIFF, and all screen devices.

The R graphics engine is fundamentally vector based, so R plots are produced very faithfully on vector-based devices. When producing output on a raster device, the quality of the result may be lower than for a vector device, but this can be ameliorated by using a higher resolution (more pixels) or by using a raster device that implements *anti-aliasing*, which helps to produce smoother lines.

In general, vector formats are superior for images that need to be viewed at a variety of scales, but raster formats will produce much smaller files if the image is very complex. For most purposes, a vector format is usually the best choice, but it is sometimes more sensible to use a raster format when a plot is visually complex, for example, if it involves a large number of data points.

It may sometimes be necessary to make further modifications to an R plot using third-party software. In such cases, another consideration is that certain modifications of an image, for example removing a particular shape, are only possible with a vector format. On the other hand, other modifications, such as making all white pixels in an image transparent, are only possible with raster formats. Because it is easy to convert a vector format to a raster version, while the reverse is very difficult if not impossible, it usually makes sense to produce a vector image from R if the image will be modified later.

PDF

PDF is a good choice of format, partly because of the widespread availability of viewing software such as Adobe Reader. It is also a very sophisticated format, so it is able to faithfully produce anything that R graphics can do.

The primary device for producing R plots in PDF format is the `pdf()` device.

The first argument is the name of the file to produce. By default, this will produce a single file, which can contain several pages of output. Section 9.5 describes how to produce a separate file for each page of output.

By default the `pdf()` device produces a seven-inch square document, but a custom physical size can be specified as the `width` and `height`, in inches. It is also possible to specify a standard paper size, e.g., `"a4"`, via the `paper` argument. However, this paper size is independent of the width and height of the actual plot *unless* the width and height are set to zero, in which case the plot expands to fit the paper size (minus 0.25 inch margins).

The default (sans-serif) font for the `pdf()` device is Helvetica, but a different default can be specified via the `family` argument. For example, `"serif"` uses a Times font, and `"mono"` produces Courier (see Section 10.4).

It is possible to use any Type 1 font in a plot with the `pdf()` device, but R must be told the location of *font metric* (`afm`) files for the font. The `pdfFonts()` function provides a list of fonts for which R already has font metrics. In order to use a font that is not on the list, it is necessary to describe the location of the font metric files using the `Type1Font()` function, then this information can be added to the *font database* using `pdfFonts()`.

In non-English locales, it may be necessary to specify an appropriate `encoding` for the file, although the `pdf()` function makes some attempt to automate this.

R also provides some support for locales with very large character sets, such as Chinese *hanzi*, Japanese *kanji*, and Korean *hanja*. For these cases, there are several predefined *CID-keyed* fonts, which are also included in the list produced by `pdfFonts()`. It is also possible to define new fonts via the `CIDFont()` function, but this does require some fairly detailed knowledge of the relevant font technology.

The `pdf()` device does *not* embed fonts within the PDF file. This is significant because PDF viewer software will *substitute* fonts if they are not embedded within a PDF file and they are not available on the system where the file is being viewed. If a non-standard font is used and font substitution occurs, the resulting plot may have missing characters or at best look quite untidy. This means that a plot should only use fonts that are known to be installed on the system where the plot is to be viewed (e.g., the default Helvetica, Times, or Courier fonts) *or* all fonts should be embedded within the PDF file using the `embedFonts()` function. In the latter case, all relevant fonts must be installed on the system that is used to create the PDF file.

In summary, any plot that makes use of the standard fonts should be fine, but any plot that makes use of more exotic fonts should call `embedFonts()` to make sure that the plot can be viewed or printed properly on any system.

When saving graphics that includes text in a PDF format, the default behavior is to use *kerning* to make small adjustments to the positioning of certain pairs of characters. For example, a lowercase 'a' beside an uppercase 'T' are placed closer together than a lowercase 'a' beside a lowercase 'o'. This facility is turned on or off via the useKerning argument.

Another special situation arises when drawing polygons that self-intersect. There are two main algorithms for determining the interior of such polygons: the *non-zero winding rule* and the *even-odd rule*. Unfortunately, the R graphics engine does not explicitly specify a *fill rule* for self-intersecting polygons, so the default is to use the non-zero winding rule. The fillOddEven argument can be used to change to the even-odd rule instead.

Another way to produce PDF output in R is to use the function that is based on the Cairo graphics library,* cairo_pdf() (on Linux or MacOS X systems). The advantage of this function is that it may provide better support for fonts, including automatic embedding of fonts, although this does depend on the installation of further software libraries.

PostScript

PostScript can be thought of as a predecessor of PDF. In some ways, PostScript is actually more sophisticated than PDF, but it does not support some of the more modern features such as semitransparent colors and hyperlinking. This means that PostScript output cannot faithfully produce everything that R graphics can do. Nevertheless, PostScript remains a very important format and is still the preferred format for some people.

The main way to produce PostScript output is using the postscript() device. This shares many features with the pdf() device as described above, including the ability to size the device, the use of Type 1 fonts, kerning, and polygon fill rules.

Device sizing is slightly different in that the paper setting is dominant over the width and height. For example, on an "a4" PostScript page, the plot will fill the page by default. The PostScript produced by R is compatible with *Encapsulated* PostScript (EPS), which is useful for including R plots within other documents (see Section 9.3), but to control the size of a plot it is necessary to specify paper="special" *as well as* an appropriate width and height. In this situation, it is usually also a good idea to specify a *portrait* orientation for the page via horizontal=FALSE. The setEPS() function is useful for setting up appropriate default settings for Encapsulated PostScript output.

---

*http://cairographics.org/.

Another difference between the PostScript device and the PDF device is that *all* fonts that are used in a PostScript plot must be "predeclared" via the `fonts` argument when the device is first opened.

One limitation with the `postscript()` device is that it does not support semi-transparency. Any attempt to draw a semitransparent color will fail with a warning. If PostScript is the required format, one avenue is to produce PDF and then convert to PostScript using third-party software such as ImageMagick.* Another option is to use the Cairo-based device `cairo_ps()` (on Linux and MacOS X systems). However, both of those options are likely to produce raster elements within the PostScript file, which means that the quality of the image may be reduced.

SVG

SVG is a format with tremendous potential because it offers an open standard vector format, as sophisticated as the PDF format, that can be embedded in web pages. Support for SVG in popular web browsers is improving and should support most static graphics that are produced in R.

On Linux systems and MacOS X, SVG output is available via the `svg()` device. On Windows, SVG output requires one of the extension packages (see Section 9.7).

Because of the limitations of the R graphics engine, it is not possible to take advantage of more advanced SVG features, such as compositing operators and animation, though some extension packages provide access to some of these features (see Section 9.7 and Section 17.4).

Windows Metafile

The Windows Metafile format is important because it is the vector format that should be most compatible with Microsoft products such as Word, Excel, and PowerPoint. This format can only be produced on Windows systems.

9.2.2 Raster formats

The raster device that users will encounter most often is the graphics window on screen. This is the quickest and simplest way to view graphical output. Screen devices are different on different operating systems: typically, a Cairo-based X Window device on Linux, a Quartz device on MacOS X, and a native

---

*`http://www.imagemagick.org/`.

Windows device on Windows. There are some differences between these devices (see Section 9.4), so R code is unlikely to produce identical results on different platforms. On Linux and MacOS X there is also an X Window device, which lacks support for some graphics features, but is faster than the Cairo-based device.

When saving graphics to a file, there are several raster formats to choose from. The PNG format is desirable because it is *lossless*, which means that it compresses the image (most raster formats compress the image to save space) in such a way that no information is lost. This means that a PNG file can be edited without reducing the quality. The JPEG format, by comparison, uses *lossy* compression so, although JPEG files will typically be smaller than PNG files, repeatedly editing a JPEG will result in a reduction in quality. Furthermore, the JPEG compression is better suited to complex images with lots of different regions (like photgraphs), whereas the PNG format does a better job with simpler images that include lines and text and large areas of constant color. Consequently, the PNG format is usually better for statistical plots, though an exception might be a very busy `image()` plot or `contour()` plot.

The JPEG format does not support semitransparency. The PNG format does, but this is only partially supported on Windows, and via the default Cairo-based devices on Linux and MacOS X.

Neither PNG nor JPEG formats support multiple pages in a document, so if a `png()` device is opened and then more than one page of output is produced, the result will be several PNG files rather than just one (by default, the file names are automatically numbered).

TIFF is a very sophisticated format that allows multiple pages of raster output within a single file. It is less well supported by web browsers, but may be the preferred format for publishers of books or journal articles.

Determining the size of a raster image is less straightforward than it is for vector formats. The `width` and `height` of a raster device are specified as a number of pixels rather than as a physical size in inches. The physical size of a raster image is then determined by the *resolution* at which it is viewed. For example, a PNG image that is 72 pixels wide will be 1 inch wide when viewed on a screen with a resolution of 72 dpi (dots per inch), but it will be only 0.75 inches wide on a screen with a resolution of 96 dpi.

It is possible to specify a fixed resolution for a raster format image via the `res` argument. However, this information will not necessarily be respected when the image is displayed. For example, web browsers tend to just use the resolution of the screen when displaying images on web pages (so the image size will still vary depending on the screen resolution). On the other hand, if a raster image is included within a LaTeX document, the resolution of the image is respected

As a general rule of thumb, if a raster image is being prepared for use on a web page, there is no point in worrying about setting the resolution, but if a raster image is being prepared for inclusion in a document that is to be typeset, such as a LaTeX or Microsoft Word document, then setting the resolution may be worthwhile, particularly if a high-quality image is required.

Because the physical size of a raster image can be ambiguous, it can be difficult to control the size of text in a raster image. The `pointsize` argument specifies the default size of text for an image, but what this means is again dependent on the resolution at which the image is displayed. The size of text is given in *big points* ($\frac{1}{72}$ inch), *relative to the* `res` *argument*. This means that the size of text is calculated as if the resolution of the image is going to be respected. The result should be as expected when a raster image is included in another document, but the result can be confusing if the image is displayed at screen resolution (e.g., when the image is used on a web page).

In summary, the physical size of text in a plot depends on the size of the text, the size of the image, the resolution of the image, and whether the image is displayed at screen resolution or at the native resolution of the image.

9.3   Including R graphics in other documents

There are two typical uses of R graphics. One is to produce basic plots on screen for exploratory data analysis, to look for patterns in the data, and the other is to produce finely tuned plots in a file format for inclusion in a larger document such as a web page or a printed report. This section deals with some issues specifically related to the latter task.

One important issue to consider is the physical size of text and the physical width of lines in a plot within the final document. Text has to be readable and lines typically need to be wide enough for print resolution so that, for example, they do not disappear when photocopied.

The default, for vector formats, is to produce a seven-inch square document, using a 12-point, sans-serif font, with lines $\frac{1}{96}$ inches wide. This is fine for viewing a plot on its own, but is much too large for a typical document, for example, when including a plot in a figure within an A4 page.

The best approach is to produce the plot at the size that it needs to be in the final document and specify the appropriate font size and line width explicitly.

9.3.1 LaTeX

Standarad vector formats such as PDF and PostScript are ideal for including within LaTeX documents. However, there is one situation where a more LaTeX-specific option may be more desirable.

One thing that LaTeX does exceptionally well is the typesetting of mathematical formulae. R's mathematical annotation facility attempts to emulate LaTeX, but it is not as good as the real thing, particularly when the fonts involved are not the TeX math fonts.

There is a special `cmsyase` font that can be used to draw mathematical formulae in R with TeX math fonts. This is available from the following web site: `http://www.stat.auckland.ac.nz/~paul/R/CM/CMR.html`.

One way to produce graphics output specifically for inclusion in a LaTeX document is to use the `pictex()` device. This produces LaTeX macros from the PiCTeX package to draw a plot. The main advantage of this is that the text in the plot will use the same font as the rest of the LaTeX document. Unfortunately, this device is very rudimentary, so is not suitable for anything other than very basic plots (it does not even support colors). See Section 9.7 for a more sophisticated alternative.

9.3.2 "Productivity" software

Microsoft software products have a tendency to play nicely with each other and with Microsoft formats, but less well with other software products and formats. This is particularly true for vector graphics formats, so possibly the best vector format for including plots in Microsoft products, such as Word and Excel, is the WMF format (Windows Meta-File). Microsoft products should cope well with the standard raster formats, though there is also the Windows-specific BMP format.

The Open Office software has better support for including PDF plots in documents and will also cope with standard raster formats.

9.3.3 Web pages

The standard way to include an image in a web page has been to use a raster format, such as PNG. With the improving support for SVG in web browsers, that format is becoming a viable vector alternative. See Section 17.4 for further discussion of this topic.

9.4　Device-specific features

Not all graphics devices are created equal. The same R code can produce slightly different graphical output depending on the graphics device format.

While the performance of vector devices should be quite consistent on all platforms (Windows, Linux, MacOS X), the performance of raster devices is much more platform dependent. On the other hand, for a specific platform, plots saved in a raster format should have the same appearance as they do on-screen.

One area where differences can become evident is in the selection of fonts. The standard set of fonts, as described in Section 10.4, should always be available, though there will be small differences in appearance on different platforms (e.g., the default `"sans"` font is Arial on Windows and Helvetica on Linux). The method for selecting fonts beyond the standard set is different on different platforms. For example, there is the `windowsFonts` function on Windows, but with Cairo-based devices on Linux and MacOS X, fonts can be specified simply by a font family name. In any case, the availability of fonts will be dependent on which fonts have been installed.

On some devices, the font size that is specified will not be honored exactly. For example, when drawing in a raw X Window window with bitmap fonts, there are only a finite set of font sizes available and this set will vary depending on which fonts are installed. For the PostScript and PDF formats, font sizes should scale appropriately to any size.

Antialiasing can dramatically improve the quality of a raster image by smoothing the appearance of lines and text. This is available by default on Linux and MacOS X, but not on Windows. If the purpose is to include a raster image in another document, then generating a high-resolution image is another way to improve quality.

The Windows screen device has less-complete support for semitransparent colors, compared to the default screen device on Linux and MacOS X.

On Linux and MacOS X, where the default screen and raster devices are Cairo based, it is also possible to produce screen output and raster formats directly via the X Window system. This typically produces a poorer quality image, for example, there is no support for semitransparent colors or antialiasing, but the rendering is faster so this option could be considered for particularly complex images.

An alternative way to produce raster format images that should produce more consistent results across platforms is to use the `bitmap()` function. The downside is that this requires the installation of additional software (Ghostscript). Section 9.7 describes some other possibilities for producing consistency across platforms.

9.5 Multiple pages of output

For a screen device, starting a new page involves clearing the window before producing more output. On Windows there is a facility for returning to previous screens of output (see the "History" menu, which is available when a graphics window has focus), but on most screen devices, the output of previous pages is lost.

If a piece of code produces several pages of plots, the `devAskNewPage()` function can be used to force a user prompt before each new page is started. This allows the user to view each page at leisure before indicating to R to move on to the next page.

For file devices, the output format dictates whether multiple pages are supported. For example, PostScript and PDF allow multiple pages, but PNG does not. It is usually possible, especially for devices that do not support multiple pages of output, to specify that each page of output produces a separate file. This is achieved by specifying the argument `onefile=FALSE` when opening a device and specifying a pattern for the file name like `file="myplot%03d"` so that the `%03d` is replaced by a three-digit number (padded with zeroes) indicating the "page number" for each file that is created.

9.6 Display lists

R maintains a *display list* for each open device, which is a record of the output on the current page of a device. This is used to redraw the output when a device is resized and can also be used to copy output from one device to another.

The function `dev.copy()` copies all output from the active device to another device. The copy may be distorted if the aspect ratio of the destination device — the ratio of the physical height and width of the device — is not the same as the aspect ratio of the active device. The function `dev.copy2eps()` is similar to `dev.copy()`, but it preserves the aspect ratio of the copy and creates a file in EPS (Encapsulated PostScript) format that is ideal for embedding in other documents (e.g., a LaTeX document). The `dev2bitmap()` function is similar in that it also tries to preserve the aspect ratio of the image, but it produces one of the output formats available via the `bitmap()` device.

The function `dev.print()` attempts to print the output on the active device. By default, this involves making a PostScript copy and then invoking the print command given by `options("printcmd")`.

The display list can consume a reasonable amount of memory if a plot is particularly complex or if there are very many devices open at the same time. For this reason it is possible to disable the display list, by typing the expression `dev.control(displaylist="inhibit")`. If the display list is disabled, output will not be redrawn when a device is resized, and output cannot be copied between devices.

There is also a `recordPlot()` function, which saves the display list to an R variable. The variable can then be passed to the `replayPlot()` function to draw the saved plot.

9.7   Extension packages

Several extension packages for R provide a number of extra graphical formats that are not provided by the **grDevices** package itself. In general, these work just like the core devices, with a function provided to open a device in the appropriate format. Additional functions may be provided for handling other features of the device, such as fonts. Table 9.2 lists some of the extension packages that provide graphics devices.

The usefulness of the **Cairo** package is that it allows Cairo-based graphics output on any platform (although it requires the Cairo graphics library to be installed first). This has two advantages: the output on a Cairo-based screen device should be very similar on all platforms and (Cairo-based) SVG output becomes available on Windows.

The **cairoDevice** package is similar, but with a focus on integration with the GTK+ GUI toolkit system (so it also requires the GTK+ libraries to be installed).

The **tikzDevice** package provides a sophisticated solution for producing graph-

Table 9.2

Graphics formats that are provided by extension packages for R and the functions that open an appropriate graphics device.

Function	Graphical Format	Package
Cairo()	Multiple formats	**Cairo**
tikz()	LaTeX PGF/TikZ file	**tikzDevice**
devSVGTips()	SVG file	**RSVGTipsDevice**
JavaGD()	Java Swing window	**JavaGD**

ical output for inclusion in LaTeX documents. The main advantages are that the fonts for text in the plot will match the fonts used in the LaTeX document and LaTeX's native mathematical formula syntax can be used for text in plots.

The **RSVGTipsDevice** package provides an alternative way to produce SVG output, with the advantage of allowing tooltips and hyperlinks to be added to the SVG file. See Section 17.4 for some other packages that take this idea further.

The **JavaGD** package allows graphical output to be included as part of a Java GUI. The **tkrplot** provides a similar facility for including R graphics in a tcltk GUI. See Section 17.3.2 for more discussion of GUI-related packages.

There is also a **canvas** package in development that produces javascript code for drawing an R plot in an HTML 5 canvas element.

For the more adventurous and developer-minded, the Omegahat Project provides an **RGraphicsDevice** package for building new R graphics devices, plus a **FlashMXML** package for generating R graphics as Adobe Flash files, among others. Another important package still being developed at the time of writing is the **Acinonyx** package. This provides an idev() function which creates a graphics device that can draw graphics *very* much faster than normal R graphics devices.

There are also several packages that do *not* produce their graphical output on R graphics devices. Instead, they create and manage their own graphics windows. Examples are the **rgl** package (see Section 16.6) and **rggobi** (see Section 17.2.2).

Chapter summary

R graphics can produce a wide variety of graphical formats. In interactive use, graphics output is drawn on screen, but it is also possible
to save graphics output in a file. A vector graphics format usually produces a better-quality result than a raster format when saving plots
to a file, but the choice of format will also depend on how the plot
will be used (e.g., included in a LaTeX document versus distributed as
part of a web page). Several extension packages provide support for
additional graphics formats.

10

Graphical Parameters

Chapter preview

This chapter describes how to specify graphical parameters, including information about specifying a single color, how to generate sets of coherent colors, information about how to specify fonts for drawing text, and information about how to produce special symbols and formatting for drawing mathematical formulae. The information in this chapter is useful for controlling the output of almost all graphics functions in R.

Graphical parameters are the arguments to functions that influence the detailed appearance of a graphical image. They apply the make-up to the basic bone structure of an image. Examples include the color and line width used to draw a line and the font used to draw text.

Despite the fact that the R graphics universe consists of two distinct graphics systems, traditional and **grid**, plus several other stand-alone systems (see, for example, Sections 17.2.2 and 16.6), the way that graphical parameters are specified is quite consistent across all of these systems.

10.1 Colors

The easiest way to specify a color in R is simply to use the color's name. For example, `"red"` can be used to specify that graphical output should be (a very bright) red. R understands a fairly large set of color names; type `colors()` (or `colours()`) to see a full list of known names.

It is also possible to specify colors using one of the standard color space descriptions. For example, the `rgb()` function allows a color to be specified as a Red-Green-Blue (RGB) triplet of intensities. Using this function, the color red is specified as `rgb(1, 0, 0)`. The function `col2rgb()` can be used to see the RGB values for a particular color name (although the resulting color channels are in the range 0 to 255 rather than 0 to 1).

```
> col2rgb("red")
```

```
       [,1]
red     255
green    0
blue     0
```

An alternative way to provide an RGB color specification is to provide a string of the form `"#RRGGBB"`, where each of the pairs RR, GG, BB consist of two hexadecimal digits giving a value in the range zero (00) to 255 (FF). In this specification, the color red is given as `"#FF0000"`.

In R, RGB color specifications are interpreted relative to the sRGB color space (IEC standard 61966).[*]

There is also an `hsv()` function for specifying a color as a Hue-Saturation-Value (HSV) triplet. The terminology of color spaces is fraught, but roughly speaking: *hue* corresponds to a position on the rainbow, from red (0), through orange, yellow, green, blue, indigo, to violet (1); *saturation* determines whether the color is dull (grayish) or bright (colorful); and *value* determines whether the color is light or dark. The HSV specification for the (very bright) color red is `hsv(0, 1, 1)`. The function `rgb2hsv()` converts a color specification from RGB to HSV.

```
> rgb2hsv(255, 0, 0)
```

```
  [,1]
h   0
s   1
v   1
```

---

[*]`http://www.color.org/chardata/rgb/srgb.xalter`.

A better alternative to either `rgb()` or `hsv()` is the `hcl()` function. Similar to `hsv()`, this function specifies colors as a hue, a *chroma* (or colorfulness, similar to saturation), and a *luminance* (or lightness, similar to value). The color `"red"` corresponds to `hcl(12, 179, 53)`.

The `hcl()` function is better than the `hsv()` function because it works in the CIE-LUV color space, in which a unit distance is close to a perceptually constant change in color, so, for example, holding chroma and luminance constant while varying only hue produces colors that are approximately similar in their visual impact on the observer.

Greyscale colors can be generated using the function `gray()` (or `gray()`). These functions take a vector of numeric values between 0 (black) and 1 (white).

One final way to specify a color is simply as an integer index into a predefined set of colors. The predefined set of colors can be viewed and modified using the `palette()` function. In the default palette, red is specified as the integer 2.

10.1.1 Semitransparent colors

All R colors are stored with an alpha transparency channel. An alpha value of 0 means fully transparent and an alpha value of 1 means fully opaque. When an alpha value is not specified, the color is opaque.

The function `rgb()` can be used to specify a color with an alpha transparency channel, simply by providing a fourth value to the function. For example, `rgb(1, 0, 0, 0.5)` specifies a semitransparent red. Alternatively, a color can be specified as a string beginning with a `"#"` and followed by *eight* hexadecimal digits. In that case, the last two hexadecimal digits specify an alpha value in the range 0 to 255. For example, `"#FF000080"` specifies a semitransparent red.

A color may also be specified as `NA`, which is usually interpreted as fully transparent (i.e., nothing is drawn). The special color name `"transparent"` can also be used to specify full transparency.

WARNING: If a graphic device does not support semitransparency, semitransparent colors are rendered as fully transparent.

10.1.2   Converting colors

There are many other ways to specify colors besides the RGB, HSV, and HCL color spaces described so far and the convertColor() function provides a mechanism for converting between different color spaces.

The following code shows an example where the color "red" is converted to the CIE-LUV color space. This can be a useful transformation because the L component of the result can be used to convert color to grayscale. The col2rgb() function is used to obtain a matrix containing the separate red, green, and blue components, those are normalized to a zero-to-one range by dividing by 255, and then the matrix is transposed so that the components are different columns. The transformation is from R's native color space, sRGB, to CIE-LUV.

```
> convertColor(t(col2rgb("red")/255), "sRGB", "Luv")
```

```
           L         u        v
[1,]  53.48418  175.3647  37.80017
```

The L component of the result corresponds to the values given for the hcl() specification of "red" on page 323. The u and v components do not correspond to the h and c components of the hcl() example because the hcl() function works in polar coordinates, whereas u and v are cartesian dimensions within the CIE-LUV color space.

Another useful tool is the adjustcolor() function, which allows the components of an existing color to be scaled. For example, the following code takes the color "red" and makes it semitransparent.

```
> adjustcolor("red", alpha.f=.5)
```

```
[1]  "#FF000080"
```

This result corresponds to the explicit color specification for semitransparent red that was given above.

Section 11.3 describes functions from add-on packages that provide further color conversions.

10.1.3   Color sets

More than one color is often required within a single plot, for example to distinguish between different groups of data symbols, and in such cases it

Table 10.1

Functions to generate color sets. R functions that can be used to generate coherent sets of colors.

Name	Description
`rainbow()`	Colors vary from red through orange, yellow, green, blue, and indigo, to violet.
`heat.colors()`	Colors vary from white, through orange, to red.
`terrain.colors()`	Colors vary from white, through brown, to green.
`topo.colors()`	Colors vary from white, through brown then green, to blue.
`cm.colors()`	Colors vary from light blue, through white, to light magenta.
`gray.colors()`	A set of shades of gray.

can be difficult to select colors that are aesthetically pleasing or are related in some way (e.g., a set of colors in which the brightness of the colors decreases in regular steps). Table 10.1 lists some functions that R provides for generating sets of colors. Each of these functions takes a single numeric argument and returns that number of colors. For example, the following code produces five colors from the `rainbow()` function.

```
> rainbow(5)
```

```
[1] "#FF0000FF" "#CCFF00FF" "#00FF66FF" "#0066FFFF"
[5] "#CC00FFFF"
```

The output of the expression `example(rainbow)` provides a nice visual summary of the color sets generated by several of these functions.

Each of the functions in Table 10.1 (apart from `gray.colors()`) selects a set of colors by taking regular steps along a path through the HSV color space. As mentioned previously, a more perceptually uniform set of colors can be obtained by working in the CIE-LUV color space. For example, the following code generates six colors from the CIE-LUV color space that vary regularly in terms of hue, but are all equally bright (the chroma component is fixed at 50) and all equally light (the luminance component is fixed at 60).

```
> hcl(seq(0, 300, 60), 50, 60)
```

```
[1] "#C87A8A" "#AC8C4E" "#6B9D59" "#00A396" "#5F96C2"
[6] "#B37EBE"
```

Section 11.3 describes functions from add-on packages that provide further tools for generating sets of colors, including functions that work in the CIE-LUV color space.

The functions `colorRamp()` and `colorRampPalette()` are a little different because they are not color set generators. Instead, they are color set *function* generators. These functions accept a set of colors and color space to work in and they interpolate a path through the color space (either joining the starting colors with straight lines or interpolating a smooth curve through the colors), then they return a function that can be called to select colors from the interpolated path.

One difference between the functions is that `colorRamp()` produces a function that can generate colors based on a sequence of values in the range 0 to 1, like `gray.colors()`, whereas `colorRampPalette()` produces a function that can generate *n* colors, like `rainbow()`.

Another difference between the functions is that `colorRamp()` returns a matrix of red, green, and blue color components, whereas `colorRampPalette()` returns a vector of colors.

The following code demonstrates `colorRampPalette()` being used to create a color set generating function that produces colors ranging from `"blue"` to `"gray"`. The function is then used to generate five colors.

```
> bluegray <- colorRampPalette(c("blue", "gray"))
> bluegray(5)
```

```
[1] "#0000FF" "#2F2FEE" "#5F5FDE" "#8E8ECE" "#BEBEBE"
```

10.1.4   Device Dependency of Color Specifications

The colors that R sends to a graphics device are sRGB colors. This should be appropriate for drawing to a screen device because most computer monitors are set up to work with sRGB. Also, colors used on web pages are typically sRGB, so raster file formats produced by R, such as PNG, should work reasonably well there too.

However, the final appearance of a color can vary considerably when it is viewed on a screen, or printed on paper, or displayed through a projector as it depends on the physical characteristics of the screen, printer ink, or projector. When an image is saved in a PDF or PostScript format, R records the fact that sRGB colors are being used so printers and viewers have some chance of producing the right result.

10.2   Line styles

It is possible to control the width of a line, the pattern used to draw the line (e.g., solid versus dashed), and the styling used for the ends and corners of a line.

10.2.1   Line widths

The width of lines is specified by a simple numeric value, e.g., lwd=3. This value is a multiple of 1/96 inch, with a lower limit of 1 pixel on some screen devices. The default value is 1.

10.2.2   Line types

R graphics supports a fixed set of predefined line types, which can be specified by name, such as "solid" or "dashed", or as an integer index (see Figure 10.1). In addition, it is possible to specify customized line types via a string of digits. In this case, each digit is a hexadecimal value that indicates a number of "units" to draw either a line or a gap. Odd digits specify line lengths and even digits specify gap lengths. For example, a dotted line is specified by lty="13", which means draw a line of length one unit then a gap of length three units. A unit corresponds to the current line width, so the result scales with line width, but is device dependent. Up to four such line-gap pairs can be specified. Figure 10.1 shows the available predefined line types and some examples of customized line types.

10.2.3   Line ends and joins

When drawing thick lines, it becomes important to select the style that is used to draw corners (joins) in the line and the style that is used to draw the ends of the line. R provides three styles for both cases: there is an lend

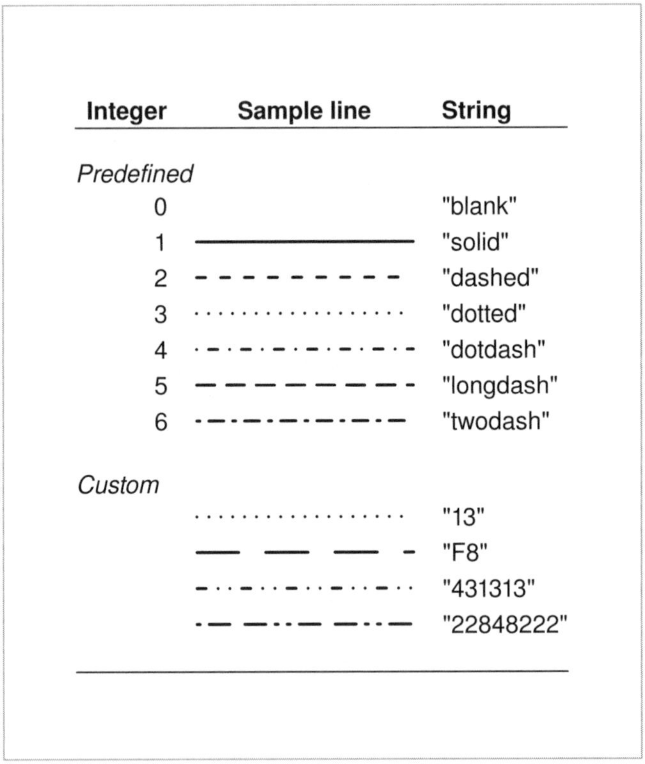

Integer	Sample line	String
Predefined		
0		"blank"
1	——————————	"solid"
2	- - - - - - - - -	"dashed"
3	· · · · · · · · · · · · ·	"dotted"
4	· — · — · — · — · ·	"dotdash"
5	— — — — — — — -	"longdash"
6	· —· —· —· —· —	"twodash"
Custom		
	· · · · · · · · · · · · ·	"13"
	— — — -	"F8"
	- · · - · · - · · - · ·	"431313"
	· — —· · — —· · —	"22848222"

Figure 10.1

Predefined and custom line types. Line type may be specified as a predefined integer, as a predefined string name, or as a string of hexadecimal characters specifying a custom line type.

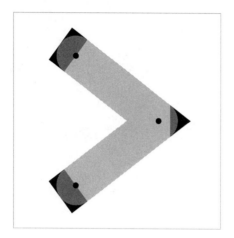

Figure 10.2
Line join and line ending styles. Three thick lines have been drawn through the same three points (indicated by black circles), but with different line end and line join styles. The black line was drawn first with `"square"` ends and `"mitre"` joins; the dark gray line was drawn on top of the black line with `"round"` ends and `"round"` joins; and the light gray line was drawn on top of that with `"butt"` ends and `"bevel"` joins.

setting to control line ends, which can be `"round"` or flat (with two variations on flat, `"square"` or `"butt"`); and there is an `ljoin` setting to control line joins, which can be `"mitre"` (pointy), `"round"`, or `"bevel"`. The differences are most easily demonstrated visually (see Figure 10.2).

When the line join style is `"mitre"`, the join style will automatically be converted to `"bevel"` if the angle at the join is too small. This is to avoid excessively pointy joins. The point at which the automatic conversion occurs is controlled by a miter limit, which specifies the ratio of the length of the miter divided by the line width. The default value is `10`, which means that the conversion occurs for joins where the angle is less than 11 degrees. Other standard values are `2`, which means that conversion occurs at angles less than 60 degrees, and `1.414`, which means that conversion occurs for angles less than 90 degrees. The minimum miter limit value is `1`.

It is important to remember that line join styles influence the corners on rectangles and polygons as well as joins in lines.

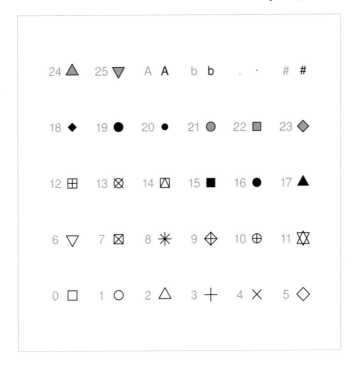

Figure 10.3
Data symbols available in R. A particular data symbol is selected by specifying an integer between 0 and 25 or a single character. In the diagram, the relevant integer or character value is shown in gray to the left of the relevant symbol.

10.3   Data symbols

The data symbol used for plotting points is specified as either an integer, which indexes one of 26 predefined data symbols (see Figure 10.3), or directly as a single character. Some of the predefined data symbols (pch between 21 and 25) allow a fill color separate from the border color.

If pch is a character then that letter is used as the plotting symbol. The character "." is treated as a special case and the device attempts to draw a very small dot.

10.4 Fonts

Specifying a font for drawing text consists of specifing a font *family*, such as Helvetica or Courier, and specifying a font *face*, such as **bold** or *italic*.

10.4.1 Font family

Every graphics device establishes a default font family, which is usually a sans-serif font such as Helvetica or Arial. A new font family can be specified using a device-independent name: `"sans"` gives a sans-serif font, like Arial; `"serif"` gives a serif font, like Times; and `"mono"` gives a monospace font, like Courier (see Table 10.2).

The device-independent font name is mapped to a device-dependent font family by individual devices. These mappings can be modified and new font names and mappings defined using functions such as `Type1Font()` to create a font description and `pdfFonts()` to register the font description with a device. On Windows, the corresponding functions are `windowsFont()` and `windowsFonts()`. The information required to define a new font is very device dependent.

The Hershey outline fonts are also distributed with R and are available for *all* output formats. The names to use with the `family` setting to obtain the different Hershey fonts are shown in Table 10.2. See the on-line help page for `Hershey` for more information on Hershey fonts.

10.4.2 Font face

The font face is usually specified as an integer value between 1 and 4. Table 10.3 shows the mapping from numbers to font faces.

The **grid** graphics system also allows the font face to be specified by name (see Table 6.4).

Table 10.2
Device-independent and Hershey font families that are distributed with R. A font family is specified as a character value.

Name	Description
Device-independent fonts	
`"serif"`	Serif variable-width font
`"sans"`	Sans-serif variable-width font
`"mono"`	Mono-spaced "typewriter" font
Hershey fonts	
`"HersheySerif"`	Serif variable-width font
`"HersheySans"`	Sans-serif variable-width font
`"HersheyScript"`	Serif "handwriting" font
`"HersheyGothicEnglish"`	Gothic script font
`"HersheyGothicGerman"`	Gothic script font
`"HersheyGothicItalian"`	Gothic script font
`"HersheySymbol"`	Serif symbol font
`"HersheySansSymbol"`	Sans-serif symbol font

Table 10.3
Possible integer font face specifications and their meanings. See Table 6.4 for font face *name* specifications. The range of valid font faces varies for different Hershey fonts, but the maximum valid value is usually 4 or less. When the font family is `"HersheySerif"`, there are a number of special font faces available.

Integer	Description
1	Roman or upright face
2	Bold face
3	Slanted or italic face
4	Bold and slanted face
5	Symbol
For the HersheySerif font family	
5	Cyrillic font
6	Slanted Cyrillic font
7	Japanese characters

10.4.3 Multi-line text

It is possible to draw text that spans several lines by inserting a new line escape sequence, `"\n"`, within a piece of text, as in the following example.

```
"first line\n second line"
```

Alternatively, simply entering a character value across several lines will produce the same result, as shown below.

```
> "first line
    second line"
```

[1] "first line\n second line"

Vertical separation can be controlled via a line height parameter, which acts as a multiplier (2 means double-spaced text).

10.4.4 Locales

R supports multibyte locales, such as UTF-8 locales and East Asian locales (Chinese, Japanese, and Korean), which means that it is possible to enter multibyte character values. There may be problems including such characters as part of graphical output on some devices. For example, Type 1 fonts on PostScript and PDF devices only work with single-byte character encodings, so an appropriate encoding may need to be specified in order to produce special characters on those devices.

10.5 Mathematical formulae

This section does not concern a graphical parameter, but it does provide important information about how to specify character values for drawing text.

Any R graphics function that draws text should accept both a normal character value, e.g., `"some text"`, and an R expression, which is typically the result of a call to the `expression()` function. If an expression is specified as the text to draw, then it is interpreted as a mathematical formula and is formatted appropriately. This section provides some simple examples of what

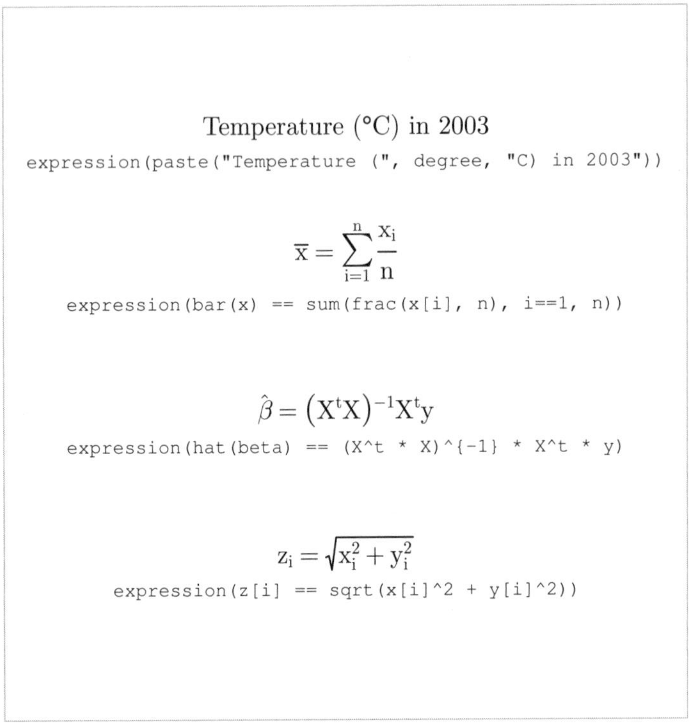

Temperature (°C) in 2003

```
expression(paste("Temperature (", degree, "C) in 2003"))
```

$$\overline{x} = \sum_{i=1}^{n} \frac{x_i}{n}$$

```
expression(bar(x) == sum(frac(x[i], n), i==1, n))
```

$$\hat{\beta} = \left(X^t X\right)^{-1} X^t y$$

```
expression(hat(beta) == (X^t * X)^{-1} * X^t * y)
```

$$z_i = \sqrt{x_i^2 + y_i^2}$$

```
expression(z[i] == sqrt(x[i]^2 + y[i]^2))
```

Figure 10.4
Mathematical formulae in plots. For each example, the output is shown in a serif font, and below that, in a typewriter font, is the R expression required to produce the output.

can be achieved. For a complete description of the available features, type `help(plotmath)` or `demo(plotmath)` in an R session.

When an R expression is provided as text to draw in graphical output, the expression is evaluated to produce a mathematical formula. This evaluation is very different from the normal evaluation of R expressions: certain names are interpreted as special mathematical symbols, e.g., `alpha` is interpreted as the Greek symbol α; certain mathematical operators are interpreted as literal symbols, e.g., a + is interpreted as a plus sign symbol; and certain functions are interpreted as mathematical operators, e.g., `sum(x, i==1, n)` is interpreted as $\sum_{i=1}^{n} x$. Figure 10.4 shows some examples of expressions and the output that they create.

In some situations, for example, when calling graphics functions from within a loop, or when calling graphics functions from within another function, the expression representing the mathematical formula must be constructed using

values within variables as well as literal symbols and constants. A variable
name within an expression will be treated as a literal symbol (i.e., the variable
name will be drawn, not the value within the variable). The solution in such
cases is to use the `substitute()` function to produce an expression. The
following code shows the use of `substitute()` to produce a label where the
year is stored in a variable.

```
> myfunction <- function(year) {
    text(0.5, 0.5, substitute(paste("Temperature (",
                                    degree, "C) in ", year),
                              list(year=year)))
  }
```

The mathematical annotation feature makes use of information about the
dimensions of individual characters to perform the formatting of the formula.
For some output formats, such information is not available, so mathematical
formulae cannot be produced.

Chapter summary

There are standard ways to specify colors, fonts, line types, and text
for virtually all graphics functions in R. There are many functions
for generating sets of colors. The CIE-LUV color space provides a
sensible foundation for generating colors in a rational fashion. Text
can be specified as an R expression, which makes it possible to draw
special characters and to produce special formatting for mathematical
formulae.

Part IV

GRAPHICS PACKAGES

11

Graphics Extensions

Chapter preview

Ths chapter describes functions from extension packages that provide additional low-level utilities for R graphics. There are sections on drawing basic shapes, labeling points and lines, generating color sets, producing fill patterns, and manipulating coordinate systems.

This part of the book is devoted to a number of graphical topics where the core statistical graphics packages in R are extended in useful ways.

This chapter describes functions from a number of packages that build useful low-level extensions on top of the core facilities of the traditional graphics system and the **grid** graphics system.

This chapter focuses mainly on *low-level* extensions of the core graphics systems: functions that provide new graphical primitives or more convenient ways to work with existing graphical primitives. Each section will focus on a certain useful graphical task and describe some functions that are available to assist with that task.

11.1 Tricks with text

This section describes functions that relate to drawing text on a graphics device.

The core graphics facilities for drawing text provide a good amount of control over the placement of an individual piece of text, but they do not provide much support for placing multiple pieces of text relative to each other or relative to other graphical output. Two important situations are common: laying out text in a strict, regular arrangement, such as a table; and laying out text in a flexible, fluid arrangement, such as labels on data points.

11.1.1 Drawing formatted text on a plot

The `textplot()` function from the **gplots** package draws a character vector as text on a graphics device.

```
> library(gplots)
```

It is designed for drawing regular R output, what R normally prints at the command line as the result of an expression. The function acts like a high-level plotting function, starting a new plot to hold the text and, by default, it automatically sizes the text to fill the width of the new plot region, so one useful application is to draw R output alongside a plot as part of a multifigure page. The function `sinkplot()` from the same package provides a convenient higher-level interface for capturing and drawing R output. The main advantage of these functions over `text()` is that they automatically draw different elements of a character vector on separate lines and use a monospace font, which makes horizontal alignment easy. The following code provides a simple example usage of `textplot()` (see Figure 11.1).

```
> par(mfrow=c(1, 2))
> plot(faithful)
> textplot(capture.output(summary(faithful)))
```

A similar function is `addtable2plot()` from the **plotrix** package.

```
> library(plotrix)
```

This function is aimed more at adding text output to an existing plot, and the text to draw is specified as a matrix or data frame (not necessarily character values). Positioning of the text output is relative to a location in the current user coordinate system, similar to the `legend()` function.

The following code adds a table showing some of the raw data values to a plot of the **pressure** data set (see Figure 11.2).

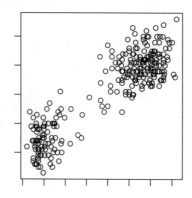

eruptions	waiting
Min. :1.600	Min. :43.0
1st Qu.:2.163	1st Qu.:58.0
Median :4.000	Median :76.0
Mean :3.488	Mean :70.9
3rd Qu.:4.454	3rd Qu.:82.0
Max. :5.100	Max. :96.0

Figure 11.1
Example output from the `textplot()` function. The plot to the left is normal output from `plot()` and the text to the right is a separate "plot" produced by `textplot()`.

```
> plot(pressure)
> addtable2plot(0, 300, pressure[13:19, ])
```

The `grid.table()` function from the **gridExtra** package provides tabular text output for **grid**-based graphics, with a large number of arguments to control borders, fonts, and justification.

```
> library(gridExtra)
```

The following code creates a table from some of the **pressure** values (see Figure 11.3).

```
> grid.table(pressure[13:19, ], show.box=TRUE,
             separator="black")
```

11.1.2 Avoiding text overlaps

A major problem that occurs when attempting to label points on a plot, or when adding a legend to a plot, is making sure that the text labels do not obscure any data values and that the text labels do not obscure each other. Often this problem is solved manually by trial and error, but this section describes some automated solutions.

To help with postioning text to avoid data symbols, the **plotrix** package provides the function `emptyspace()` to determine the largest empty region

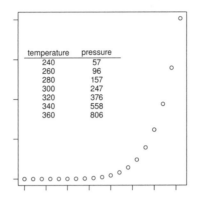

Figure 11.2

Example output from the `addtable2plot()` function.

	temperature	pressure
13	240	57
14	260	96
15	280	157
16	300	247
17	320	376
18	340	558
19	360	806

Figure 11.3

Example output from the `grid.table()` function.

within a plot, though it only returns the central location of that region and makes no guarantees about what sort of output will fit into the region. The `largest.empty()` function in **Hmisc** is similar, except that it has arguments `width` and `height` to specify a minimum size for the empty region (in user coordinates).

```
> library(Hmisc)
```

The following code uses these two functions to find empty space in a scatter-plot. First, a scatterplot of random points is drawn.

```
> x <- rnorm(20)
> y <- rnorm(20)
> plot(x, y, pch=16, col="gray")
```

Now the `emptyspace()` function is called to determine the central point within the largest empty space in the plot. This location is used to draw a text label (see Figure 11.4).

```
> xy <- emptyspace(x, y)
> text(xy, label="largest\nempty\nregion")
```

Finally, the `largest.empty()` function is used to request an empty space that is length one on each side (in user coordinates) and a rectangle of the requested size is drawn (see Figure 11.4).

```
> xy2 <- largest.empty(x, y, 1, 1)
> rect(xy2$x - .5, xy2$y - .5,
        xy2$x + .5, xy2$y + .5)
```

Switching now to the problem of placing text values so that they do not overlap with each other, the **plotrix** package has two functions for automatically positioning text labels relative to data points. The `spread.labels()` function is designed for a situation where the points to be labeled form a vertical or horizontal band, which leaves plenty of room for labels either above and below or to either side. The function shifts the labels away from the points and alternates the "side" that the label is drawn on. It also draws lines between each label and its corresponding point. The following code provides a simple example using random data, where the points form a vertical band so labels can be drawn to either side (see Figure 11.5).

```
> x <- runif(10)
> y <- rnorm(10)
```

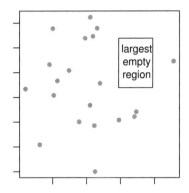

Figure 11.4
Finding empty space in a plot with the `emptyspace()` function (shown by the text)
and the `largest.empty()` function (shown by the rectangle).

```
> plot(x, y, pch=21, bg="gray", ylim=c(-3, 3), asp=1)
> spread.labels(x, y, labels=1:10)
```

The `thigmophobe.labels()` function places the labels right next to the points,
but attempts to avoid overlaps by choosing which "side" to draw the label on
(below, to the left, above, or to the right). The following code demonstrates
the use of this function (see Figure 11.5).

```
> plot(x, y, pch=21, bg="gray",
        ylim=c(-2, 3), xlim=c(-.5, 1.5))
> thigmophobe.labels(x, y, labels=1:10)
```

An alternative approach is provided by the `spread.labs()` function from the
TeachingDemos package.

```
> library(TeachingDemos)
```

This is similar to the `spread.labels()` function, but it places all labels along
a single dimension, adjusting the space between them to avoid overlaps. This
function does not draw the labels; it just returns modified locations for the
labels in one direction. The following code uses this function to generate a new
set of values, `adjy`, from the original y-values of the data points, then draws
the labels using those new values. It also uses the new values a second time
to draw line segments between the labels and the points (see Figure 11.5).

```
> plot(x, y, pch=21, bg="gray", ylim=c(-3, 3), asp=1)
> adjy <- spread.labs(y, strheight("10", cex=1.5))
> text(-0.5, adjy, labels=1:10, pos=2)
> segments(-0.5, adjy, x, y)
```

Yet another option is the pointLabel() function from the **maptools** package.

```
> library(maptools)
```

This function considers eight possible positions for each label relative to its data point (below, below and to the left, to the left, above and to the left, etc) and attempts to find an optimal placement of all labels. This is similar to thigmophobe.labels(), but the algorithm is more sophisticated and the labels are drawn much closer to the points. The following code demonstrates the use of this function (see Figure 11.5).

```
> plot(x, y, pch=16, col="gray", ylim=c(-2, 3), xlim=c(-.5, 1.5))
> pointLabel(x, y, labels=as.character(1:10))
```

The final function considered here is labcurve() from the **Hmisc** package. This function is a little different because it is designed for labeling *lines* on a plot rather than points. The complete set of lines to be labeled must be provided as a list of lists, then the function determines locations for the labels where there is large separation between the curves (so that it is obvious to which curve each label belongs). The following code constructs a list of four curves by permuting the x- and y-values from the previous examples.

```
> sx <- sort(x)
> sy <- sort(y)
> lines <- list(A=list(x=sx, y=y, lty=1),
                B=list(x=sx, y=sy, lty=2),
                C=list(x=sx, y=rev(y), lty=3),
                D=list(x=sx, y=rev(sy), lty=4))
```

The next code starts a new plot, draws the curves, then calls labcurve() to label each curve (see Figure 11.5).

```
> plot(x, y, type="n", ylim=c(-3, 3))
> lapply(lines, function(l) do.call("lines", l))
> labcurve(lines)
```

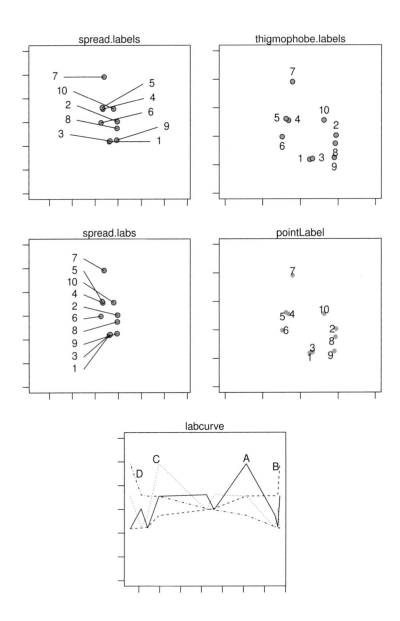

Figure 11.5

Drawing non-overlapping labels in a plot: top-left is the `spread.labels()` function; top-right is the `thigmophobe.labels()` function; middle-left is `spread.labs()`; middle-right is `pointLabel()`; and bottom is `labcurve()`.

With all of these functions, it may be necessary to manually adjust the axis limits in order to accommodate all labels. All of these functions are also only compatible with traditional graphics; the **directlabels** package, still in development at the time of writing, provides the `direct.label()` function for a general labeling paradigm with **grid**-based packages.

11.2 Peculiar primitives

A number of functions provide variations and embellishments on the standard set of graphical shapes, such as lines, rectangles, polygons, and text, that are provided by the core graphics systems.

The **plotrix** package provides circles and arc segments via `draw.circle()` and `draw.arc()`. There is also an `arctext()` function to draw text along an arc. For example, the following code draws a series of concentric circles and arcs, plus a piece of text on a curve (see Figure 11.6).

```
> draw.circle(.1, .9, radius=1:5/100)
> draw.arc(.3, .9, radius=1:5/100,
           deg1=45, deg2=seq(360, 160, -50))
> arctext("arctext", center=c(.5, .85), radius=.05,
           stretch=1.2)
```

The **plotrix** package also provides the `boxed.labels()` function, which draws text, like the traditional function `text()`, but adds a background rectangle, which can be filled. The rectangle is automatically sized to enclose the text. The `textbox()` function takes a different approach, drawing a rectangle with a given width and then breaking text into multiple lines so that it fits into the rectangle width. The height of the rectangle then depends on how many lines of text are produced. The following code draws an example of both type of boxed text (see Figure 11.6).

```
> boxed.labels(.7, .85, "boxed.labels", bg="gray90")
> textbox(c(.85, 1), .9, "this is a textbox .")
```

Three functions in the **plotrix** package produce different fill patterns within rectangles. The `gradient.rect()` function produces a gradient fill effect by drawing many thin slices of different colors within a rectangle and the `cylindrect()` function is a special case where the gradient fill is produced so as to create the illusion of a 3D cylinder. The `rectFill()` function produces fill patterns by drawing multiple plotting symbols within a rectangle. The following code draws examples of these fill patterns (see Figure 11.6).

```
> gradient.rect(.05, .5, .15, .7, col=gray(0:20/21))
> cylindrect(.25, .5, .35, .7, "black")
> rectFill(.45, .5, .55, .7, pch=16)
```

One further **plotrix** graphical primitive is provided by the `polygon.shadow()` function, which produces a fake shadow effect. This can be used to create the illusion of three dimensions, with a graphical shape appearing to hover slightly above the surface of the page. The following code shows how this function can be combined with the standard `polygon()` function (see Figure 11.6).

```
> x <- c(.65, .65, .75, .75)
> y <- c(.5, .7, .7, .5)
> polygon.shadow(x, y, offset=c(2/100, -2/100))
> polygon(x, y, col="white")
```

A similar function, `shadowtext()`, is provided by **TeachingDemos** package. This draws a background or border around text to make it stand out from its surroundings. The following code draws a simple example (see Figure 11.6).

```
> shadowtext(.9, .6, "shadowtext")
```

The **TeachingDemos** package also provides the `my.symbols()` function. This allows an arbitrary shape to be used as the plotting symbol, similar to the function `symbols()`, except that the user has complete control over what the shape looks like. The user defines the custom symbol as a matrix or list of x- and y-values, or as a function that creates a matrix or list or draws the symbol directly. The coordinate system for expressing the custom symbol is a square region with values ranging from −1 to 1 in both dimensions. The following code provides a simple example which draws predefined `ms.male` and `ms.female` symbols (see 11.6).

```
> my.symbols(seq(.3, .7, .2), .3,
             ms.male, inches=.2)
> my.symbols(c(.4, .6), .3,
             ms.female, inches=.2)
```

The **TeachingDemos** package also has a `panel.my.symbols()` function for drawing custom data symbols on **lattice** plots.

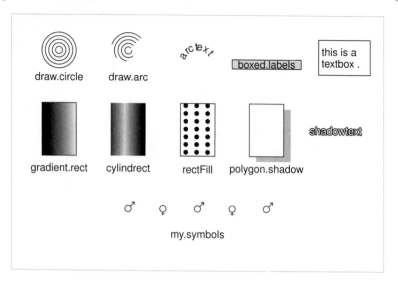

Figure 11.6
Graphical primitives produced by the functions `draw.circle()`, `draw.arc()`, `arctext()`, `boxed.labels()`, `textbox()`, `gradient.rect()`, `cylindrect()`, `rectFill()`, `polygon.shadow()`, `shadowtext()`, and `my.symbols()`.

The **gridExtra** package provides several new primitives for **grid**-based graphics: the `grid.ellipse()` function draws an ellipse, specified by `size`, `ar` (aspect ratio), and `angle`; `grid.pattern()` fills a rectangle with one of six possible fill patterns; and `grid.barbed()` draws points and lines for specified x and y locations in the `type="b"` style of the traditional graphics `plot()` function.

The following code demonstrates the use of these functions (see Figure 11.7).

```
> grid.ellipse(x=1:6/7, y=rep(.8, 6), size=.1,
               default.units="npc", size.unit="npc",
               ar=1:6, angle=1:6*15/180*pi)


> grid.pattern(x=1:6/7, y=.5, width=unit(.1, "npc"),
               height=unit(.1, "npc"), pattern=1:6,
               motif.cex=.7, gp=gpar(fill="gray80"))


> grid.barbed(1:6/7, y=rep(c(.15, .25), 3),
              size=unit(.05, "snpc"),
              pch=21, gp=gpar(fill="gray"))
```

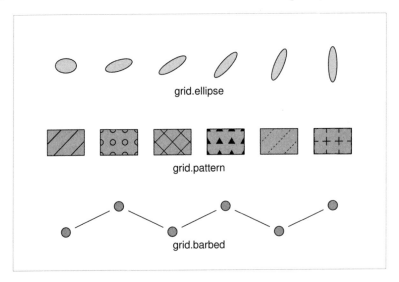

Figure 11.7
Graphical primitives produced by the functions `grid.ellipse()`, `grid.pattern()`,
and `grid.barbed()`.

Another package that produces a variety of geometric shapes and fill patterns
is the **shape** package (see Section 15.4).

11.2.1   Confidence bars

A common extension of a standard scatterplot involves adding confidence bars
to the data symbols. This effect can be achieved relatively easily with the
low-level `arrows()` function, but a number of packages provide convenience
functions.

There are in fact so many solutions that several different packages, including
gplots and **plotrix**, all provide a function called `plotCI()` and **Hmisc** is not
the only package that provides an `errbar()` function for the same purpose.
The **gplots** package includes an even higher-level function, `plotmeans()` that
calculates the ranges of the area bars as well. The following code uses that
function to plot the average fuel consumption for cars with different numbers
of cylinders, with confidence intervals represented by vertical bars (see Figure
11.8).

```
> plotmeans(mpg ~ cyl, mtcars,
          barcol="black", n.label=FALSE, connect=FALSE)
```

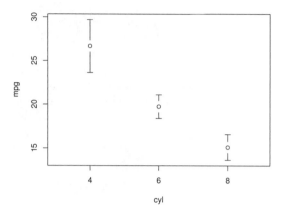

Figure 11.8

A scatterplot of average fuel consumption for cars with different numbers of engine cylinders, with confidence represented by vertical bars.

In the **lattice** world, confidence bars are one feature of the `xYplot()` function from the **Hmisc** package.

The **ggplot2** package provides several geoms that can be used for this purpose, including the `geom_errorbar()`, `geom_pointrange()`, and `geom_ribbon()` functions.

11.3   Calculations on colors

A number of packages provide additional functions for generating sets of colors. This section describes a small selection of these packages that have a specific focus on color.

11.3.1   The colorspace package

The **colorspace** package provides functions to create colors in a variety of color spaces, plus functions to convert between color spaces. There is some overlap with the `convertColor()` function from the **grDevices** package (see Section 10.1).

```
> library(colorspace)
```

In terms of generating color sets, the **colorspace** package provides analogs of some functions from **grDevices** that work in the (superior) CIE-LUV color space: `rainbow_hcl()`, `heat_hcl()`, and `terrain_hcl()`. In addition, there are functions for generating sequential and diverging color sets (in CIE-LUV). The `sequential_hcl()` function interpolates a sequence of n colors of a given hue, between a starting value for chroma and luminance and an end value for chroma and luminance. The following code draws a series of circles filled with 10 grayscale colors from a luminance of 20 to a luminance of 90 (see Figure 11.9).

```
> grid.rect(1:10/11, .75, width=1/15, height=1/3,
           gp=gpar(col=NA,
             fill=sequential_hcl(10, 0, 0, c(20, 90))))
```

The `diverge_hcl()` function interpolates a sequence of colors from a starting hue, chroma, and luminance, through a neutral color, given by a second luminance (and chroma of zero), to an end hue with the starting luminance. The following code draws a series of circles filled with 10 grayscale colors from a luminance of 20 through a luminance of 90 and back to a luminance of 20. The output from this code and the previous code is shown in Figure 11.9.

```
> grid.rect(1:10/11, .25, width=1/15, height=1/3,
           gp=gpar(col=NA,
             fill=diverge_hcl(10, 0, 0, c(20, 90))))
```

Both `sequential_hcl()` and `diverge_hcl()` have a `power` argument to allow for non-linear interpolation between the start and end colors.

11.3.2   The RColorBrewer package

The ColorBrewer web site provides carefully designed palettes of colors for use in coloring regions of maps.* The **RColorBrewer** package provides the `brewer.pal()` function for selecting one of these palettes, by specifying the name and size of the palette. For example, the following code selects the `"Pastel1"` palette containing five colors.

```
> library(RColorBrewer)
```

---

*`http://colorbrewer2.org/`.

Figure 11.9
Two sets of grayscale colors generated by `sequential_hcl()` (top) and `diverge_hcl()` (bottom).

```
> brewer.pal(5, "Pastel1")

[1] "#FBB4AE" "#B3CDE3" "#CCEBC5" "#DECBE4" "#FED9A6"
```

The `display.brewer.all()` functions produces a convenient display of all of the ColorBrewer palettes.

11.3.3 The munsell package

The Munsell color space is a perceptually based system for specifying colors as a hue, value, and chroma triplet. Munsell colors are specified with an idiosyncratic syntax and the **munsell** package provides a `mnsl2hex()` function to convert such a specification to an R color. For example, the following code converts a Munsell specification for a colorful, medium-lightness, purple to an R color.

```
> library(munsell)
> mnsl2hex("5P 5/10")

[1] "#9060A8"
```

The package also has functions for selecting and manipulating colors within the Munsell color space.

11.3.4 The dichromat package

A significant percentage of males suffer from some form of color blindness, for example, an inability to distinguish between reds and greens. The **dichromat** package provides a list containing a number of palettes that are suitable for people with this sort of vision deficiency. For example, the following code produces a set of 10 divergent colors ranging from a dark brown to light brown then light blue to dark blue.

```
> library(dichromat)
> colorschemes$BrowntoBlue.10

 [1] "#663000" "#996136" "#CC9B7A" "#D9AF98" "#F2DACE"
 [6] "#CCFDFF" "#99F8FF" "#66F0FF" "#33E4FF" "#00AACC"
```

11.4 Custom coordinates

One problem with positioning output within a traditional plot, using functions like text() and lines(), is that locations are given in user coordinates (relative to the scales on the plot axes). Simple conceptual locations, such as "top-left corner," are awkward to specify, so several packages provide functions to make this sort of task a bit easier.

11.4.1 Converting between traditional coordinate systems

The smartlegend() function from the **gplots** package allows a legend, as drawn by the legend() function, to be positioned using "left", "center", "right", "bottom", and "top". Similarly, corner.label() from the **plotrix** package will locate text, as drawn by text(), based on a position that is specified from −1 (left) to 1 (right) in both horizontal and vertical dimensions.

These functions provide a small amount of convenience over the traditional functions grconvertX() and grconvertY(), which support conversions between a much wider variety of traditional coordinate systems (see Section 3.1.1).

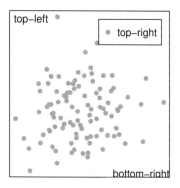

Figure 11.10
Example output from the functions `corner.label()`, `smartlegend()`, and `grconvertX()`.

The following code draws a scatterplot of random data.

```
> plot(rnorm(100), rnorm(100), pch=16, col="gray",
        ann=FALSE, axes=FALSE)
> box()
```

The next code uses `corner.label()` to put a label in the top-left corner of a plot, `smartlegend()` to draw a legend at top-right, and `grconvertX()` and `grconvertY()` to draw a label at bottom-right (see Figure 11.10).

```
> corner.label("top-left", x=-1, y=1)

> smartlegend(x="right", y="top",
            legend="top-right", pch=16,
            col="gray", bg="white")

> text(grconvertX(1, "npc"), grconvertY(0, "npc"),
        adj=c(1, 0), labels="bottom-right")
```

In **grid**-based graphics, these sorts of conversions are supported by the core system via the concept of units (see Section 6.3).

11.4.2 Subplots

This section concerns the problem of drawing a small plot *within* a larger plot.

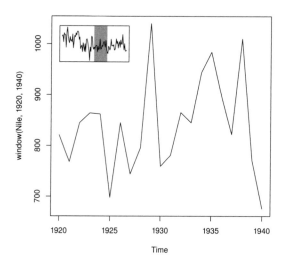

Figure 11.11

An example of drawing a small plot within a larger plot with the subplot() function.

The traditional graphics system has numerous graphical state settings that allow control over the location and size of plot regions (see Section 3.2.6), but again these can require a bit of effort to use. The subplot() function in the **TeachingDemos** package provides a convenient front end to these settings, so that a subplot can be drawn in a single call.

For example, the following code draws a time series plot for a subset of the Nile data set, with a small plot embedded in the top-left corner. The small plot shows the complete Nile time series, with a gray region to indicate the extent of the main plot (see Figure 11.11). The small plot is positioned within the main plot using x and y, the size is given in inches, and the horizontal justification is controlled by hadj.

```
> plot(window(Nile, 1920, 1940))
> subplot({ plot(Nile, axes=FALSE, ann=FALSE)
            rect(1920, 0, 1940, 2000, border=NA, col="gray")
            box()
            lines(Nile) },
          x=1920, y=1000, size=c(1.5, .75), hadj=0)
```

In **grid**-based systems, this sort of problem is solved using viewports (see Section 6.5).

11.5 Atypical axes

The **plotrix** package provides two functions that provide special effects for axis labeling.

The first of these addresses the issue of overlapping axis tick labels. The default behavior of the `axis()` function is to only draw axis tick labels that do not overlap. The `staxlab()` function provides an alternative solution by drawing the labels that would overlap at a different offset from the axis. The following code provides a simple example by drawing lots of tick labels along the x-axis (see Figure 11.12).

```
> with(pressure,
     {
         plot(temperature, pressure, axes=FALSE)
         axis(2)
         box()
         staxlab(1, at=temperature, cex=.7)
     })
```

Alternatively, this function can be used to draw the tick labels at an angle to avoid overlaps.

The **plotrix** package also provides the `axis.break()` function. This adds a visual "break" to an axis, for example, in the following code a break is put on the x-axis as a visual cue that the left end of the axis does *not* represent 0 (see Figure 11.12).

```
> kelvin <- pressure$temperature + 273.15

> with(pressure,
     {
         plot(kelvin, pressure, xlim=c(250, 650))
         axis.break(1)
     })
```

Another task that can be awkward in traditional plots is modifying the scales on an existing plot (see Section 3.4.5). The **plotrix** package provides one convenience function for this task, `revaxis()`. This function draws a complete plot and reverses the direction of the x- or y-axis (or both) so that, for example, y-values increase *down* the page rather than up the page. The following code shows a simple example (see Figure 11.13).

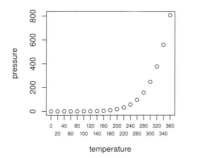 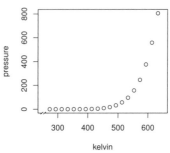

Figure 11.12
Examples of axis annotation from the `staxlab()` (left) and `axis.break()` (right)
functions.

```
> with(pressure,
        revaxis(temperature, pressure))
```

The **TeachingDemos** package also provides functions in this area. The
`updateusr()` function allows the user coordinate system to be modified by
mapping a pair of points in each dimension to a new pair of points. In the
following code, the **pressure** data are plotted with temperature on the Kelvin
scale. The `updateusr()` function is used to transform the x-axis scale to be
in Celsius (0 kelvin maps to −273.15 celsius) and the y-axis scale is left alone.
A vertical line at the boiling point of water is then added to the plot using
the new Celsius scale (see Figure 11.13).

```
> plot(kelvin, pressure$pressure)
> updateusr(c(0, 1), 0:1, c(-273.15, -272.15), 0:1)
> abline(v=100)
> text(x=100, y=700, " water boils", adj=0)
```

Another function provided by **TeachingDemos** is the `zoomplot()` function.
This function resets the axis scale limits *and* redraws the plot, creating the
effect of zooming in or out. For example, the following code draws a plot of
the **pressure** data set then zooms in on the bottom-left section of the plot
(the result is shown in Figure 11.13).

```
> plot(pressure)
> zoomplot(c(0, 150), c(0, 3))
```

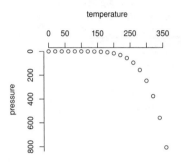

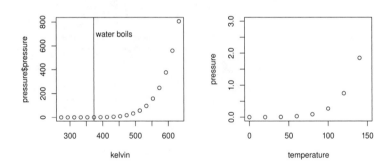

Figure 11.13
Examples of modifying the scales on the axes using **revaxis()** (top) to reverse the
direction of the y-axis, **updateusr()** (bottom-left) to transform the x-axis scale from
Kelvin to Celsius, and **zoomplot()** (bottom-right) to reset the axis limits.

Chapter summary

There are many functions in a range of extension packages that provide
low-level extensions to the core R graphics drawing tools for positioning text, drawing shapes, drawing fill patterns, manipulating plotting
coordinates, and generating colors.

12

Plot Extensions

Chapter preview

This chapter describes functions from extension packages that produce complete plots that are either reasonably common or well known or arise from a common type of data, but are not catered for by high-level functions in the core R graphics packages.

An important feature of the various R graphics systems is that they are generally built for extensibility. If there is not a function that already produces a particular image, there are functions that can be put together to produce that image, given enough time and patience.

However, it can be much faster to get a result if someone else has already done at least some of the work, so it is often worthwhile using an existing function if one can be found.

The standard plot types—scatterplots, barplots, histograms, and boxplots—are all well catered for by existing functions, but there are many well-known types of plots that are not provided by the core R graphics packages.

This chapter looks at a selection of these types of plots and the packages that provide functions for drawing them.

12.1   Venn diagrams

A number of packages provide functions for producing Venn diagrams, two of
which will be demonstrated in this section.

In both cases, the data can be specified as a data frame with a column of
logical values for each set. If row i of column j is TRUE then case i is a
member of set j.

The following code creates a data frame from the Titanic data set. There
are four sets representing passengers (as opposed to crew), adults, males, and
survivors.

```
> TitanicDF <- as.data.frame(Titanic)
> TitanicList <- lapply(TitanicDF[1:4], rep, TitanicDF$Freq)
> TitanicSets <-
      data.frame(passenger=TitanicList$Class != "Crew",
                 adult=TitanicList$Age == "Adult",
                 male=TitanicList$Sex == "Male",
                 survivor=TitanicList$Survived == "Yes")
> head(TitanicSets)
```

```
  passenger adult male survivor
1      TRUE FALSE TRUE    FALSE
2      TRUE FALSE TRUE    FALSE
3      TRUE FALSE TRUE    FALSE
4      TRUE FALSE TRUE    FALSE
5      TRUE FALSE TRUE    FALSE
6      TRUE FALSE TRUE    FALSE
```

The **gplots** package provides the venn() function for drawing Venn diagrams.

```
> library(gplots)
```

It can represent up to five sets at once. Two or three sets are drawn using
overlapping circles, with no attempt made to represent degree of intersection
by amount of overlap (see Figure 12.1). These plots are drawn using the **grid**
graphics system.

```
> venn(TitanicSets[1:2])
```

```
> venn(TitanicSets[1:3])
```

With four or five sets, the diagram is drawn using ellipses, again with no attempt made to represent degree of intersection by amount of overlap (see Figure 12.1). This plot is drawn using the traditional graphics system.

```
> venn(TitanicSets)
```

An alternative is provided by the **venneuler** package.

```
> library(venneuler)
```

This package has the `venneuler()` function, which generates an object describing the location and size of overlapping circles for a Venn diagram, where the overlap of the circles approximates the amount of intersection between the sets. There is a `plot()` method for drawing the result.

The following code produces **venneuler** versions of the previous three Venn diagrams (see Figure 12.1).

```
> plot(venneuler(TitanicSets[1:2]),
        col=hcl(0, 0, c(60, 80), .5),
        alpha=NA, border="black")

> plot(venneuler(TitanicSets[1:3]),
        col=hcl(0, 0, seq(40, 80, 20), .5),
        alpha=NA, border="black")

> plot(venneuler(TitanicSets[1:4]),
        col=hcl(0, 0, seq(20, 80, 20), .5),
        alpha=NA, border="black")
```

This package is based on **Java** code and requires the **rJava** package. The drawing is based on the traditional graphics system.

12.2 Chernoff faces

One approach to visualizing multivariate data is to produce a small plot for each case in the data set (see Section 2.5). One of the more entertaining

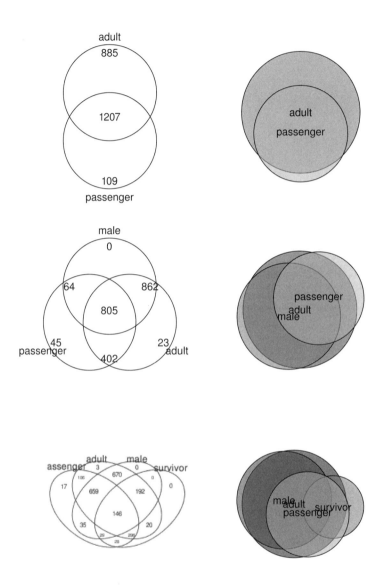

Figure 12.1

Venn diagrams from the **gplots** package and the **venneuler** package.

variations on this idea is the Chernoff face, which draws a cartoon human head with the size and shape of different facial features depending on the values of different variables.

The **TeachingDemos** package provides two functions for drawing Chernoff faces: `faces()` and `faces2()`.

```
> library(TeachingDemos)
```

These functions differ in which features are controlled by variables in the data set, so they produce different styles of faces. The following code draws Chernoff faces based on the first five judges in the `USJudegRatings` data set (lawyers' ratings of US Supreme Court judges). The result is shown in Figure 12.2.

```
> faces(USJudgeRatings[1:5, ], nrow=1, ncol=5)
```

```
> faces2(USJudgeRatings[1:5, ], nrow=1, ncol=5, scale="all")
```

Another variation is provided by the `symbol()` function from the **symbols** package (see Figure 12.2).

```
> library(symbols)
```

```
> symbol(USJudgeRatings[1:5, ], type="face")
```

This function can also be used to draw other sorts of small plots, including stars (like the `stars()` function), profile plots, and a "stick figure" plot.

Both of these packages are based on the traditional graphics system.

12.3 Ternary plots

A ternary plot can be useful for plotting *compositional* data, where three values sum to a constant value. An example is *soil texture* data, which consists of three proportions that sum to one: the proportions of clay, silt, and sand in a soil sample.

This section will use as an example a set of 13 soil samples from different regions of Peru (derived from the `soil` data set in the **agricolae** package).

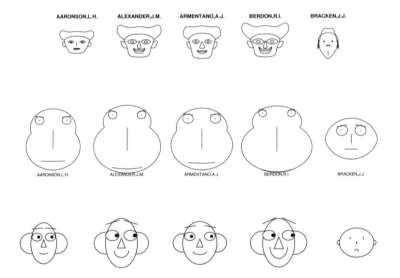

Figure 12.2
Four implementations of Chernoff faces: `faces()` and `faces2()` from the **Teach-ingDemos** package; `faces()` from the **aplpack** package; and `symbol()` from the **symbols** package.

```
> head(soils)
```

```
   sand silt clay
1    68   26    6
2    52   28   20
3    48   38   14
4    36   36   28
5    50   28   22
6    44   46   10
```

The **vcd** package provides the `ternaryplot()` function.

```
> library(vcd)
```

This function expects the data in three columns and each row should sum to the same number. The plot resulting from the following code is shown in Figure 12.3.

```
> ternaryplot(soils, col="black",
              grid_color="black", labels_color="black")
```

The `ternaryplot()` function is based on the **grid** graphics system, but all other examples in this section are based on traditional graphics.

The **plotrix** package also provides a ternary plot, which just has a different labeling style.

```
> library(plotrix)
```

The function is called `triax.plot()` and the data are supplied as before (see Figure 12.3).

```
> triax.plot(soils,   cex.ticks=.5)
```

Another option is provided by the **compositions** package.

```
> library(compositions)
```

This package has an `rcomp()` function for creating a `composition` object and there is an appropriate `plot()` method. This package also provides some lower-level annotation support. For example, the following code plots the soil data on a ternary plot, then adds a horizontal line representing the FAO "fine" and "very fine" soil classification criteria (more than 35% clay and more than 60% clay).

```
> plot(rcomp(soils))
> lines(rcomp(rbind(c(.4, 0, .6),
                    c(0, .4, .6))))
> lines(rcomp(rbind(c(.65, 0, .35),
                    c(0, .65, .35))))
```

The **compositions** package provides a more general approach to compositional data. For example, the data can be composed from more than three values, in which case a matrix of ternary plots is produced. The package also implements more sophisticated approaches to compositional data, which can result in different visualizations.

12.3.1   Soil texture diagrams

For the specific case of soil texture data, the **soiltexture** package provides a comprehensive set of functions for drawing, annotating, and customizing soil texture diagrams, complete with classification boundaries.

```
> library(soiltexture)
```

The following code uses the `TT.plot()` function to draw a soil texture diagram for the Peru soil samples. This function requires the data in a very specific format.

```
> TTsoils <- soils
> names(TTsoils) <- c("SAND", "SILT", "CLAY")
> TT.plot(tri.data=TTsoils)
```

12.4   Polar plots

The pie chart is a much-maligned graphical device for displaying counts or proportions as slices of a circle. The problem is that, in general, people are not good at perceiving the absolute or relative sizes of angles, compared to judging lengths or positions in normal cartesian coordinates.

However, pie charts are not the only way to present data on a polar coordinate system. It is possible to produce polar plots that represent values by lengths (of bars extending from the center) rather than angles and there are certain

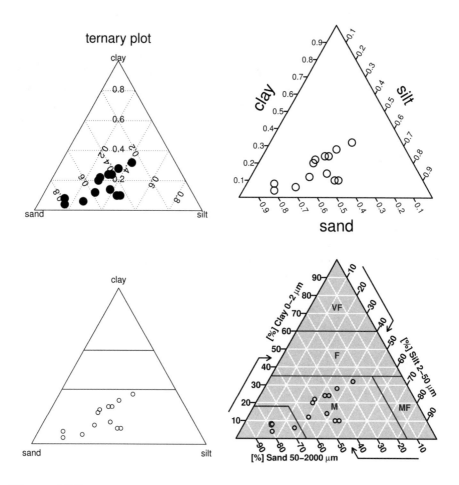

Figure 12.3

Four implementations of ternary plots: the `ternaryplot()` function from **vcd** (top-left); the `triax.plot()` function from **plotrix** (top-right); the `plot()` method for `"rcomp"` objects from the **compositions** package (bottom-left); and `TT.plot()` from **soiltexture** (bottom-right).

sorts of data, such as wind direction, that are naturally represented as an angle. For data that have a repeating cycle, such as hourly measurements over several days, a polar representation can also be useful for revealing periodic features.

This section looks at some packages that provide functions for producing plots using polar coordinates.

The data used in this section are wind measurements from the New Zealand National Climate Database (`http://cliflo.niwa.co.nz/`). The data consist of wind speed and direction from daily (9am) observation for approximately two years (September 2008 to September 2010). Observations at each time point are taken from 11 different weather stations scattered around the Auckland region.

```
> head(wind9am)
```

```
    Station              Date     Speed Dir
1     12325 2008-09-26 09:00:00 3.194444 211
2     12326 2008-09-26 09:00:00 3.194444 221
3     12327 2008-09-26 09:00:00 1.805556 212
4     12328 2008-09-26 09:00:00 1.611111 225
5     18195 2008-09-26 09:00:00 1.611111 215
6     22164 2008-09-26 09:00:00 2.805556 196
```

There is also a smaller data set consisting of hourly average wind speed based on data from 11 weather stations. There are hourly readings every day for one month (September 2010).

```
> head(hourlySpeed)
```

```
  hour day    Speed
1    0 237 1.626263
2    1 237 1.575758
3    2 237 1.618687
4    3 237 1.489899
5    4 237 1.138889
6    5 237 1.343434
```

The first function, `polar.plot()`, is from the **plotrix** package. This can be used to produce a variety of plots, but the code below is just a polar scatterplot of individual daily wind observations, with points plotted around the circle based on the wind direction and the distance from the center of the circle based on wind speed. The first two arguments provide the radius variable and the angle variable, and `rp.type` is specified here to plot data symbols at each point. The function normally draws its polar grid on top of the data symbols, so the code makes use of the standard painters model to first draw an empty plot with a grid and then draw another plot over the top to draw the actual data symbols. The data symbols are very dense at the center of the plot so the color of the data symbols is made semitransparent, with the level of transparency varying according to distance from the center. This allows some of the density structure to be seen at the center of the plot (see Figure 12.4).

```
> with(wind9am,
       {
           polar.plot(Speed, Dir, rp.type="s",
                      start=90, clockwise=TRUE,
                      point.col=NA)
           polar.plot(Speed, Dir, rp.type="s",
                      start=90, clockwise=TRUE,
                      add=TRUE,
                      point.symbols=16,
                      point.col=rgb(0,0,0, .3*Speed/max(Speed)),
                      show.grid=FALSE, show.radial.grid=FALSE)
       })
```

An important consideration with all polar plots is where to place the angle origin and which direction angles increase (clockwise or anti-clockwise), so all functions in this section provide some way to control these features. In this case, the `start` argument controls the origin and `clockwise` controls the direction.

Another way to view the distribution of these observations is provided by the `polarFreq()` function from the **openair** package.

```
> library(openair)
```

This function draws a polar image plot, with the color of each annulus sector determined by the number of points within that region.

The data must be provided to this function with variables named `ws` for the radius variable and `wd` for the angle variable. There should also be a `date` variable, even though that is not always used.

The following code demonstrates the use of this function and the output is shown in Figure 12.4.

```
> with(wind9am,
        polarFreq(data.frame(ws=Speed, wd=Dir, date=Date),
                  cols=gray(10:1/11), border.col="black"))
```

The functions in the **openair** package assume that the angle origin is pointing up (north) and that angles proceed clockwise, so there is no need to explicitly set the origin or angle direction in this case.

It should also be obvious that the **openair** package is built on **lattice**. In other words, the drawing is based on the **grid** graphics system. There are more examples from this package later in this section.

The previous two examples have demonstrated the use of points and areas within a polar plot. The next example shows one use of line segments, again using the `polar.plot()` function from **plotrix**. This plot is based on an aggregated form of the hourly wind data. The values are averaged across days to leave only 24 values, one for each hour of the day.

```
> hourSpeed <- aggregate(hourlySpeed["Speed"],
                         list(hour=hourlySpeed$hour),
                         mean)
> head(hourSpeed)
```

	hour	Speed
1	0	2.404572
2	1	2.387312
3	2	2.318133
4	3	2.231682
5	4	2.330435
6	5	2.398445

The angles on this plot represent hours of the day (0 to 23), but distance from the center still representing wind speed. As the following code demonstrates, there are numerous arguments that allow control over the placement of the grid lines and labels that provide the axes in a polar plot. It is also possible to control the range of radial values.

```
> polar.plot(hourSpeed$Speed, hourSpeed$hour * 15,
             start=90, clockwise=TRUE, lwd=5,
             label.pos=seq(15, 360, 15), labels=1:24,
             radial.lim=c(0, 4.5))
```

The **circular** package also provides functions that produce polar plots, although this package is focused on *circular* data, which is to say that only

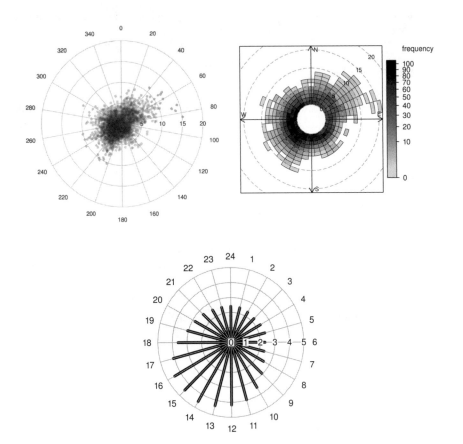

Figure 12.4

Polar plots produced by `polar.plot()` from the **plotrix** package (left and bottom) and `polarFreq()` from the **openair** package (right). These are the polar-coordinates analogs of a scatterplot (top-left), a barplot (bottom), and an image plot (top-right).

the angle variable is of interest. The focus of these plots is mostly on the distribution of the angle values.

```
> library(circular)
```

The next few examples will work with the daily wind data from just one of the weather stations in the wind9am data set.

```
> station22254dir <- with(wind9am, Dir[Station == 22254])
```

The first step is to create a circular object using the circular() function. Notice that the object itself contains information about the origin and direction of angles.

```
> station22254 <- circular(station22254dir,
                           units="degrees",
                           zero=pi/2, rotation="clock")
```

The following code draws the individual data values as data symbols on the circumference of a circle, using an appropriate plot() method (see Figure 12.5).

```
> plot(station22254, stack=TRUE, sep=.06)
```

This next code calculates, then plots, a density estimate for the wind directions at this station (see Figure 12.5).

```
> plot(density(station22254, bw=45),
        main="", xlab="", ylab="")
```

The rose.diag() function plots a polar histogram from the wind direction values, which provides yet another view of the distribution of the angles (see Figure 12.5).

```
> rose.diag(station22254, bins=36, prop=3)
```

One useful feature of the **circular** package is that it provides low-level functions for annotating a polar plot. The next example works with the hourly wind data (there is a wind speed value for each hour of each day for approximately one month). The directions for these circular data represent hours.

```
> windHours <- circular(hourlySpeed$hour,
                        units="hours",
                        zero=pi/2, rotation="clock")
```

In the following code, a bare plot is created. The `lines()` method for `circular` ojects is then used to add lines to the plot. The lines show how wind speeds vary over the day. Finally, the `axis()` method for `circular` objects is used to add labeling to the plot (see Figure 12.5).

```
> plot(windHours, col=NA, shrink=1.2, axes=FALSE)
> lines(windHours,
        0.5*hourlySpeed$Speed/max(hourlySpeed$Speed),
        nosort=TRUE, lty="dotted", join=FALSE)
> axis.circular(template="clock24")
```

12.4.1 Wind roses

In the specific case of plotting wind data, a specific style of plot called a *wind rose* is popular. This plots wedges (or "paddles") for a small set of angle ranges to represent how common each wind direction is. The wind speed is represented by breaking the wedge into separate bands, similar to a stacked bar chart.

The **openair** package provides the `windRose()` function. As mentioned previously, this package is built on **lattice** so it is capable of multipanel wind roses. The following code produces a wind rose of the daily wind data for each weather station in the data set (see Figure 12.6).

```
> with(wind9am,
        windRose(data.frame(ws=Speed, wd=Dir,
                            date=Date, station=factor(Station)),
                 paddle=FALSE,
                 type="station", cols=gray(4:1/6), width=2))
```

The **ggplot2** package provides a general mechanism for drawing in polar coordinates via the `coord_polar()` function (see Section 5.9).

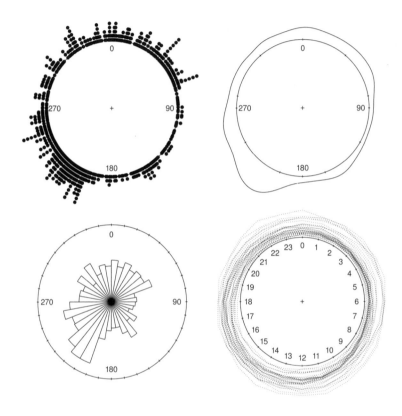

Figure 12.5
Plots of circular data produced by the functions plot.circular() (top-left), plot.density.circular() (top-right), rose.diag() (bottom-left), and lines.circular() and axis.circular() (bottom-right). All of these functions are from the **openair** package.

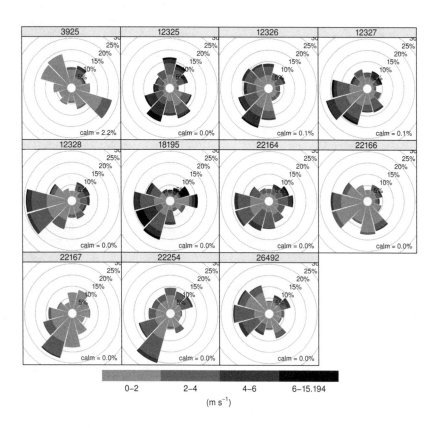

Figure 12.6

Wind rose plots produced by `windRose()` from the **openair** package. The different panels show wind data from different weather stations.

12.5   Hexagonal binning

This section describes another option for avoiding overlapping data symbols
when plotting a very large number of points on a scatterplot (see Section 2.4).
This solution involves hexagonal binning and drawing a representation of the
density of points in different regions of the plot rather than the points themselves. The **hexbin** package provides functions for performing the binning
operation and drawing the resulting hexagons.

```
> library(hexbin)
```

For example, the following code draws a normal scatterplot of `Serum.Iron`
against `Transferin` from the `NHANES` data set (part of the **hexbin** package)
and the main body of points consists of a dark blob (see Figure 12.7).

```
> data(NHANES)
> plot(Serum.Iron ~ Transferin, NHANES)
```

A hexagonal binning plot allows some of the structure of the main body of
points to be seen. The following code uses the `hexbinplot()` function from
the **hexbin** package (see Figure 12.7).

```
> hexbinplot(Serum.Iron ~ Transferin, NHANES)
```

The `hexbinplot()` function provides a **lattice** style interface so that multipanel conditioning and multiple groups of data within panels are also possible
(see Figure 12.7).

```
> hexbinplot(Serum.Iron ~ Transferin | Sex, NHANES)
```

There is also a `hexplom()` function for drawing hexagonal binning plot matrices.

The `geom_hex()` and `stat_binhex()` functions can be used to produce this
sort of plot in **ggplot2**.

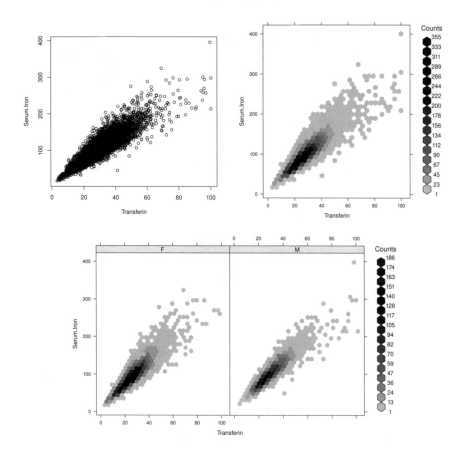

Figure 12.7
At top-left is a scatterplot with many overlapping points. The hexagonal binning
plot at top-right plots point densities in hexagonal regions to reveal some of the
distributional structure of the main body of points. The bottom plot is a multipanel
hexagonal binning plot.

Chapter summary

Many extension packages provide high-level functions for producing types of plots that are not available in the core R graphics system. This chapter described some packages and functions for drawing Venn diagrams, Chernoff faces, polar coordinate plots, ternary plots, and hexagonal binning plots.

13

Graphics for Categorical Data

Chapter preview

This chapter describes plots for categorical data. Some of these plots can be produced using traditional, **lattice**, or **ggplot2** functions, but the **vcd** package provides both specializations and generalizations that go beyond the facilities provided by the core graphics packages.

This chapter focuses on graphics for visualizing categorical data. Although categorical data are extremely common, the techniques for visualizing categorical data, especially in a multivariate setting, are much less well known than the corresponding plot types for continuous variables. For example, the mosaic plot for displaying the relationship between two categorical variables is much less common than the ubiquitous scatterplot for two continuous variables. One purpose of this chapter is to describe functions that can draw plots for multivariate categorical data sets.

13.1 The vcd package

The main package described in this chapter is **vcd**, a package that was originally created as an implementation of the ideas in Michael Friendly's book *Visualizing Categorical Data*. This package is built on the **grid** graphics system and that provides the second main purpose of this chapter, which is to describe **grid**-based functions for displaying categorical data. This leads to features such as **lattice**-like multipanel displays for categorical data and sophisticated support for customization of categorical plots.

13.2   XMM-Newton

Data from the X-ray Multi-Mirror space telescope (XMM) will be used to provide examples in this chapter.

XMM was launched by the European Space Agency in December 1999 to help study exotic astronomical objects such as black holes and pulsars. Researchers must apply for a time slot to use the telescope in a competitive process and this chapter will make use of public information about the successful applications in this process from 2007 (see http://xmm.esac.esa.int/). Proposals are classified by the following variables:

Category: This describes what sort of object is being studied. There are seven categories, labeled A to G.

Priority: This describes the assessed importance of the study, from A (highest) to C (lowest).

Schedule: This is a binary variable indicating whether the study needs to be carried out at a particular point in time (fixed) or whether it can be carried out anytime (free).

nObs: The number of "pointings" involved in the study (the number of times the telescope needs to be focused upon a particular region of the sky), as a categorical variable with levels single and multiple.

Duration: The total proposed observation duration, in seconds.

```
> head(xmm)
```

	Category	Priority	Schedule	Duration	nObs
1	D	C	free	56000	single
2	A	A	fixed	13000	multiple
3	A	A	fixed	13000	multiple
4	A	A	fixed	18000	multiple
5	A	B	free	10000	single
6	A	B	free	18000	single

13.3   Plots of categorical data

For a single categorical variable, the standard visualization approach is a barplot, though a more modern approach acknowledges that the widths of the bars are redundant and a dotplot is considered to be more appropriate. Figure 13.1 shows some simple examples of these plots, which are produced using the traditional graphics functions `barplot()` and `dotchart()`. These types of plots can also easily be reproduced in either **lattice** or **ggplot2** (also shown in Figure 13.1).

There are also standard plots for visualizing the dependence of one or more continuous variables on a single categorical variable. This leads to multiple boxplots or, using the categorical variable as a grouping variable, a scatterplot with multiple data series. These sorts of plots are again easy to produce in either traditional graphics, **lattice**, or **ggplot2**.

However, plots that show a categorical variable as the *dependent* variable, or plots that show the relationship between several categorical variables, are much less common.

13.4   Categorical data on the y-axis

Two plots that show the distribution of a categorical variable dependent on a continuous variable are the spinogram and the conditional density plot. Figure 13.2 shows examples of these sorts of plots where the dependent variable is the `Priority` assigned to XMM proposals and the independent variable is the proposal `Duration`. In both cases, the shading in the vertical direction represents the proportion of proposals in each `Priority` category, conditional on the `Duration`.

These plots can be produced in traditional graphics using the `spineplot()` and `cdplot()` functions. The **vcd** package provides **grid**-based versions via `cd_plot()` and `spine()`.

```
> library(vcd)
```

The following code demonstrates the use of these functions (see Figure 13.2).

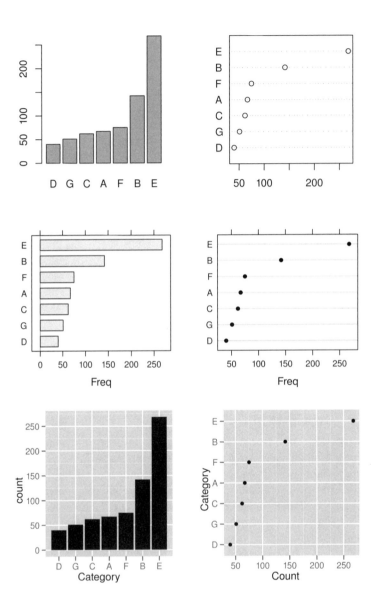

Figure 13.1

Barplots (left) and dotplots (right) of the frequency of XMM proposals in each
scientific category. The top row shows traditional graphics versions, the plots in
the middle row are produced using **lattice**, and the bottom row shows **ggplot2**
versions. The counts have been deliberately ordered from smallest to largest.

```
> spine(Priority ~ Duration, xmm)

> cd_plot(Priority ~ Duration, xmm)

> durn <- xmm$Duration/1000
> cd_plot(Priority ~ durn, xmm, xlab="Duration (1000s)")
```

13.5   Visualizing contingency tables

One approach to visualizing several categorical variables together is to extend the simple barplot to a stacked barplot or a side-by-side barplot. These sorts of plots can be produced with `barplot()` in traditional graphics and are also straightforward in **lattice** and **ggplot2**. Figure 13.3 shows examples of these types of plots to show the distribution of priority ratings for XMM proposals with `free` versus `fixed` timings.

A more general form of the stacked barplot is the mosaic plot. In this sort of plot, a rectangle is divided into rows, based on the proportion of observations in each category of a variable. Each row is then subdivided into columns, based on the proportions or counts in a second categorical variable. Each column can then be further subdivided into rows according to a third variable and so on. The areas of the resulting rectangles represent the counts in the corresponding cell of a multiway contingency table.

In traditional graphics, mosaic plots can be generated with the `mosaicplot()` function. The **vcd** package provides a **grid**-based version via the `mosaic()` function. Figure 13.4 shows an example for the XMM proposal data, which is equivalent to the stacked barplots in Figure 13.3. The code is shown below.

```
> mosaic(Priority ~ Schedule, xmm)
```

One advantage of the mosaic plot is that it presents information on the relative frequencies of both variables at once. For example, the barplots in Figure 13.3 do not reflect the fact that most proposals have `free` timings. Another advantage is that mosaic plots generalize to more than two variables. The following code uses `mosaic()` to draw a mosaic plot showing the relative frequencies of different proposal timings, the distribution of priorities within each timing category, and the distribution of `single` versus `multiple` observation proposals within each combination of timing and priority (see the bottom plot of Figure 13.4).

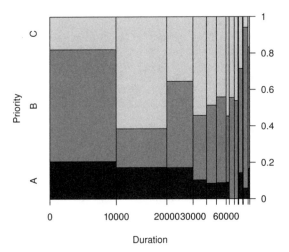

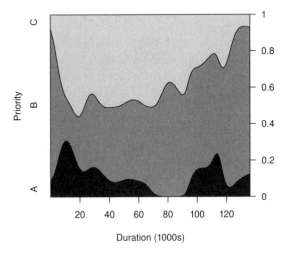

Figure 13.2

A spinogram (top) and a conditional density plot (bottom) showing the proportion
of proposals assigned to each priority category, conditional on the proposal duration.
Notice the non-linear x-axis on the spinogram.

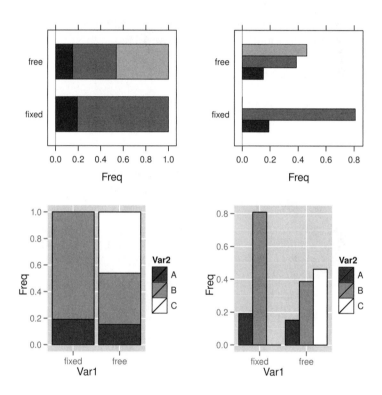

Figure 13.3
Barplots of two categorical variables, either stacked (left) or side by side (right), from
lattice (top) and **ggplot2** (bottom), showing the proportion of proposals assigned
to each priority category for proposals with `fixed` versus `free` timing.

```
> mosaic(nObs ~ Schedule + Priority, xmm)
```

The order in which variables are used in a mosaic plot has an enormous impact on what sorts of comparisons can be made and this leads to a number of further generalizations of the idea of a grid of rectangular regions.

The `tile()` function in the **vcd** package produces a grid of rectangles where the area of each rectangle represents the proportion of observations in each combination of categories. This sort of plot is called a *tile display* or a *fluctuation diagram*. The following code produces a tile display version of the bottom mosaic plot in Figure 13.4 (see Figure 13.5).

```
> tile(nObs ~ Schedule + Priority, xmm)
```

Another variation on the mosaic plot is the double-decker plot. In this case, a rectangle is divided into columns based on frequencies of one categorical variable and then each column is subdivided into more columns based on another categorical variable. This breakdown continues until the last (binary) variable, which is represented by dividing each column into shaded rows. This plot is specifically aimed at visualizing the final (binary) variable conditional on the levels of all other variables.

The **vcd** package can provides the `doubledecker()` function for producing this type of plot. The following code uses this function to produce another variation on the mosaic plot in Figure 13.4 (see Figure 13.5).

```
> doubledecker(nObs ~ Schedule + Priority, xmm)
```

The **vcd** package also provides support for model-based displays of categorical data. For example, mosaic displays can be shaded and colored to represent residuals from an independence model. There are also functions that provide **grid**-based versions of assocation plots, `assoc()`, and fourfold plots, `fourfold()`.

13.6   Categorical plot matrices

The **vcd** package also provides a categorical plot matrix analog to the scatterplot matrix. This is provided as a `"table"` method for the `pairs()` function.

The **vcd** package provides the `structable()` function for conveniently tabulating data from a formula. As an example, the following code generates a

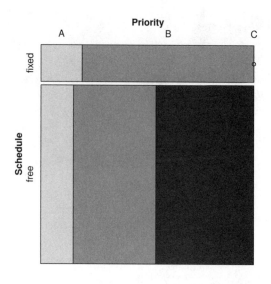

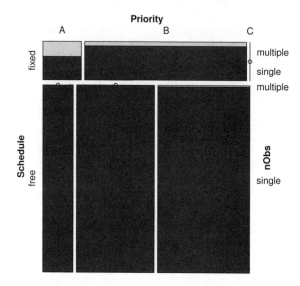

Figure 13.4
Mosaic plots for two variables (top) and three variables (bottom), produced by the **vcd** package.

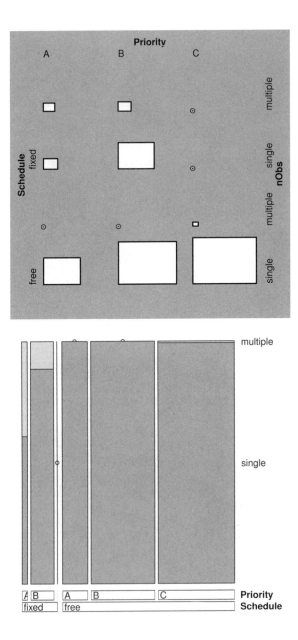

Figure 13.5
A tile display (top) and a double-decker plot (bottom) of three categorical variables,
produced by the **vcd** package.

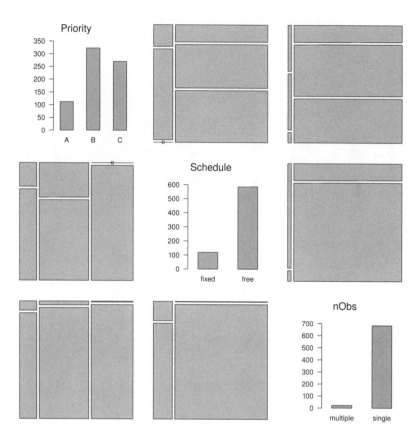

Figure 13.6
A categorical plot matrix, with a mosaic plot for each pair of variables and barplots
on the diagonal.

three-way table from the `Priority`, `Schedule`, and `nObs` variables. This is
then passed to the `pairs()` function to produce a matrix of mosaic plots, with
barplots on the diagonal (see Figure 13.6).

```
> pairs(structable(nObs ~ Priority + Schedule, xmm),
        space=.15)
```

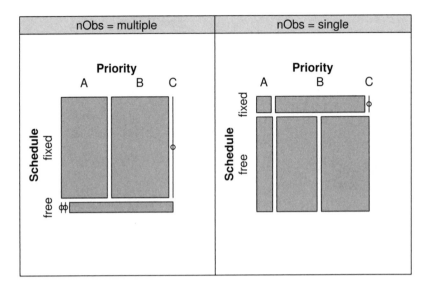

Figure 13.7
Multipanel conditioning mosaic plots, with a mosaic plot panel for each level of a
categorical variable.

13.7   Multipanel categorical plots

Another way of incorporating a categorical variable into a display is to use it
as a conditioning variable, so that individual subplots, or panels, are produced
for each level of the variable.

It is quite simple to create multipanel conditioned barplots with **lattice**, or
facetted barplots with **ggplot2**, but the **vcd** package adds the ability to generate multpanel conditioning plots for the mosaic plot and its variations. The
cotabplot() produces output much like the **lattice** high-level functions, but
allows **vcd** plots as panel functions.

The following code generates a mosaic plot of XMM proposal Schedule versus
Priority for both single and multiple pointing proposals (see Figure 13.7).

13.8   Customizing categorical plots

The **vcd** package is designed to allow a great deal of customization of the
plots that it creates. One of the difficulties in producing complex plots of
multivariate categorical data is that there is no obvious algorithm for placing
category labels that will always avoid overlapping the text of adjacent labels.
The number of labels will also vary from plot to plot, and there are typically
many labels, so it is not possible to provide a separate argument to control
the detailed position and appearance of each individual label.

This problem is addressed in **vcd** by providing a general mechanism for specifying not only details for labels, but also details for shading the rectangles
within the plots and details for the spacing between the rectangles.

The plots in Figures 13.4 and 13.5 made use of these customization features
and this section will demonstrate the real code that was used to generate those
figures.

The following code was used to produce the bottom plot in Figure 13.4.

```
> mosaic(nObs ~ Schedule + Priority, xmm,
         labeling_args=list(rot_labels=c(right=0),
           offset_labels=c(right=-.5),
           just_labels=c(right="left")),
         margin=c(right=4))
```

The default mosaic plot has labels written vertically on both the left and
right of the plot, but in this case that would mean that the labels overlap.
Instead this code specifies that the labels on the right of the plot should be
horizontal and left-aligned and that they should be offset from the plot by
half a line less than the default. The right margin for the plot is made slightly
larger than the default to accommodate the horizontal labels. This code uses
named vectors as the argument values to avoid having to specify values for
the bottom, left, and top of the plot as well; the unspecified parameters just
retain their defaults.

The key insight that is embodied in the **vcd** package is that most of the
plots of multidimensional tables of counts are just a two-dimensional array of
rectangles. The calculation of the placement and sizes of the rectangles varies,
and the appearance of the rectangles varies, and the labeling of the rectangles
varies, but there is a common underlying structure.

This means that a wide range of plots can be produced from a single function, with arguments to allow some details to vary. The above example demonstrates the `labeling_args` argument, which allows for small deviations from the default behavior. There is also a `labeling` argument, which allows larger customizations by specifying a completely different labeling *function*. Furthermore, there are similar arguments for controlling shading of rectangles and spacing between rectangles: `shading_args` and `shading`, and `spacing_args` and `spacing`.

The code below shows a very simple use of the `shade` argument. This code was used to produce the top plot in Figure 13.5.

```
> tile(nObs ~ Schedule + Priority, xmm,
       tile_type="area",
       shade=TRUE,
       gp=gpar(lwd=2, fill="white"),
       pos_labels=c(left="left", top="left", right="left"),
       just_labels=c(left="left", top="left", right="left"),
       pop=FALSE, newpage=FALSE)
```

The positioning of the labels has been customized again, but the new feature is that the rectangles in the plot are forced to have a gray fill and thicker lines than the default. This is to ensure that the small rectangles in this plot are still visible. The argument `shade=TRUE` means that the rectangles are filled and the value of the `gp` argument provides the graphical parameters that dictate how that fill occurs.

Even finer control can be obtained by taking advantage of the fact that **vcd** is built on top of the **grid** graphics system, plus the fact that **vcd** provides useful names for many of the viewports and grobs that it creates.

The top plot in Figure 13.5 also provides an example of this flexibility. The default fluctuation diagram draws just a very small dot when the observed count in a cell is zero. In order to make these cells more visible, and to distinguish them from cells that have a small count, in Figure 13.5 a circle is drawn around each small dot that represents a zero. The code below shows how this is achieved for one of the zero cells.

```
> downViewport("cell:Schedule=fixed,Priority=C,nObs=multiple")
> grid.circle(0, 0, r=unit(1, "mm"))
> upViewport(0)
```

The point is that the `tile()` function uses **grid** viewports to arrange its drawing *and* it names those viewports in a rational manner. By specifying `pop=FALSE` in the call to `tile()`, those viewports were retained so that, in the code above, each viewport corresponding to a zero cell can be revisited and a circle added at the bottom-left corner.

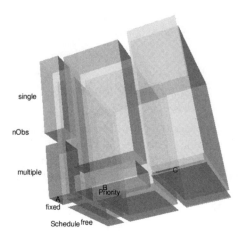

Figure 13.8
A three-dimensional mosaic plot produced by the `mosaic3d()` function from the
vcdExtra package.

13.9 The vcdExtra package

The **vcdExtra** package mostly extends the model-based features of **vcd**, by
providing support for displaying diagnostics from a wider range of categorical
data analysis models.

```
> library(vcdExtra)
```

However, **vcdExtra** also provides a `mosaic3d()` function for producing three-
dimensional mosaic plots, with help from the **rgl** package (see Section 16.6).
Figure 13.8 shows the output from the following call to `mosaic3d()`.

```
> mosaic3d(structable(~ Priority + Schedule + nObs, xmm))
```

Chapter summary

Simple plots of one or two categorical variables can be produced using either traditional graphics, **lattice**, or **ggplot2**. The **vcd** package provides **grid**-based versions of some less common categorical plots, plus a number of more complex, multivariate plots for categorical data. The **vcd** package allows for a great deal of customization of its plots.

14

Maps

Chapter preview

This chapter describes how to draw geographical maps in R. Drawing maps requires specialized extension packages because there are
unique file formats for map data and drawing maps involves special
transformations that are not provided by the core graphics systems.

From a purely graphical perspective, drawing a map appears to be simply a
matter of rendering one or more polygons that represent the borders of one or
more countries, states, or regions. For example, it is not immediately obvious
whether Figure 14.1 is just a rectangle, or a map of the State of Colorado.

However, a number of issues combine to ensure that producing a map is often
not a trivial task.

Figure 14.1
A map of the State of Colorado. This could be mistaken for a simple rectangle
because a simple map is just a set of polygonal shapes.

Map data: The polygons that describe regions on a map can be very detailed and complex, so it is usual to obtain information about map boundaries from external files, which often have a special format. This leads to two problems: locating map data in the first place and then finding functions that can be used to read the map data into R.

Lakes in islands on lakes: One classic example of complexity in a map is the existence of a *hole* within a geographical region, such as a country that contains a lake. The situation gets worse if the lake contains an island and worse still if the island has its own lake, particularly if the goal is to fill areas of land with a color, but leave areas of water empty. Functions that draw maps must be aware of this problem.

Projections: Locations and regions on a map are often described in terms of *geographic coordinates*, longitude and latitude, but these are locations and regions on the surface of a three-dimensional sphere whereas map display typically occurs on a two-dimensional page or computer screen. Longitude and latitude are often converted to two-dimensional *projected coordinates*, which can be drawn in a straightforward manner, but there are many such projections to choose from. Drawing a map requires both knowledge of whether the locations and regions have been projected and, if not, possibly some way of applying a projection to the polygons or other graphical shapes that are to be drawn.

Aspect ratio: The locations of map polygons typically relate to a physical scale, for example, a set of locations on the Earth's surface. In order to faithfully represent these locations, it is important to control the (ratio of the) physical dimensions of the polygons as they are drawn. Drawing a map requires the ability to control the *aspect ratio* of the drawing region.

Annotation: A map rarely consists of just the border outline of a geographical region. For example, additional lines may be added to represent other features such as cities, roads, railways, and rivers. More often, the map is just providing a context for displaying the values of one or more variables, such as fill colors to indicate disease incidence within different regions, data symbols to represent the locations of events, or an overlay of weather patterns such as wind speed and direction. Any drawing that consists of multiple *layers* of *annotation* like this must ensure that any projection is consistent across all layers. Also, there needs to be some way to reliably match a data value with the appropriate polygon within a map.

The following sections elaborate on these issues and describe some of the packages and functions in R that provide solutions.

This is a very simplified and brief introduction to some of the mapping tools that are available for R. A far more thorough discussion of the ideas, tools, and techniques is provided in the book *Applied Spatial Data Analysis with R* by Roger Bivand, Edzer Pebesma, and Virgilio Gómez-Rubio.

14.1   Map data

The simplest way to produce a map in R is with the **maps** package, because **maps** provides several of its own sets of map information.

14.1.1   The maps package

The **maps** package provides the function map() for drawing maps. The first argument to this function specifies the name of a "database" that contains the map information. The default is "world", which is a set of low-resolution country boundaries. Other options are: "france", which includes boundaries of the *departments* of France; "italy", which includes boundaries of the Italian *provinces*; "state" and "county", which provide state and county borders in the USA; and "nz", which provides the outline of the main islands of New Zealand.

By default, all polygons in a given database are drawn, but the region argument can be used to specify (by name) just a subset of polygons, and the xlim and ylim arguments may also be used to limit the drawing to just a particular range of longitude and latitude.

The following code shows a simple example using the default "world" database, but only drawing the polygon corresponding to Brazil (see Figure 14.2).

```
> library(maps)

> map(regions="Brazil", fill=TRUE, col="gray")
```

The map() function uses the traditional graphics system to draw the map. The **gmaps** package can use the **maps** package map databases, but render the map with **grid**.

The map produced with the default "world" database is quite coarse and the **mapdata** package has a higher-resolution map database. For example, the following code produces a more detailed version of Figure 14.2.

Figure 14.2
A simple map of Brazil produced using the **maps** package.

```
> library(mapdata)

> map(database="worldHires", regions="Brazil",
      fill=TRUE, col="gray")
```

However, the map data from both the **maps** and the **mapdata** package may not be sufficiently accurate or up-to-date for some purposes. For more modern, accurate, and detailed maps there are more sophisticated sources of map data and more sophisticated packages for working with them.

14.1.2　Shapefiles

A very common format for storing geographical information is the *shapefile* format. More accurately, such files are called *ESRI shapefiles* because the format is controlled by the ESRI company that produces GIS software. A shapefile is actually a collection of files; the geographical information is stored in several files with a common name stem and different suffixes (e.g., .shp, .shx, and .dbx).

Shapefiles are available from many places on the internet, including various governmental agencies. Files were obtained from the GSHHS (Global Self-consistent, Hierarchical, High-resolution Shoreline) database and the Natural Earth Project for the figures in this chapter.*

The **maptools** package provides several functions to read shapefiles, the most general of which is the readShapeSpatial() function. The following code uses this function to read a shape file that contains state boundaries for Brazil.

```
> library(maptools)
```

```
> brazil <-
      readShapeSpatial(system.file("extra", "10m-brazil.shp",
                                   package="RGraphics"))
```

The **maptools** package depends on the **sp** package, which is automatically loaded when **maptools** is loaded. The **sp** package provides data structures and functions for working with spatial data in R and this is what gets created by the call to readShapeSpatial(). The variable brazil is a special "Spatial" object that combines the polygons to draw the map, plus extra information about the regions that those polygons represent. The extra information will be used in Section 14.2; for now, the map itself can be drawn with a call to plot(), as shown below, because the **sp** package defines a method for "Spatial" objects (see Figure 14.3).

```
> plot(brazil, col="gray")
```

The **maptools** package also provides functions for reading data in other formats, for example, there is the Rgshhs() function for reading map files from the GSHHS database (see Section 14.3). There are also functions for converting map databases from the **maps** package to "Spatial" objects, for example, map2SpatialPolygons().

14.2 Map annotation

Up to this point, a map has only been considered as a set of polygons representing regions. However, other sorts of shapes are also relevant for drawing maps: lines to represent roads or rivers; and points to represent specific locations of interest such as cities or mountain peaks.

---

*ftp://ftp.soest.hawaii.edu/pwessel/gshhs; http://www.naturalearthdata.com/.

Figure 14.3
A map of Brazil showing the different states. Made with Natural Earth: Free vector
and raster map data @ `naturalearthdata.com`.

Furthermore, the region boundaries on a map are useful only really as a context for other sorts of data. For example, regions may be filled with different
colors to indicate population size and points may be labeled with city names.

This is where the `"Spatial"` data structures from the **sp** package become
very useful because they contain not only the shapes to draw a map, but
also additional information such as place names and region names that are
associated with the shapes.

For example, the `"Spatial"` object `brazil` contains polygons to draw each
of the Brazilian states, *plus* it contains the name of the region that each state
belongs to. The extra information can be accessed just like accessing variables
in an R data frame.

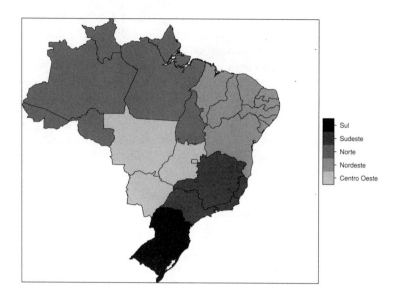

Figure 14.4
A map of Brazil showing the different state boundaries, with states filled according
to which region they come from.

```
> brazil$Regions
```

```
 [1] Norte          Norte          Nordeste      Norte
 [5] Norte          Norte          Centro Oeste  Centro Oeste
 [9] Sudeste        Centro Oeste   Sul           Sul
[13] Sul            Nordeste       Nordeste      Nordeste
[17] Nordeste       Sudeste        Nordeste      Sudeste
[21] Nordeste       Nordeste       Norte         Norte
[25] Centro Oeste   Sudeste        Nordeste
Levels: Centro Oeste Nordeste Norte Sudeste Sul
```

The following code draws a map of Brazil with the states shaded according
to region (see Figure 14.4).

```
> spplot(brazil, "Regions", col.regions=gray(5:1/6))
```

This example introduces another function from the **sp** package for drawing
maps, the spplot() function, which provides several points of difference from
the plot() method used previously.

The major advantage that has been demonstrated here is that it makes it easy to annotate the fill regions on the map according to the levels of a variable simply by naming the relevant variable. Another difference is that the spplot() function calls **lattice** functions to draw the map. This means that it is possible to produce multipanel map displays (by specifying more than one variable as the second argument to spplot()) and, because the output is **grid** based, it is possible to perform detailed customizations.

The next example demonstrates a more complex annotation. In this case, lines, points, and text are added to the original map to indicate the boundaries of regions and the locations and names of the state capitals (see Figure 14.5).

The data for the region boundaries and the data for the locations and names of the state capitals in this example come from separate shapefiles.

```
> brazilRegions <-
      readShapeSpatial(system.file("extra",
                                    "10m_brazil_regions.shp",
                                    package="RGraphics"))
```

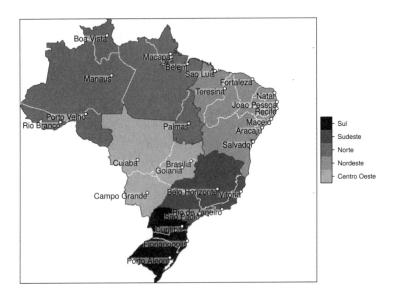

Figure 14.5
A map of Brazil showing the different state boundaries, with states filled according to which region they come from and state capitals shown.

```
> brazilCapitals <-
      readShapeSpatial(system.file("extra",
                              "10m_brazil_capitals.shp",
                              package="RGraphics"))
```

The following code draws a map with each state colored according to its region similar to before, but with white borders for the regions. The important change is that this time a custom panel function is specified, which adds darker borders around each region via sp.lines(), points for each state capital via sp.points(), and labels for each capital via sp.text(). There are also semi-transparent rectangles behind each label to aid visibility. Those are added using low-level **grid** calls.

```
> spplot(brazil, "Regions",
         col.regions=gray.colors(5, 0.8, 0.3),
         col="white",
         panel=function(...) {
             panel.polygonsplot(...)
             sp.lines(brazilRegions, col="gray40")
             labels <- brazilCapitals$Name
             w <- stringWidth(labels)
             h <- stringHeight(labels)
             locs <- coordinates(brazilCapitals)
             grid.rect(unit(locs[, 1], "native"),
                       unit(locs[, 2], "native"),
                       w, h, just=c("right", "top"),
                       gp=gpar(col=NA, fill=rgb(1, 1, 1, .5)))
             sp.text(locs, labels, adj=c(1, 1))
             sp.points(brazilCapitals, pch=21,
                       col="black", fill="white")
         })
```

14.3 Complex polygons

The busy-looking region in the North of the map of Brazil (see, for example, Figure 14.3) represents the mouth of the Amazon river. At the heart of that river mouth lies the island of Marajo, which is large enough to contain its own lakes.

The shape files used in previous figures, from the Natural Earth project, do

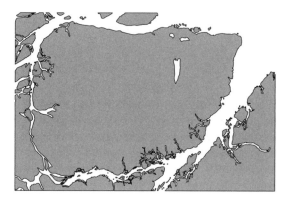

Figure 14.6
A map of Marajo Island in the mouth of the Amazon River. This is the largest island that is completely surrounded by fresh water in the world. The map data come from GSHHS.

not have sufficient detail to show these lakes, but the GSHHS database does provide the required resolution. The following code reads in a shapefile that was generated from the GSHHS data.

```
> marajo <-
      readShapeSpatial(system.file("extra", "marajo.shp",
                                   package="RGraphics"))
```

The following code draws Marajo Island and its surrounds. The important features are the three lakes toward the north-east of the island. These have been filled white by specifying the **pbg** argument.

```
> plot(marajo, col="gray", pbg="white")
```

The main point is that the **sp** functions for drawing maps can detect and handle these situations.

14.4   Map projections

All geographic locations can be specified by a longitude (east-west of greenwich) and latitude (north-south of the equator). However, these are locations

Figure 14.7
A map of Iceland drawn without specifying any projection information. The result
is very distorted because a degree of longitude at northerly climes is a much smaller
distance than a degree of latitude.

on a three-dimensional (approximate) sphere and maps are typically drawn
on a two-dimensional page or screen.

A very simple *projection* of longitude and latitude to a two-dimensional plotting surface involves assigning longitude to the x-axis and latitude to the
y-axis, but there are many other possibilities.

In the worst case, no information is known about the map projection. In this
situation, the **sp** functions for drawing maps will set the plot aspect ratio to 1
(a unit in the x-direction is the same physical size as a unit in the y-direction),
but that will not always produce a nice result.

For example, the following code reads in and draws the counties of Iceland
without supplying any projection information (see Figure 14.7).

```
> iceland <-
      readShapeSpatial(system.file("extra", "10m-iceland.shp",
                                   package="RGraphics"))

> plot(iceland, col="gray")
```

The map data are in geographic coordinates and plotting a country like Iceland by simply mapping longitude and latitude to x and y, with an aspect
ratio of 1, creates severe distortion because, at northerly or southerly latitudes, a single degree of longitude spans a much smaller distance than it does
at the equator. The following code adds projection information, using the
CRS() and `proj4string()` functions so that R knows that the map is in geographic coordinates. The projection value is a PROJ.4 specification[*] and this

---

[*]`http://trac.osgeo.org/proj/`.

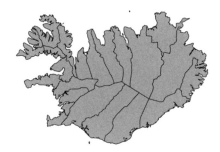

Figure 14.8

A map of Iceland drawn with an aspect ratio adjustment made for the fact that the map is in geometric coordinates. The result is much less distorted than Figure 14.7.

information should be obtained from the supplier of the shapefile or in some cases this will be part of a shapefile in the form of a file with a `.prj` suffix.

```
> proj4string(iceland) <- CRS("+proj=longlat +ellps=WGS84")
```

Now the map can be drawn again (see Figure 14.8). With the projection information specified, the `plot()` method calculates an aspect ratio for the plot to account for the displacement from the equator and this produces a better result.

```
> plot(iceland, col="gray")
```

However, the regions for this map of Iceland are still in geographic coordinates. A more sophisticated solution would be to use a proper map projection to produce projected coordinates. This requires loading the package **rgdal**.

```
> library(rgdal)
```

The choice of projection depends on the use for the map because different projections preserve different map features. For example Google Maps uses a Mercator projection because this preserves angles (so that, for example, streets that meet at a cross road are drawn at right angles). The following code projects the map of Iceland using a Mercator projection with the `spTransform()` function (and then draws the map; see Figure 14.9).

```
> icelandMercator <- spTransform(iceland,
                         CRS("+proj=merc +ellps=GRS80"))
```

Figure 14.9
A map of Iceland drawn with a Mercator projection. The result is very similar to the adjusted map of geometric coordinates (which is drawn in gray with white borders in the background).

```
> plot(icelandMercator)
```

The result is very similar to the previous map, but with this projection the map has known useful properties, rather than just a rough adjustment using the plot aspect ratio.

The following code demonstrates another type of projection. First, projection information is added for the Brazil map used in previous sections.

```
> proj4string(brazil) <- CRS("+proj=longlat +ellps=WGS84")
```

Now the map is transformed using an orthographic projection, which shows what Brazil would look like if viewed from orbit at a location above longitude zero and latitude zero (see Figure 14.10). Grid lines have been added to the plot, using the `gridlines()` function, to help understand the projection.

```
> brazilOrtho <- spTransform(brazil, CRS("+proj=ortho"))
> plot(brazilOrtho, col="gray")
```

14.5 Raster maps

The examples so far have demonstrated *vector* data (polygonal regions) and *point* data (city locations). A third main type of information that can be used in a map is *raster* data, which consists of a regular grid of values (pixels).

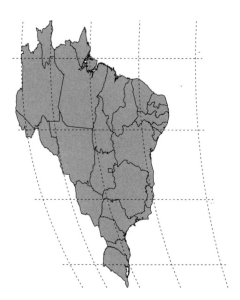

Figure 14.10
A map of Brazil using an orthographic projection. This is what Brazil would look
like from space (hovering above the equator directly south of Greenwich).

Reading and rendering raster data is quite straightforward (see also Chapter
18). The main problem that arises is aligning raster data and vector (or
point) data on a map together. The **raster** package provides support for
reading and manipulating raster data, including combining raster and vector
data for drawing maps.

```
> library(raster)
```

The following code uses the **raster()** function from the **raster** package to
read in a raster image from the Natural Earth project and sets the projection information using the **projection()** function. This image provides a
grayscale shaded relief of the entire Earth.

```
> worldRelief <- raster("SR_LR.tif")
> projection(worldRelief) <- CRS("+proj=longlat +ellps=WGS84")
```

The result is a special **"RasterLayer"** object and the following code uses the
crop() function to reduce that very large image down to just that part of
the image that corresponds to the extent of the vector map of Brazil from

Figure 14.11
A map of Brazil with a shaded relief background to indicate elevation. Made with
Natural Earth. Free vector and raster map data @ naturalearthdata.com.

previous examples. The two objects now correspond to each other and have
the same projection, so they can be drawn together.

```
> brazilRelief <- crop(worldRelief, brazil)
```

The raster image can be rendered using a "RasterLayer" method for the
image() function and then the vector state borders are added using the
plot() method for "Spatial" objects, with add=TRUE (see Figure 14.11).

```
> image(brazilRelief, col=gray(0:255/255), maxpixels=1e6)
> plot(brazil, add=TRUE)
```

14.6   Other packages

Several other packages provide further mapping extensions or completely alternative paradigms for mapping in R.

For example, the **PBSmapping** package provides its own self-contained set of map data files, data structures for working with the map data, and functions for manipulating and rendering the map data. The **rworldmap** package also contains its own map data plus convenient high-level functions for producing world maps of coarse geographic data sets that contain one value per country.

The **RgoogleMaps** package provides functions for downloading Google Maps tiles for use as raster map data. Going the other way, there are functions in **maptools** for writing out map data as KML files for use in Google Earth and Google Maps.

The **latticeExtra** package provides a `mapplot()` function for **lattice** and **ggplot2** has the `borders()` function, plus a few others, to support drawing maps from the **maps** package.

Chapter summary

Drawing maps requires specialized graphics functions to handle special file formats and rendering details such as map projections and lakes within islands. Simple maps can be drawn using the **maps** package and its built-in map data. More sophisticated results can be obtained using the **maptools** package to read in map data, the **rgdal** package to handle projections, and the **sp** and **raster** packages for map data manipulation and map drawing.

15

Node-and-edge Graphs

Chapter preview

This chapter describes how to produce node-and-edge graphs with R. Node-and-edge graphs are a special case because producing a graph visualization usually requires special algorithms to determine a useful arrangement of the nodes on a page. Several packages are described that provide this facility, notably **Rgraphviz** and **igraph**. This chapter also describes how to draw more regular node-and-edge diagrams in R.

The main graphics systems in R, traditional, **lattice**, and **ggplot2**, are all focused on producing graphs in the sense of statistical *plots*. Another common meaning of the term "graph" is a set of nodes with edges connecting them. This chapter describes packages that are focused on producing images of this sort of node-and-edge graph.

There are three important steps involved in producing an image of a node-and-edge graph:

1. An object representing the graph must be created.

2. A layout must be generated for the graph, which provides a description of where all of the nodes and edges should be drawn.

3. The graph needs to be rendered, which involves drawing the information in the graph representation, such as the names of the nodes, at the locations specified in the layout.

15.1   Creating graphs

A graph consists of a set of nodes and a set of edges, where each edge is
a connection between any two nodes. This information about a graph can
be represented in R in any number of ways and different packages provide
different solutions. However, many of the packages that work with graphs,
particularly those that produce graph visualizations, depend on graph repre-
sentations provided by the **graph** package.

15.1.1   The graph package

The **graph** package provides two simple ways to specify a graph.

```
> library(graph)
```

The graph can be described in terms of explicit node names and an explicit
list of edges between nodes, a `"graphNEL"` object, or in terms of an adjacency
matrix, a `"graphAM"` object, where there is an edge between node i and node
j if element (i, j) of the matrix has the value 1.

The following code shows how to create a simple graph from a vector of node
names and a list of edges. This graph shows the relationships between the
five core graphics packages in R. A visualization of this simple graph is shown
in Figure 15.1. Notice that the graph is *directed*, so edges go only one way
between two nodes.

```
> nodes <- c("grDevices", "graphics", "grid",
             "lattice", "ggplot2")
> edgeList <-
     list(grDevices=list(edges=c("graphics", "grid")),
          graphics=list(),
          grid=list(edges=c("lattice", "ggplot2")),
          lattice=list(),
          ggplot2=list())
> simpleGNEL <- new("graphNEL",
                    nodes=nodes,
                    edgeL=edgeList,
                    edgemode="directed")
```

The following code shows an equivalent graph defined via an adjacency matrix.

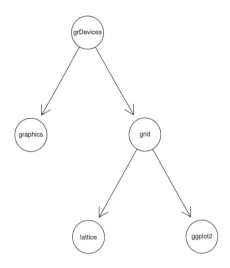

Figure 15.1
A simple node-and-edge graph consisting of five nodes. This shows the relationship between the core graphics functions in R. This graph has been rendered by the **Rgraphviz** package.

```
> adjMat <- rbind(grDevices=c(0, 1, 1, 0, 0),
                  graphics=rep(0, 5),
                  grid=c(0, 0, 0, 1, 1),
                  lattice=rep(0, 5),
                  ggplot2=rep(0, 5))
> simpleGAM <- new("graphAM", adjMat, edgemode="directed")
```

For small graphs, it is feasible to specify the nodes and edges by hand as in the above examples, but for larger graphs the set of nodes and the edge list, or the adjacency matrix, can be generated programmatically.

Alternatively, an external description of a graph in the GXL* format can be read into R, as a `"graphNEL"` object, using the `fromGXL()` function. For example, Figure 15.2 shows a file, `simplegraph.gxl`, containing GXL code for the simple graph example above and the following code creates a `"graphNEL"` object from that file.

```
> simpleGNEL <- fromGXL(file("simplegraph.gxl"))
```

---

*http://www.gupro.de/GXL/.

```
<?xml version="1.0"?>
<gxl:gxl xmlns:gxl="http://www.gupro.de/GXL/gxl-1.1.dtd">
 <gxl:graph id="graphNEL" edgemode="directed">
  <gxl:node id="grDevices" />
  <gxl:node id="graphics" />
  <gxl:node id="grid" />
  <gxl:node id="lattice" />
  <gxl:node id="ggplot2" />
  <gxl:edge id="1" from="grDevices" to="graphics"/>
  <gxl:edge id="2" from="grDevices" to="grid"/>
  <gxl:edge id="3" from="grid" to="lattice"/>
  <gxl:edge id="4" from="grid" to="ggplot2"/>
 </gxl:graph>
</gxl:gxl>
```

Figure 15.2
A simple GXL file that describes a graph consisting of five nodes. This is the graph
that is drawn in Figure 15.1.

The **graph** package also provides functions for manipulating graphs. For
example, the subGraph() function can be used to extract a subgraph from
a larger graph and the leaves() function can be used to determine the leaf
nodes of a graph. In the following code, subGraph() is used to extract just
the **grid**-related nodes from simpleGNEL and leaves() is used to find the leaf
nodes of simpleGNEL.

```
> smallGNEL <- subGraph(c("grid", "lattice", "ggplot2"),
                        simpleGNEL)
> smallGNEL


A graphNEL graph with directed edges
Number of Nodes = 3
Number of Edges = 2


> leaves(simpleGNEL, "out")


[1] "graphics" "lattice"  "ggplot2"
```

15.2 Graph layout and rendering

Having generated a representation of a graph, drawing the graph requires deciding where to draw nodes and edges, the *layout* step, and then drawing the nodes and edges at those locations, the *rendering* step.

For very simple graphs, it may be feasible or desirable to position the nodes and edges by hand. That scenario is dealt with in Section 15.4. This section addresses the more complex problem of positioning graphs with numerous nodes and edges. In this case, some sort of *layout algorithm* must be employed.

Several packages implement graph layout algorithms, but the focus in this section is on the **Rgraphviz** package, which provides an interface to the graphviz software library.*

15.2.1 The Rgraphviz package

The **Rgraphviz** package is part of the Bioconductor project. This package provides both layout and rendering facilities for graphs that have been created using the **graph** package.

```
> library(Rgraphviz)
```

Rendering a graph is as simple as calling the generic `plot()` function with a `"graphNEL"` or `"graphAM"` object as the argument. The following code renders the simple graph introduced earlier to produce the result shown in Figure 15.1.

```
> plot(simpleGNEL)
```

This rendering uses the default layout algorithm for **Rgraphviz**, which is called dot. This algorithm places nodes in a hierarchical layout consisting of horizontal layers, tries to keep edges short, and tries to avoid edge crossings.

The second argument to this `plot()` method is the algorithm to use for laying out the graph. There are several options; the following code shows how to select a neato layout, which treats edges as if they are springs and finds a layout that balances the tension of the springs. Figure 15.3 shows the resulting layout for the simple graph.

---

*http://www.graphviz.org/.

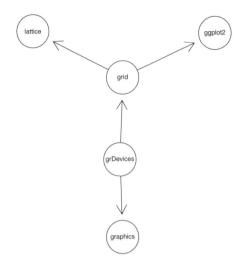

Figure 15.3
Rendering a graph with **Rgraphviz**, using the neato layout algorithm. This should
be compared with the graph layout in Figure 15.1, which used the default dot layout
algorithm.

```
> plot(simpleGNEL, "neato")
```

The help page for `"GraphvizLayouts"` provides more information on the layout algorithms.

15.2.2   Graph attributes

It is also possible to supply graph *attributes*, which affect the rendering of
the graph. The following code demonstrates one way to do this, by supplying
additional arguments to the call to `plot()`. In this case, the `nodeAttrs` and

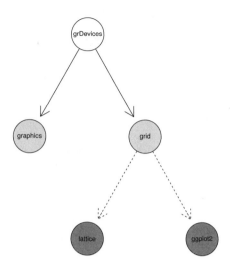

Figure 15.4

A simple graph rendered with **Rgraphviz**, but using graph attributes to modify the appearance of nodes and edges.

`edgeAttrs` arguments are used to modify the appearance of individual nodes and edges. The resulting graph is shown in Figure 15.4.

```
> plot(simpleGNEL,
       edgeAttrs=list(lty=c(`grDevices~graphics`="solid",
                            `grDevices~grid`="solid",
                            `grid~lattice`="dashed",
                            `grid~ggplot2`="dashed")),
       nodeAttrs=list(fillcolor=c(grDevices="white",
                          graphics="gray90", grid="gray90",
                          lattice="gray60", ggplot2="gray60")))
```

The help page for `"GraphvizAttributes"` provides a list of available attributes and their meanings. It may also help to refer to the documentation on the graphviz web site itself.[*]

---

[*]http://www.graphviz.org/doc/info/attrs.html.

15.2.3   Customization

The layout and rendering of graphs can be performed as separate steps. One way to do this, using the `layoutGraph()` and `renderGraph()` functions, is shown in the following code. The result is the same as Figure 15.1.

```
> layoutGNEL <- layoutGraph(simpleGNEL)
> renderGraph(layoutGNEL)
```

The usefulness of separating the steps like this is that an intermediate object, here `layoutGNEL`, is created with information about the arrangement of the nodes and edges.

This leads to an alternative method of customizing the appearance of nodes and edges using functions like `nodeRenderInfo()` and `edgeRenderInfo()`. The following code modifies the background color for some nodes and the line style for some edges to create the same result as shown in Figure 15.4.

```
> nodeRenderInfo(layoutGNEL) <-
      list(fill=c(graphics="gray90", grid="gray90",
              lattice="gray60", ggplot2="gray60"))
> edgeRenderInfo(layoutGNEL) <-
      list(lty=c(`grid~lattice`="dashed",
              `grid~ggplot2`="dashed"))
```

Another option is to make use of the layout information in the intermediate object to add further drawing to a graph or even to take control of drawing the graph itself. For example, the `plot()` method for graph objects and the `renderInfo()` function are based on traditional graphics. Figure 15.5 shows an example where `layoutGraph()` was used to calculate node positions, then the resulting layout was rendered using **grid**. The code for this example is available on the book web site.

15.2.4   Output formats

By rendering a graph using R, it is possible to produce output in any of the graphics formats that the R engine supports (see Chapter 9). However, those formats are purely for displaying a graph.

Another option is to save a graph in a format that facilitates further editing of the graph (using other software), for example in graphviz's native dot format or the native format of a diagram editor such as xfig or dia.* The function

---

*http://www.xfig.org/; http://projects.gnome.org/dia/.

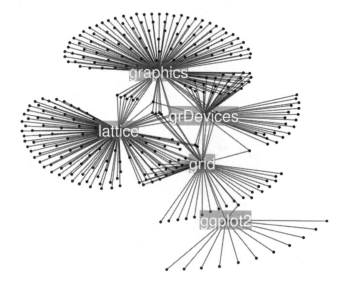

Figure 15.5
A more complex graph laid out using the neato layout algorithm. This graph has been created by using **Rgraphviz** to do the layout, but then **grid** to do the rendering. The graph has a node for all packages that directly depend on or import one of the core graphics packages in R (based on the state of CRAN on March 5, 2010).

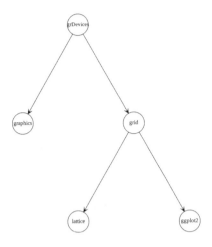

Figure 15.6

A simple graph that was created in R, but laid out *and rendered* by graphviz.

toFile() provides several extra formats of this sort.

Yet another way to work is to use the toFile() function to call graphviz to perform not only the graph layout, but also the graph rendering, which in some cases may produce a higher-quality result compared to a rendering in R. For example, the following code produces a PDF file that has been both laid out and rendered by graphviz (see Figure 15.6). In this case, the agopen() function is used to lay out the graph and provide toFile() with the correct sort of R object that it requires.

```
> toFile(agopen(simpleGNEL, ""),
         filename="Figures/graph-graphvizrender.pdf",
         fileType="pdf")
```

15.2.5   Hypergraphs

The **graph** package only supports standard directed or undirected graphs, where an edge connects exactly two nodes. In a *hypergraph*, an edge can connect more than two nodes. The **hypergraph** and **hyperdraw** packages from

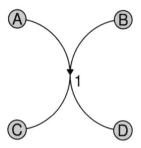

Figure 15.7
A simple hypergraph consisting of one hyperedge that connects one pair of nodes
to another pair of nodes. This hypergraph has been rendered by the **hyperdraw**
package.

the Bioconductor project provide some facilities for creating and rendering
hypergraphs.

```
> library(hyperdraw)
```

The following code provides a simple example. A hypergraph is constructed
using functions from the **hypergraph** package and then the graph is plotted
using graphBPH() and a hypergraph plot() method from the **hyperdraw**
package. The result is shown in Figure 15.7.

```
> dh <- DirectedHyperedge(c("A", "B"), c("C", "D"))
> hg <- Hypergraph(LETTERS[1:4], list(dh))
> plot(graphBPH(hg))
```

15.3 Other packages

The combination of **graph** and **Rgraphviz** provides only one possible approach to drawing node-and-edge graphs in R. This section desribes several
other packages that provide functions for creating and rendering graphs.

15.3.1   The igraph package

The **igraph** package provides a large set of functions both for creating graphs and for laying out and rendering graphs.

```
> library(igraph)
```

This package provides several convenient features for creating graphs. On one hand, there are functions that provide simple interfaces for creating the graph structure. For example, the graph() function accepts a numeric vector where each pair of values describes an edge. The following code creates the simple graph structure from Section 15.1.1.

```
> simpleIgraph <- graph(c(0, 1, 0, 2, 2, 3, 2, 4))
```

Another interface is provided by the graph.formula() function, which allows the edges to be specified using a special syntax. The following code creates the simple graph from Section 15.1.1.

```
> formulaIgraph <- graph.formula(grDevices -+ graphics,
                                  grDevices -+ grid,
                                  grid -+ lattice,
                                  grid -+ ggplot2)
```

The **igraph** package also has a number of functions that generate regular or well-known graphs. For example, the graph.tree() function produces regular hierarchical graphs and the graph.full() function produces regular fully connected graphs (see Figure 15.8).

```
> treeIgraph <- graph.tree(10)
> fullIgraph <- graph.full(10)
```

There is also the graph.famous() function for well-known "named" graphs, the graph.atlas() function to create one of the 1253 graphs from the book *An Atlas of Graphs*, and many more.

The **igraph** package also offers a wide variety of graph layout algorithms. For example, the layout.reingold.tilford() function performs a hierarchical layout similar to the **dot** algorithm of **graphviz** and the layout.spring() is in a similar spirit to the **neato** algorithm. In addition, the **igraph** package offers several more variations on the spring or force layout algorithm and layout.circle() to place all nodes on the circumference of a circle.

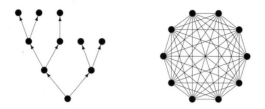

Figure 15.8
Two examples of regular graphs: a tree graph (left) and a fully connected graph (right).

The sizing and labeling of graphs in the rendered output is less automated in **igraph**, but it is possible to control these features via functions such as `set.vertex.attribute()` and `set.edge.attribute()`.

The existence of the `igraph.to.graphNEL()` function means that one fruitful approach is to make use of the **igraph** package to generate a graph and then convert it to something that **Rgraphviz** can render.

The **igraph** package has several other distinctive features. The `tkplot()` function provides an interactive editor, which can be used to click and drag individual nodes to fine tune the layout of a graph. There is also the function `read.graph()` to read a graph description from an external file in a variety of formats, plus the `write.graph()` function to save a graph in one of those formats.

15.3.2   The network package

The **network** package is part of the statnet suite of software packages for network analysis.* It provides the basic visualization functions for network objects.

This package is notable for supporting a very general concept of a graph. For example, it can cope with *hypergraphs*, where a single edge can connect more than two nodes, in addition to the standard graph where an edge connects exactly two nodes.

A graph may be created via the `network()` function, supplying the number of nodes and the graph edges as an adjacency matrix or as an "edge list" (actually

---

*`http://www.statnetproject.org/`.

a matrix, where each row specifies an edge). The following code creates the
simple directed graph from previous sections.

```
> library(network)
```

```
> simpleNetwork <-
      network(rbind(c(1, 2),
                    c(1, 3),
                    c(3, 4),
                    c(3, 5)),
              vertex.attr=list(vertex.names=nodes))
```

The **network** package can also lay out and render graphs, though there are
only a few layout algorithms available and the rendering style is different again
from **Rgraphviz** and **igraph**. For example, node labels are drawn adjacent
to nodes rather than within nodes.

A `plot()` method for `"network"` objects performs the layout and rendering.
This function has many parameters to allow control over the appearance of
the rendered graph, including `mode`, which controls the layout algorithm. The
following code draws the `simpleNetwork` object (see Figure 15.9).

```
> plot(simpleNetwork, mode="fruchtermanreingold",
       vertex.col=1, displaylabels=TRUE)
```

One advantage of this package is that it does not depend on any third-party
software to perform the graph layout.

Many other packages provide rendering of graphs or trees for particular areas
of application. For example, the **ape** package provides a range of layout styles
and flexible facilities for labeling nodes and edges of phylogenetic trees.

15.4 Diagrams

This section looks at drawing arrangements of nodes and edges when the
positioning is more deliberate or does not require automating, such as in the
production of flow charts.

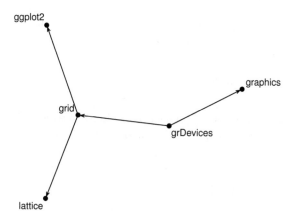

Figure 15.9
A simple network consisting of three nodes, with edge between nodes 1 and 2, between nodes 2 and 3, and between nodes 3 and 1. This network has been rendered by the **network** package.

15.4.1   The diagram and shape packages

The **shape** package provides functions for drawing a variety of geometric shapes and arrowheads and the **diagram** package provides functions for positioning shapes and drawing lines or curves between them. Together, these packages provide convenient functions for producing simple diagrams consisting of nodes and edges.

```
> library(diagram)
```

The function `coordinates()` provides a convenient way to calculate locations (on a zero-to-one scale) for a simple arrangement of nodes. Given a vector of n integers, this will calculate positions for nodes arranged in n rows, where each integer describes how many nodes are placed in each row. The following code calculates locations for eight nodes arranged two per row in four rows.

```
> nodePos <- coordinates(c(2, 2, 2, 2))
```

The locations are all on a normalized coordinate system, so a simple call to `plot.new()` will create a plot region within which these coordinates can be used.

```
> plot.new()
```

The function `straightarrow()` draws a line with an arrowhead on it. The following code shows an example using the node positions calculated to draw a line between node position 1 and node position 3. There are also functions for drawing curved lines or lines that travel between points in a city-block fashion.

```
> straightarrow(nodePos[1, ], nodePos[3,])
```

The function `textrect()` draws a piece of text within a rectangle (with a drop shadow). For example, the following code draws the label `"start"` within a rectangle at node position 1. Arguments to the function allow the rectangle and the text to be sized appropriately.

```
> textrect(nodePos[1, ], .05, .025, lab="start")
```

There are also functions for drawing text labels within ellipses, or diamonds, or with no surround at all.

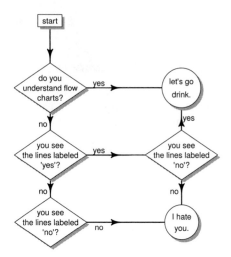

Figure 15.10
A flow chart about understanding flow charts produced using the **diagram** package (based on the xkcd comic strip `http://xkcd.com/518/`).

The flow chart in Figure 15.10 was created from the code above, plus several other similar calls to draw further lines and text labels. The full code is available from the book web site.

The **diagram** package also provides convenience functions for producing simple arrangements of networks of nodes and edges in a single function call, for example the `plotmat()` function.

Output from the **diagram** package is produced using traditional graphics; Section 7.5.2 describes some features of **grid** graphics that can be used to produce similar results based on **grid**.

Chapter summary

Node-and-edge graphs can be created using the **graph** package and laid out and rendered using **Rgraphviz**. The **igraph** package provides a complete alternative. The **diagram** package provides tools for producing more regular arrangements of nodes and edges, such as a flow chart, where the layout is determined by the user.

16

3D Graphics

Chapter preview

This chapter describes several approaches to drawing 3D images in R. Drawing in 3D requires several new concepts compared to drawing in 2D, so there is a brief introduction to those concepts. The core R graphics system only provides a 2D graphics engine, but several functions are described that provide limited 3D support based on that. Graphics based on a genuine 3D graphics engine is provided by the **rgl** package.

The core R graphics system provides a 2D graphics engine. All drawing occurs on a two-dimensional plane, usually based on simple (x, y) cartesian coordinates. This chapter looks at packages and functions that allow drawing in three dimensions, where locations are in terms of (x, y, z) triplets, it is possible to work with shapes that represent volumes rather than just areas, and a number of special effects are possible.

16.1 3D graphics concepts

The journey from a pair of data values to the location of a data symbol on a scatterplot, while not entirely trivial, is sufficiently straightforward that it can be taught to primary school children. It is still convenient, and more accurate, to have graphics software take care of the details of the drawing, but because the concepts are simple, it is simple to control the way that software draws

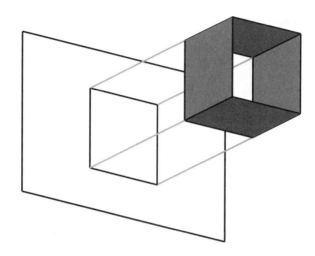

Figure 16.1

A diagram showing a 3D object being projected onto a 2D surface.

a two-dimensional plot (see, for example, the discussion of plot regions in Section 3.1 and the concept of viewports in Section 6.5).

The journey from a *triplet* of data values to the location of a data symbol in a *three-dimensional* plot is, unfortunately, a little more difficult to grasp. Although there is graphics software that can perform the necessary calculations to produce the drawing, because the concepts are more complex, a little more learning is required to be able to control the software that draws a three-dimensional plot.

Projections: The main step that has to occur is a projection from 3D space into 2D space (a page or a computer screen). A simple way to think about this is to imagine a 2D plane, like a sheet of paper, being placed into the 3D space. Each point in the 3D space can be projected onto the 2D plane by drawing a perpendicular line from the plane to the point (see Figure 16.1).

This simple projection is called an *orthogonal* projection. Different views of a 3D object are obtained by placing the 2D plane at different locations within the 3D space.

Viewpoints: An orthogonal projection has the advantage of making it easier to compare distances between locations in 3D space, but it does not provide good depth cues, so it may be hard for the viewer to perceive the image as three dimensional, and it does not produce a realistic-looking

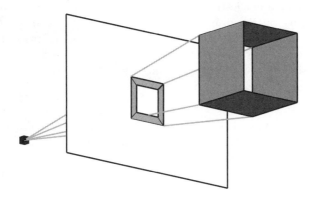

Figure 16.2
A diagram showing a 3D object being projected onto a 2D surface from a 3D viewpoint.

image, which can be a problem when drawing real-world objects. A more realistic three-dimensional image is obtained by using a *perspective* projection.

In this case, a 2D plane is located in 3D space as before, but then a viewing location in 3D space, called a *viewpoint*, is also defined. Now the 3D space is projected onto the 2D plane by drawing at the intersection of a line from each point in 3D space to the viewpoint (see Figure 16.2).

The result from a perspective projection creates a much more effective illusion of three dimensions. The amount of perspective distortion and the amount of the scene that is visible are determined by how close the viewpoint is to the viewing plane, how big the viewing plane is, and how close the viewing plane is to the 3D scene.

Typically, the viewpoint, the viewing direction, and the viewing plane are determined automatically so that the entire 3D scene is visible, but it may be necessary to modify their positions in order to obtain a more interesting view. Each of the systems encountered in this chapter will provide some way of specifying the three-dimensional view.

Aspect ratio: In a two-dimensional plot, the aspect ratio controls whether data are plotted within a square region or within a rectangle.

In a three-dimensional plot, the aspect ratio controls whether the data are plotted within a cube or within a cuboid (or rectangular prism). It is important to control the aspect ratio whenever any of the three data dimensions have different scales.

Transformation matrices: The transformation from a location in three dimensions to a location on a two-dimensional page or screen can be summarized by a single transformation matrix. This matrix can be useful for adding further drawing to a 3D plot because it allows new 3D locations to be projected into 2D in exactly the same was as the original plot. Each of the packages that are described in this chapter will provide some way of obtaining and working with this matrix.

Drawing primitives: Just like in two dimensions, a three-dimensional plot is composed mainly of points, lines, polygons (areas), and text. The difference in three dimensions is that a polygon is a two-dimensional *surface* with an orientation in 3D space. For example, a cube in 3D is composed of six identical polygons with different locations *and orientations*. It will usually be possible to add extra low-level output to a 3D plot just as extra output can be added to 2D plots in R.

An additional concept in three dimensions is that of a *sprite*. This is a two-dimensional image that *always* faces toward the viewpoint no matter where the three-dimensional scene is being viewed from. An example use of sprites is to draw text so that it can always be read no matter where a 3D plot is viewed from.

Parameters: The standard graphical parameters are available for controlling the appearance of shapes in three dimensions, for example, line widths, text size, and fill colors.

The situation can be more complex than in two dimensions though because the *lighting* of the 3D scene may be taken into account. Two surfaces that have an identical fill color but have different orientations will be shaded differently by the same light source. Using a light source can be useful for producing a more realistic-looking three-dimensional surface, though it involves the additional work of specifying a location and direction for the light source. Some functions that draw 3D surfaces provide a way to specify a light source for the scene.

Axes: Annotating a three-dimensional plot, particularly drawing axes, is harder than annotating two-dimensional plots because it is harder to automatically determine the layout and position of items like axes so that they do not obscure and are not obscured by other objects in the 3D scene. This means that default labeling and axes tend to be less useful in a three-dimensional plot and there is generally less support for customizing axes.

16.2 The Canterbury earthquake

This section describes a data set that will be used for examples throughout the chapter.

On September 4, 2010, the Canterbury region of the South Island experienced New Zealand's worst earthquake for 80 years. The main quake, which caused the majority of the damage, registered 7.1 on the Richter scale. This was followed by hundreds of smaller quakes over the following weeks.

The New Zealand GeoNet project[*] provides earthquake data for the whole of New Zealand, including location (longitude and latitude), depth (kilometers underground), and magnitude. Earthquakes occurring on September 4 and the following few days are available in the data frame NZquakes.

```
> head(NZquakes)

        LAT      LONG    MAG     DEPTH
1 -41.80190 174.2286 2.652   15.3622
2 -38.87300 175.8345 4.296  116.4915
3 -37.88906 176.9134 3.029    5.0000
4 -39.73341 176.9492 2.674   38.1103
5 -39.15628 174.9325 2.230   31.2663
6 -37.90366 176.9359 2.886    0.6285
```

16.3 Traditional graphics

The only traditional graphics function that draws a three-dimensional plot is the persp() function. Given a regular grid of x- and y-locations, plus z-values at each location, this function produces a three-dimensional surface from the z-values.

As an example, the following code generates a two-dimensional kernel density estimate for the geographic distribution of earthquakes in the Canterbury region.

---

[*]http://www.geonet.org.nz/.

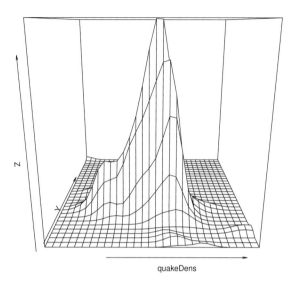

quakeDens

Figure 16.3
A 3D surface showing the two-dimensional density estimate of the locations of earth-quakes in the region of Darfield, New Zealand, following the 7.1 magnitude earth-quake on September 4, 2010.

```
> cantyQuakes <- quakes[quakes$LAT < -42.4 & quakes$LAT > -44 &
                quakes$LONG > 171 & quakes$LONG < 173.5, ]
> library(MASS)
> quakeDens <- kde2d(cantyQuakes$LONG, cantyQuakes$LAT, n=30)
```

This density estimate is drawn as a three-dimensional surface with the follow-ing code (see Figure 16.3).

```
> persp(quakeDens)
```

As this code demonstrates, the data to plot may be given as a single list with x, y, and z components instead of three separate arguments. The x- and y-values should be in ascending order and just provide the locations of the grid lines (they are *vectors* not matrices). The z argument provides a matrix of *heights* for the surface.

This example also demonstrates several of the difficulties in obtaining a useful view of a three-dimensional plot. The following code improves the view in several ways (see Figure 16.4).

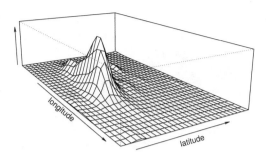

Figure 16.4
An improved view of the 3D surface showing the two-dimensional density estimate
of the locations of earthquakes in the region of Darfield, New Zealand, following the
7.1 magnitude earthquake on September 4, 2010.

```
> persp(quakeDens, scale=FALSE, expand=0.02,
        theta=60, d=.1, r=.1,
        xlab="longitude", ylab="latitude", zlab="")
```

The x-values in this plot are longitudes and the y-values are latitudes. By
setting `scales=FALSE`, the plot is created inside a cuboid with dimensions
that reflect the geographic range of the data (wider in an east-west direction
than in the north-south direction). The z-values are densities, with a much
wider range than either x or y, so `expand=0.02` is used to scale the z values
(to avoid having a very tall and thin cuboid).

The main feature of the density estimate is a ridge running east-to-west. This
is hard to see in the default view in Figure 16.3, so `theta=60` is used to specify
a different view point. The `theta` argument specifies an angle of rotation
about the z-dimension and there is also a `phi` argument to specify a rotation
up or down. The default view has `phi=15`, which provides a view looking
"down" on the drawn surface. With the change in aspect ratio, described
above, there is very little perspective distortion in the view, so `r=.1` is used
to shift the viewpoint closer to the viewing plane and `d=.1` is used to shift the
viewing plane closer to the scene, thereby restoring the perspective effect.

The axis labels have also been corrected via `xlab`, `ylab`, and `zlab`.

The `persp()` function also provides a basic lighting model via a `shade` argument (not used here). Specifying a value greater than 0 for the `shade`
argument generates a light source (the larger the value, the less diffuse the
light), which can be positioned via `ltheta` and `lphi`. The result is a different
shading on each surface to represent incident light.

Another option is to specify a color for each individual polygon of the surface,

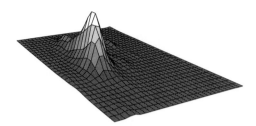

Figure 16.5

A 3D surface showing the two-dimensional density estimate of the locations of earth-quakes in the region of Darfield, New Zealand, following the 7.1 magnitude earth-quake on September 4, 2010. Each face on the surface is colored according to its (interpolated) z value (high points are white, low areas are dark gray).

for example, coloring polygons according to their z-value. This is slightly complicated by the fact that there are fewer polygons to color than there are z-values because the z-values give the locations of the polygon *vertices*. The following code shows one way to generate a set of z-based colors so that "higher" areas of the surface are a lighter color than "lower" areas (see Figure 16.5). This code also removes all axes and the bounding box via the **axes** and **box** arguments.

```
> zinterp <- with(quakeDens,
                  z[-1, -1] + z[-1, -ncol(z)] +
                  z[-nrow(z), -1] + z[-nrow(z), -ncol(z)])
> persp(quakeDens, scale=FALSE, expand=0.02,
        theta=60, d=.1, r=.1, axes=FALSE, box=FALSE,
        col=gray(.4 + 1:6/10)[cut(zinterp, 6)])
```

The return value from **persp()** is the transformation matrix for the plot. This can be used, in combination with the **trans3d()** function and functions such as **lines()**, **points()**, and **text()**, to add further lines, points, and text to the plot. An example was given in Section 3.4.6.

The major drawback to the **persp()** function is that it has a very simple approach to *hidden surface removal*—deciding which parts of a 3D scene are visible. This means that, except for very simple cases, it is not possible to produce more than one surface on a plot, and even drawing additional points and lines may not be possible. For more complex cases, a more sophisticated 3D system is required (see Section 16.6).

16.4   lattice graphics

The **lattice** package provides two functions for drawing three-dimensional plots: `wireframe()` and `cloud()`.

The `wireframe()` function produces a three-dimensional surface similar to `persp()`, with the additional benefit of the standard **lattice** multipanel conditioning feature. The data to plot may be provided simply as a matrix of z values or as a formula of the form z ~ x + y (where x and y should be locations on a regular grid).

The `cloud()` function produces a three-dimensional scatterplot, which consists of a set of points within a three-dimensional scene. In this case, the x- and y-values no longer have to reside on a regular grid.

In order to demonstrate the `cloud()` function, the following code further subsets the Canterbury earthquake data to retain only earthquakes that occurred closer than 20 km to the Earth's surface.

```
> shallowCantyQuakes <- subset(cantyQuakes, DEPTH < 20)
```

For both `cloud()` and `wireframe()`, the aspect ratio of the plot is controlled via the `aspect` argument. This consists of two numeric values: the first gives the ratio of the x-dimension to the y-dimension and the second gives the ratio of the z-dimension to the x-dimension. Alternatively, the plot region can be left as a cube and the scales can be manipulated via `xlim`, `ylim`, and `zlim`.

The viewpoint and the amount of perspective is controlled via the `screen` and `distance` arguments. The latter specifies the distance of the viewing plane from the 3D scene (0 to 1) and the former controls the view by rotating the *data*. The value of `screen` should be a list with components x, y, and z to specify a rotation of the data about the corresponding dimension. The viewpoint is along the z-dimension by default, so to view the z-values from side on, a rotation about x or y is necessary. The default `screen` is `list(z = 40, x = -60)`.

The following code produces several different views of the shallow earthquake data (see Figure 16.6).

```
> for (i in seq(40, 80, 20)) {
     print(cloud(-DEPTH ~ LONG + LAT, shallowCantyQuakes,
              xlim=c(171, 173), ylim=c(-44.5, -42.5),
              pch=16, col=rgb(0, 0, 0, .5),
              screen=list(z=i, x=-70)))
  }
```

It is much harder to perceive three-dimensional structures in a 3D scatterplot compared to a 3D surface plot, but presenting a series of views like this can help to provide further visual cues. In this case, the data are arranged in a tall thin cloud that runs east-to-west (along the longitude dimension). It is also possible to see horizontal bands (these are around 5 km and 12 km), which turn out to be artifacts of the process used to determine earthquake location and depth.

16.5   The scatterplot3d package

The **scatterplot3d** package provides a single function, scatterplot3d(), which produces a three-dimensional scatterplot. The distinction between this and the **lattice** cloud() function is that the scatterplot3d() output uses the traditional graphics system, whereas cloud() is based on **grid**.

The xyz.coords() function is used to transform any reasonable form of input into a set of x, y, and z values to plot. The data can be given as: a formula of the form z ~ x + y; three separate numeric vectors, or a list of three vectors; a matrix or data frame, in which case the first three columns are used.

This package only provides an orthogonal projection, but allows fine control over the detailed appearance of the plot, including axis labeling.

There is limited control over the direction from which the plot is viewed, via the **angle** argument. This specifies the angle between the x- and y-axes (from 0 to 180), where this is the angle between these axes *after* projection (i.e., in the two-dimensional viewing plane).

There is also some control over the aspect ratio via the **scale.y** argument, which controls the ratio of the y-axis to both x and z (the latter two are always equal length).

The return value from scatterplot3d() is a list of functions, which can be used to add further output to the plot. The xyz.convert component of this list is a function to convert three-dimensional coordinates into two dimensions

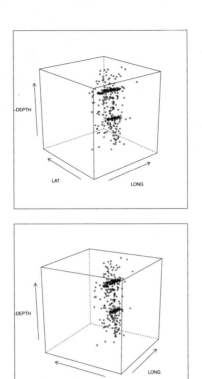

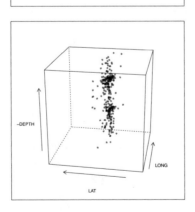

Figure 16.6

A 3D scatterplot showing the locations and depths of shallow earthquakes (less than 20km deep) in the region of Darfield, New Zealand, following the 7.1 magnitude earthquake on September 4, 2010.

so that the standard lines(), points(), and text() functions can be used
to add further output. The points3d, plane3d, and box3d components of
the list are functions that provide convenient special cases to add extra points
(or lines), a plane, or a bounding box to an existing 3D scatterplot.

The following code demonstrates working with the return value from the
scatterplot3d() function. The first step is to set up a three-dimensional
scatterplot of the shallow earthquake data using scatterplot3d() (see Figure 16.7). Because type="n", no points are drawn. Instead, the return value
is captured so that further output can be added to the plot.

```
> library(scatterplot3d)


> s3d <- with(shallowCantyQuakes,
            scatterplot3d(-DEPTH ~ LONG + LAT,
                          angle=30, scale.y=0.45, type="n",
                          pch=16, color=rgb(0, 0, 0, .5),
                          x.ticklabs=pretty(LONG, 3),
                          grid=FALSE, zlim=c(-20, 0)))
```

The next code adds a contour plot to the lower plane of the three-dimensional
scatterplot. This makes use of contourLines() to generate contour lines
from a two-dimensional density estimate of the shallow earthquake locations
(similar to the calculation in Section 16.3). Each of these contour lines is
upgraded to a three-dimensional line, by setting the z-value to -20, then
the three-dimensional lines are converted to two-dimensional lines using the
xyz.coords component of the result from scatterplot3d(). Each contour
is then drawn using the polygon() function, with a gray fill color based on
the contour level.

```
> quakeDensXY <- kde2d(shallowCantyQuakes$LONG,
                       shallowCantyQuakes$LAT, n=30)
> lapply(contourLines(quakeDensXY, nlevels=8),
        function(cl) {
            polygon(s3d$xyz.convert(cl$x, cl$y,
                                   rep(-20, length(cl$x))),
                    lwd=.5, col=gray(.8 - cl$level/20),
                    border=NA)
        })
```

Similar calculations and drawing are also performed to show the joint distribution of longitude and depth, which is drawn on the back plane of the plot,
and to show the joint distribution of latitude and depth (code not shown).

As a final step, the shallow earthquake points are drawn. These are drawn last so that the contour lines do not obscure any data points.

```
> with(shallowCantyQuakes,
        s3d$points3d(-DEPTH ~ LONG + LAT, pch=16,
                     col=rgb(0, 0, 0, .3)))
```

16.6   The rgl package

All of the previous approaches to three-dimensional plots only provide partial implementations of a three-dimensional graphics system. For example, none of them can properly perform hidden-surface removal to determine which parts of a three-dimensional scene are visible.

The topic of this section, the **rgl** package, has no such inhibitions. This package provides access to the OpenGL graphics system, which is a fully featured 3D graphics engine.

```
> library(rgl)
```

The **rgl** package does require a software implementation of OpenGL to be installed. On Windows and MacOS X systems, something appropriate should be installed by default, but users on Linux systems may need to install the Mesa 3D graphics library.*

The **rgl** package provides both high-level plotting functions that can produce entire three-dimensional plots, and low-level functions that can be used to construct a more general 3D scene from basic primitives.

The output from **rgl** drawing functions is produced in a special **rgl** graphics device. Only **rgl** functions can draw to these devices, but the resulting image is interactive (the view can be manipulated using the mouse) and it is possible to produce 3D images and effects that go well beyond what is possible with functions based on traditional graphics or **grid** graphics.

An example of a high-level **rgl** function is the `persp3d()` function, which produces a three-dimensional surface, similar to `persp()`. The following code

---

*http://www.mesa3d.org/.

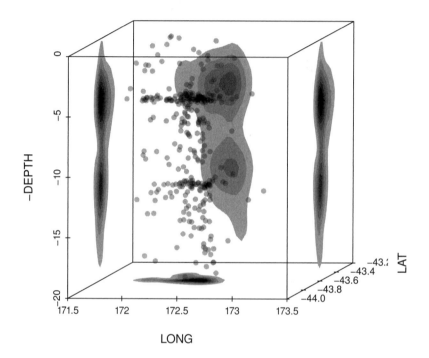

Figure 16.7

A 3D scatterplot showing the locations and depths of shallow earthquakes (less than 20 km deep) in the region of Darfield, New Zealand, following the 7.1 magnitude earthquake on September 4, 2010. Several contours of two-dimensional density estimates of the point cloud have been added.

Figure 16.8

A 3D surface plot of earthquakes.

draws a representation of the two-dimensional density estimate of Canterbury earthquake locations from Section 16.3 (see Figure 16.8).

```
> persp3d(quakeDens$x, quakeDens$y, quakeDens$z,
          aspect=c(1, 0.55, .2), col="white", box=FALSE,
          axes=FALSE, xlab="", ylab="", zlab="")
```

This code demonstrates the use of the `aspect` argument to control the aspect ratio of the resulting plot. This consists of three values, which specify the ratios for the three plot dimensions, relative to a cube with sides of length 1. The default axes and labels have all been turned off with standard arguments.

One important feature of this plot, compared to the shaded `persp()` plot in Figure 16.4, is that the surface shading is very smooth. This is an example of the effect of the more sophisticated underlying OpenGL graphics engine. The color of the surface is specified at each *vertex* and the system interpolates across each face of the surface to produce a smooth result. All of these features are optional, for example, a non-smooth shading is still possible by setting the `smooth` argument to `FALSE`.

Another important feature of the plot produced by `persp3d()` is that it is interactive. The view can be manipulated by clicking and dragging with the mouse. This removes the need to explicitly set a 3D viewpoint and being able to rapidly change the view makes it much easier to perceive the three-dimensional features of a plot. Section 17.2.2 describes another package that provides this sort of facility.

On the downside, because the plot is not drawn in a normal R graphics device, some facilities are not available. For example, it is not possible to draw mathematical formulae using R expressions (see Section 10.5) and there is no built-in support for multiple plots let alone the multipanel conditioning offered by **lattice** and **ggplot2**. It is also not possible to save the plot in the usual array of file formats (see Section 9.2), though the **rgl** package does provide the snapshot3d() function for generating a raster image (e.g., a PNG file) and the rgl.postscript() function to save the scene to a PostScript file.

In addition to the persp3d() function, there is a plot3d() function for producing a 3D scatterplot, including the option of drawing 3D spheres for data points.

The **rgl** package also provides a number of low-level functions that add three-dimensional objects to an existing 3D scene, such as lines3d(), points3d(), and surface3d(). The text3d() function adds text that can be positioned in 3D space, but is always oriented toward the viewing plane.

Importantly, the **rgl** package is capable of hidden surface removal, so additional 3D objects can be added to a plot without having to worry about which objects obscure each other. This means that there is no need to work directly with a transformation matrix. That complexity is all handled by the underlying OpenGL system.

In order to demonstrate this, the following code makes use of the **misc3d** package, which builds on *rgl* and offers functions for drawing three-dimensional contour surfaces. The following code uses the kde3d() function to generate a three-dimensional density estimate for the locations of the shallow earthquakes.

```
> library(misc3d)
```

```
> d <- with(shallowCantyQuakes,
            {
                kde3d(LONG, LAT, -DEPTH,
                    h=c(.1, .1, 2), n = 30)
            })
```

The next code draws a 3D scatterplot of the shallow earthquakes and then adds two semitransparent 3D contours to the plot (see Figure 16.9).

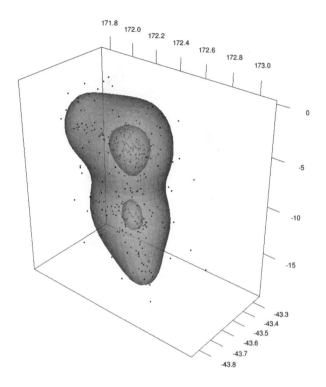

Figure 16.9

A 3D contour plot of earthquakes.

```
> with(shallowCantyQuakes,
      {
          plot3d(LONG, LAT, -DEPTH,
                aspect=c(1, 0.55, 1),
                axes=TRUE, box=FALSE,
                xlab="", ylab="", zlab="")
          contour3d(d$d, c(.4, .1), d$x, d$y, d$z,
                  color=rep("gray80", 2),
                  color2="gray", specular="black",
                  engine="rgl", add=TRUE, alpha=.5)
      })
```

The **rgl** package can also be used at a lower level, which allows a scene to
be constructed by creating and positioning 3D objects, plus there are more
advanced 3D effects such as lighting and surface reflectance that can be con-

Figure 16.10
A 3D image that is not a statistical plot. Any resemblance to persons either real or fictional is purely coincidental.

trolled.

It is possible to generate basic three-dimensional shapes with functions like `ellipse3d()` and `cylinder3d()`, position the shapes through `rotate3d()` and `translate3d()`, then view them with `shade3d()`. The viewpoint can be explicitly set using `par3d()`, light sources can be added with `light3d()`, and the material properties of surfaces controlled with `material3d()`. It is also possible to specify a bitmap image as a *texture* for a 3D surface.

An example of an image constructed using these lower-level **rgl** functions is shown in Figure 16.10. The code is available from the book web site.

16.7   The vrmlgen package

Another approach to producing 3D images is provided by the **vrmlgen** package. Rather than drawing to screen, this package saves 3D images in two file formats that can then be viewed in a web browser: VRML and LiveGraphics3D.*

```
> library(vrmlgen)
```

The package provides both high-level functions for producing complete plots, such as `cloud3d()`, and low-level functions for building 3D scenes from simple primitives, such as `points3d()`, `lines3d()`, and `text3d()`.

The following code produces a VRML file containing a 3D scatterplot of the Canterbury earthquake data and Figure 16.11 shows the file being viewed in a web browser.

```
> with(shallowCantyQuakes,
        cloud3d(LONG, LAT, -DEPTH,
                filename="vrmlgen.wrl",
                cols="white"))
```

Like **rgl**, one benefit of the **vrmlgen** package is that it is possible to interact with the resulting 3D image. One additional benefit with **vrmlgen** is that the resulting image can be viewed with just a web browser — R is not needed to view and interact with the image — so it is easy to share the files widely.

---

*http://www.web3d.org/x3d/specifications/vrml/;
http://www.vis.uni-stuttgart.de/~kraus/LiveGraphics3D/.

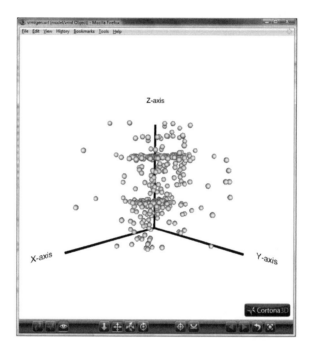

Figure 16.11
A 3D image that has been saved in a VRML format. The image is being viewed in
a web browser.

Chapter summary

It is possible to produce basic 3D scatterplots and 3D surfaces in traditional graphics, with the `persp()` function or the **scatterplot3d** package. The `cloud()` and `wireframe()` functions provide similar facilities in **lattice** graphics. For sophisticated 3D images and effects, there is the **rgl** package.

17

Dynamic and Interactive Graphics

Chapter preview

This chapter describes some functions and packages for producing dynamic and interactive graphics. The core R graphics system is designed for producing static graphics, so this chapter introduces some new graphical concepts as well as a number of extension packages. As well as packages that provide ready-made interactive graphics, there is a section on packages that provide tools for developing new interactive graphics.

The strength of the core R graphics engine lies in the production of complex static plots with flexible control of fine details. It is possible to create some simple dynamic and interactive effects, but for anything remotely complex one of the extension packages described in this chapter is required.

17.1 Dynamic graphics

A dynamic plot is one which changes over time, such as a plot of a stock index that is constantly updated. The difference between a static plot and a dynamic plot is the difference between a photograph and a video.

The simplest approach to producing a dynamic plot with R is to generate a sequence of plots and show them rapidly one after the other. In simple cases, this can be achieved with a normal graphics window. For example, the

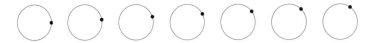

Figure 17.1
Seven frames from a simple animation of a dot traveling in a circle.

following code produces a dot traveling around the circumference of a circle. On some platforms, the display may flicker badly as each new frame is drawn. Figure 17.1 shows a sequence of the frames in this animation.

```
> n <- 40
> t <- seq(0, 2*pi, length=n)
> x <- cos(t)
> y <- sin(t)

> for (i in 1:n) {
      plot.new()
      plot.window(c(-1, 1), c(-1, 1))
      lines(x, y)
      points(x[i], y[i], pch=16, cex=2)
      Sys.sleep(.05)
  }
```

An alternative approach is to save each frame of the animation to a separate file and then use third-party software, such as **ffmpeg**,* to combine the frames together into a movie format. The **animation** package provides several convenient functions that implement variations on this approach.

17.1.1   The animation package

The **animation** package can produce animation files in a variety of formats, for example, an animated GIF, an Adobe Flash animation, or a web page with the animation embedded in it.

In order to demonstrate each of these, the following code defines a function, `orbit()`, which will do the drawing of the animation frames.

---

*http://ffmpeg.org/.

```
> orbit <- function() {
      par(pty="s", mar=rep(1, 4))
      for (i in 1:n) {
          plot.new()
          plot.window(c(-1, 1), c(-1, 1))
          lines(x, y)
          points(x[i], y[i], pch=16, cex=2)
      }
  }
```

Producing an animated GIF is now simply a matter of calling the `saveMovie()` function, passing a call to the function `orbit()` as the first argument. Other arguments specify the delay between frames, where to put the files that are generated, and what to call them.

```
> library(animation)
```

```
> saveMovie(orbit(), interval=0.05, moviename="orbit.gif")
```

This function uses ImageMagick to create the final file, so that software must also be installed.

Producing an Adobe Flash animation is also very straightforward, this time using the function saveSWF(). This function requires that the SWFTools software is installed.[*]

```
> saveSWF(orbit(), interval=0.05, swf.name="orbit.swf")
```

Generating a web page with an embedded animation is only slightly more effort. The first step is to call the `ani.options()` function, which sets up basic parameters of the animation. Next, the `ani.start()` function is called to initialize some files for the web page. The third step is to call the function that draws the animation frames, `orbit()` in this case, and the final step is to call the `ani.stop()` function, which finishes off the creation of the web page. Figure 17.2 shows the resulting web page.

```
> ani.options(interval=0.05, outdir="orbitImages",
              filename="orbit.html")
> ani.start()
> orbit()
> ani.stop()
```

---

[*]`http://www.swftools.org/`.

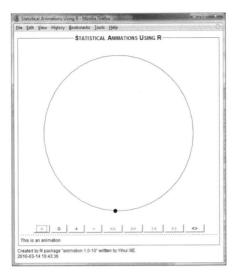

Figure 17.2

An animation embedded in a web page, as produced by the **animation** package.

17.2   Interactive graphics

An interactive plot is dynamic in the sense that it can change rapidly, but the changes occur as a result of user input. The difference between a dynamic plot and an interactive plot is the difference between a video and a video game.

In order to demonstrate the usefulness of interactive graphics, it is useful to consider the following static graphics example.

Figure 17.3 shows two simple *static* scatterplots of miles-per-gallon versus engine displacement for the mtcars data. In both plots, a subset of the points has been highlighted: cars with three-speed gearboxes on the left and cars with four-speed gearboxes on the right. Figure 17.4 shows a similar plot, using **lattice** to draw separate panels for different gearboxes.

These graphs allow us to inspect the relationship between two continuous variables (mpg and disp) and one categorical variable (gear) all at once by presenting two different *views* of the same data.

If the categorical variable had more categories, more plots (or panels) would be required. Similarly, another categorical variable of interest (e.g., number of cylinders in the engine) could be accommodated by producing even more

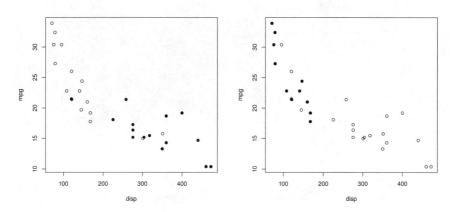

Figure 17.3
Two static plots of miles-per-gallon versus engine displacement for the mtcars data.
In the left-hand plot, points corresponding to cars with a three-speed gearbox have
been highlighted and, in the right-hand plot, points have been highlighted for four-
speed gearboxes.

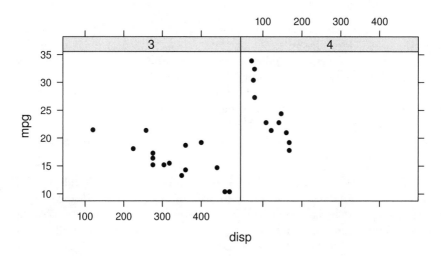

Figure 17.4
A static **lattice** plot showing miles-per-gallon versus engine displacement for the
mtcars data, with separate panels for cars with a three-speed gearbox and cars with
a four-speed gearbox.

Figure 17.5

Two interactive plots: an interactive scatterplot showing miles-per-gallon versus engine displacment for the `mtcars` data and a separate barchart representing the frequencies of cars with different numbers of gears. The bar for three-speed gearboxes has been selected in the barchart.

plots. However, at some point the number of plots will exceed the space available on the page (or the size of each plot will become too small to see).

Figure 17.5 shows an interactive graphics approach to this situation. In this case, there are two different types of plots: a scatterplot of miles-per-gallon versus engine displacement and a barchart of number of gears. The plots are interactive so that, for example, if the user clicks within a bar in the barchart, that bar is highlighted. Furthermore, the plots are *linked* so that, for example, if a bar is highlighted in the barchart, the corresponding points in the scatterplot are also highlighted. In Figure 17.5, the bar representing cars with three-speed gearboxes has been selected and the corresponding points in the scatterplot are also highlighted.

Through interacting with these plots, it is possible to inspect the relationship between three variables, just like in Figures 17.3 and 17.4. However, this interactive approach scales much better to larger numbers of categories and a greater number of variables and it provides much greater flexibility. For example, Figure 17.6 shows the same interactive plots as Figure 17.5, but this time a selection has been made in the scatterplot. Now, when points in the scatterplot are selected, corresponding regions of the bars in the barplot are also highlighted.

The important point is that it is possible to ask many more questions of the data by interacting with these plots compared to the static plots and the

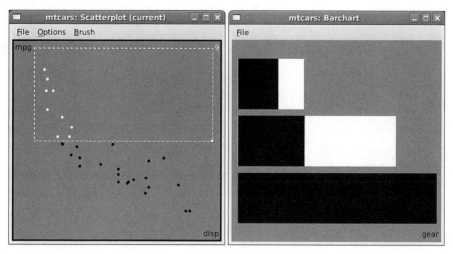

Figure 17.6

Two linked plots: an interactive scatterplot showing miles-per-gallon versus engine displacment for the `mtcars` data and a separate barchart representing the frequencies of cars with different numbers of gears. The points for cars with a miles-per-gallon greater than 20 have been selected in the scatterplot.

answers to these questions are generated very rapidly.

The importance of interactive statistical graphics is that it allows *many different views* of the data to be produced and explored *very rapidly.*

There are of course some negatives to interactive graphics, for example, it is harder to generate and save a record of the actions taken when using interactive graphics (because it is harder to save and share mouse movements than it is to save and share code) and plots tend to be less ornate because the graphics systems are focused more on speed than on fine control of the details of a plot.

17.2.1 Tools and techniques

There are many examples of interactive graphics software, but there are many interactive features that they all have in common. This section describes some of the standard sorts of interactions that are provided by interactive statistical graphics systems. Some specific software tools are described in the next sections. The book *Interactive Graphics for Data Analysis* by Martin Theus and Simon Urbanek provides a more in-depth discussion.

Brushing and linking: The idea of *brushing* is that the user can click or drag regions with the mouse to select subsets of the data. This may

just serve to highlight a particular subset, but more often it is combined
with *linking*, whereby selections in one plot also highlight corresponding
points in a separate, linked plot.

The main idea is that the view of the data is altered by highlighting
only a subset of the plots so that they are visually distinguishable from
the parts of the plot that are not selected.

There are many variations on the basic idea: the "brush" can be a single
shape, a rectangular region, or a free-form shape (a "lasso"); the highlighting that is produced by brushing may be persistent or transient;
and, when highlighting is persistent, it may be possible to combine one
brush operation with another (in effect specifying more than one subsetting condition).

Furthermore, different sorts of plots present different graphical elements
for selection. For example, a barchart offers rectangular regions, while a
scatterplot offers only point symbols, and it makes sense to select either
edges or nodes in a node-and-edge graph.

Identification: The idea of *identification* is that the user can hover the mouse
over or click the mouse on a graphical element in a plot in order to obtain
more information about the underlying data. The main idea is that the
view of the data is augmented by presenting more detailed information
about the selected element.

The additional information may be drawn next to the selected item or it
could be presented in a *tooltip* (a separate small window of information
that is temporarily shown on top of the plot). In the extreme case, a
mouse click may generate a new display entirely (e.g., a hyperlink or a
"drill-down").

Zooming: The idea of *zooming* is to change the scale on a plot so that only
a subset of the data is visible. A classic example is a time series that
extends over several years and we want to be able to view only one year
at a time or one month at a time as well as viewing the entire time
series. The main idea is that it is possible to rapidly alter the subset of
the data that is being viewed.

Dragging: The idea of *dragging* is to click (to select) and then drag elements
of the plot using the mouse. One way to view this technique is that
the interaction can be used to make changes to the actual data. This
may be a dangerous thing to do, but it has applications in teaching and
sensitivity analysis. A simple example might involve moving a single
data point on a scatterplot to see the impact on a fitted regression line.

Alternatively, dragging may be applied to some feature of the display.
Rather than altering the data values, the interaction modifies the manner in which the data are being viewed. A good example of this sort of

interaction is the ability to rearrange the order of bars within a barplot or the order of variables within a parallel coordinates plot. This sort of interaction can also be used with node-and-edge graphs to manually fine-tune an automated graph layout.

It is possible to achieve some of these effects with the core R graphics system. For example, the `identify()` function in traditional graphics allows interactive labeling of points in a scatterplot. The `panel.identify()` function in **lattice** provides a similar feature and there is a lower-level `grid.locator()` in **grid**. However, for sophisticated interactive graphics, a more complete solution is required.

The following sections describe two packages that provide these interactive graphics features. In both cases, the packages provide their own graphics systems, independent of core R graphics.

17.2.2   The rggobi package

The **rggobi** package provides an interface between R and the GGobi software for dynamic and interactive graphics.* This section provides a very brief introduction to the **rggobi** package. The book *Interactive and Dynamic Graphics for Data Analysis* by Di Cook and Deborah Swayne provides a much more complete description of the package and the underlying GGobi system.

The **rggobi** package requires GGobi to be installed, which in turn requires the GTK+ software library.†

```
> library(rggobi)
```

In the most basic case, this package just simplifies the task of getting data from R into GGobi.

```
> gg <- ggobi(mtcars)
```

The `ggobi()` function starts up a GGobi session using the supplied data frame as the data set to explore. Figure 17.7 shows the initial GGobi window that is created by the code above.

At this point, it is possible to leave R altogether and simply interact with the GGobi system through its windows and dialogs. For example, the D̲isplay

---

*http://www.ggobi.org/.
†http://www.gtk.org/.

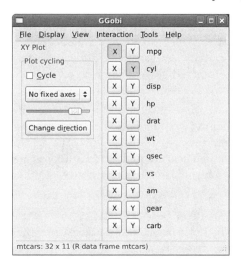

Figure 17.7

The GGobi main window after starting GGobi from R with the `mtcars` data set.

menu can be used to produce new plots and the buttons on the main GGobi window can be used to select which variables are used within each plot.

Alternatively, some of these operations can be performed from R. For example, the `display()` function can be used to generate new plots within the GGobi session. The following code creates the scatterplot and a barchart from Figure 17.5.

```
> display(gg[1], vars=list(X="disp", Y="mpg"))
> display(gg[1], "Barchart", vars=list(X="gear"))
```

GGobi can produce scatterplots, barcharts, scatterplot matrices, parallel coordinate plots, and time series. A scatterplot encompasses both the traditional two-dimensional scatterplot and a one-dimensional, density view of a single variable. The latter is, by default, an average shifted histogram, but it can also be viewed as a jittered strip plot. A barchart may also be viewed as a spineplot.

GGobi provides many ways to interact with the data. For example, with the barchart from Figure 17.5 active, Brush interaction can be selected from the Interaction menu on the main GGobi window. This produces a small square brush cursor in the barchart window; clicking in one of the bars in the barchart causes that bar to be highlighted *and* the corresponding points in the scatterplot to be highlighted (as in Figure 17.5). The documentation for GGobi describes what other sorts of interactions are possible.

Again, something similar to the interaction just described can be produced by writing code from R rather than via the menus and dialogs of the GGobi system. The following code uses the `colorscheme()` function to set the default GGobi color scheme to a set of three distinct hues and the `glyph_color()` function to assign different hues to different levels of the `gear` factor.

```
> colorscheme(gg) <- "Accent 3"
> mtcarsGG <- gg["mtcars"]
> glyph_color(mtcarsGG) <- mtcars$gear
```

This sort of interactive graphics with GGobi via R code is not very effective because it removes the main point of interactivity, which is to be able to generate multiple views of the data very rapidly. However, there are several reasons why it makes sense to interact with GGobi via R code rather than via GGobi's native menus and dialogs: writing code produces a record of the interaction; GGobi draws plots much faster than R graphics so a programmed series of plots can be viewed more comfortably; and ultimately, R code can be used to augment the capabilities of GGobi, for example, by providing data analysis techniques and even for programming new types of interactive plots.

It is important to note that the GGobi plots are produced in separate windows that are under the control of the GGobi system. It is not possible to add any output to those plots using core R graphics functions. On the other hand, GGobi does provide a Save Display Description option in its Tools menu, which produces a file that can be read into R and drawn with the **DescribeDisplay** package.

Tours

One of the main uses of GGobi is to visualize multidimensional data and, besides the interactive techniques, the main tool that GGobi provides for this purpose is the *tour*. This consists of a series of projections of many dimensions into only one or two dimensions, typically with the purpose of identifying "interesting" projections, such as separate clusters of points in the data. This feature combines both dynamic and interactive graphics because a series of different projections is generated and rapidly displayed, while at the same time interactions such as brushing, possibly of linked plots, can be combined to explore a high-dimensional data set.

17.2.3   The iplots package

Another option for interactive graphics is provided by the **iplots** package. This package provides functions for creating plots that are similar to the

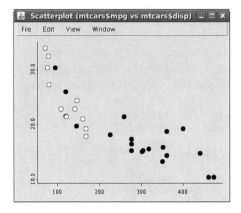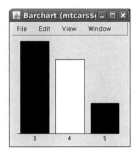

Figure 17.8
Two **iplots** interactive plots showing miles-per-gallon versus engine displacement for
the `mtcars` data and a separate barchart representing the frequencies of cars with
different numbers of gears. The bar for four-speed gearboxes has been selected in
the barchart.

standard plot types, but which respond to user interaction and are automatically linked. For example, the following code uses the `iplot()` and `ibar()`
functions to produce a scatterplot and a linked barchart from the `mtcars` data
frame, similar to those shown in Figure 17.5.

```
> library(iplots)
```

```
> iplot(mtcars$disp, mtcars$mpg)
> ibar(mtcars$gear)
```

A click in one of the bars of the barchart will highlight that bar *and* highlight
corresponding points in the scatterplot. Figure 17.8 shows the result of clicking
in the bar that represents cars that have a four-speed gear box. A selection
can also be made by clicking and dragging to specify a rectangular subregion
of a plot.

One point of interest with the **iplots** package is that it adds interactivity
to several additional plot types. The `ibox()` function produces interactive
boxplots, the `imap()` function produces interactive maps, and there is an
`imosaic()` for producing interactive mosaic plots for exploring multivariate
categorical data.

Like **rggobi**, plots produced by the **iplots** package appear in separate graphics
windows and normal R graphics functions cannot be used to add output to
those plots. However, there are **iplots** functions for adding annotations to
iplots plots: the `iabline()` function for straight lines, `ilines()` for paths

and polygons, and `itext()` for text labels. For example, the following code adds the car names to the interactive scatterplot that was produced above. The call to the function `iplot.set()` is necessary to make the scatterplot the active plot. The **iplots** package provides functions for navigating between multiple plot windows similar to those provided for normal R graphics devices.

```
> iplot.set(1)
> itext(mtcars$disp, mtcars$mpg, rownames(mtcars))
```

In addition to manual interaction with **iplots** plots, it is possible to programmatically affect the plots. For example, the `iset.select()` function can be used to highlight a subset of the data; the following code highlights cars with four-speed gear boxes, in both the scatterplot and the barchart.

```
> iset.select(mtcars$gear == 4)
```

This function also allows the specified subset to be combined with a preexisting selection, either by union or intersection. Conversely, there is a `iset.selected()` function that can be used to obtain the indices of selected observations following a manual selection.

Developing new interactive plots

Another interesting feature of the **iplots** package is that it provides facilities for extending the interactive behavior of a plot. This is currently based on the `ievent.wait()` function, which makes it possible to write R code that will respond to user interaction on a plot.

The following code provides a simple demonstration. First of all, the text labels that had been added above are removed from the scatterplot. This can be done using the `iobj.rm()` function.

```
> iplot.set(1)
> iobj.rm()
```

The next code adds new text labels to the interactive scatterplot, but each of these labels is only visible if the corresponding data point is currently selected. Figure 17.9 shows the result

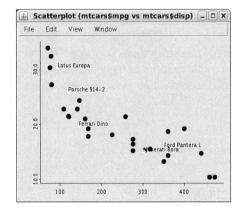

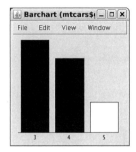

Figure 17.9
Two **iplots** interactive plots showing miles-per-gallon versus engine displacment for
the mtcars data and a separate barchart representing the frequencies of cars with
different numbers of gears. The scatterplot has been customized so that labels are
drawn for selected points (and the bar for five-speed gearboxes has been selected in
the barchart).

```
> labels <- mapply("itext",
                   mtcars$disp, mtcars$mpg, rownames(mtcars),
                   MoreArgs=list(visible=FALSE), SIMPLIFY=FALSE)
> olds <- NULL
> while (!is.null(ievent.wait())) {
      if (iset.sel.changed()) {
          s <- iset.selected()
          if (length(s) > 1) {
              lapply(labels[s], iobj.opt, visible = TRUE)
          }
          if (length(olds) > 1) {
              lapply(labels[olds], iobj.opt, visible = FALSE)
          }
          olds <- s
      }
  }
```

The ievent.wait() function is similar to the locator() function in traditional graphics in that it blocks the R console, so it is only possible to interact
with **iplots** windows—the R command line is unresponsive until the above
code is terminated (by selecting Break from the File menu of an **iplots** window).

The **iplots** package is built on top of the **rJava** package so Java must be
installed for this package to work.

At the time of writing, a new version of **iplots**, called **Acinonyx**, is being developed to offer greater speed and flexibility. This may become the preferred package in the future.

17.3 Graphics GUIs

This section is similar to the previous section on interactive graphics, because it also deals with graphics that change in response to user interaction. However, where Section 17.2 was more concerned with the user interacting with elements of the graphical image itself, such as lines, points, and axes, in this section, the interaction is typically with components that are separate from the image, such as menus, dialogs, buttons, and sliders.

Another distinction is that, in this section, graphics is usually produced on a standard R graphics device, rather than using external software to do the drawing, so all normal R graphics functions can be used to produce graphics output.

17.3.1 GUIs for R

On Windows and MacOS X, the standard interface for R consists of menus and dialogs in addition to the basic R command-line. These interfaces allow some simple graphical operations to be performed via the mouse, such as printing a plot or saving it in a different format, but do not provide interactive graphics in the sense used in this chapter.

A number of extension packages provide alternative GUIs for R. Some of these provide provide further convenience for *creating* plots, for example the **Rcmdr** package provides dialogs for the user to fill in the relevant parameters for the plot. Figure 17.10 shows the dialog box from the **Rcmdr** package that is used to create a scatterplot.

A slightly more advanced example is the **pmg** (Poor Man's GUI) package, which in addition to providing a similar dialog interface, has a "Lattice explorer" window (available from the Plots menu). This allows variables to be dragged onto a canvas to create plots. Figure 17.11 shows a **lattice** xyplot() that was created using this drag-and-drop feature.

Another quite sophisticated example of the plot dialog approach is the **latticist** package. The latticist() function from this package creates a window

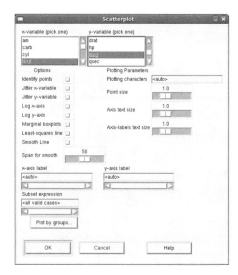

Figure 17.10
An **Rcmdr** dialog box for creating a scatterplot. This dialog is being used to create
a scatterplot of miles-per-gallon as a function of engine displacement for the mtcars
data frame.

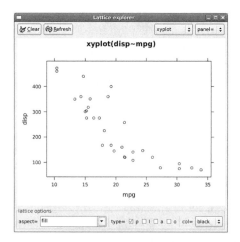

Figure 17.11
The Lattice Explorer from the **pmg** package. This plot has been created by dragging
the variables disp and mpg from a spreadsheet-like view of the mtcars data frame
onto the Lattice Explorer window.

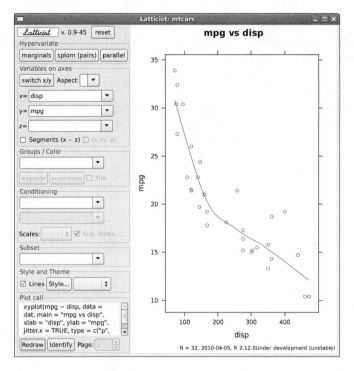

Figure 17.12
A **latticist** window showing a scatterplot of miles-per-gallon versus engine displacement for the `mtcars` data frame. The plot may be interactively modified by manipulating the menus and buttons to the left of the plot.

containing a **lattice** plot, with a large selection of interactive elements alongside that allow the plot to be modified in many ways, including conditioning the data into multiple panels. The following code creates a **lattice** scatterplot from the `mtcars` data frame. Figure 17.12 shows the resulting window.

```
> library(latticist)
> latticist(mtcars, list(xvar="disp", yvar="mpg"),
            use.playwith=FALSE)
```

The **playwith** package differs from the other R graphics GUIs because it also provides direct interaction with the elements of the plot, such as brushing data points. However, unlike the packages in Section 17.2, this interaction is with a normal R graphics device, so the plots have all of the variety and fine detail of standard R plots.

As an example, the following code uses the `playwith()` function to create two windows that are linked so that brushing points in one window highlights the relevant points in *both* windows (see Figure 17.13). In this example, the three cars with the greatest weight (`wt`) are shown to be the three cars with the largest engines (`disp`).

```
> library(playwith)
> playwith(xyplot(mpg ~ disp, mtcars))
> playwith(xyplot(qsec ~ wt, mtcars),
            new=TRUE, link.to=playDevCur())
```

It is also possible to interactively rescale or zoom the plots in the **playwith** window.

The interactivity provided by this package is not as flexible as that provided by the packages in Section 17.2. For example, brushing a scatterplot has no effect on a linked barchart, but it is nevertheless an impressive augmentation of R's normally static-only graphics.

The **latticist** package is integrated with **playwith** by default (which explains the `use.playwith=FALSE` in the previous code example).

17.3.2   GUI toolkits

This section looks at packages that work at a lower level than the packages in the previous section. The packages in this section provide the basic interactive tools that can be used to create GUIs like **Rcmdr** and its ilk.

This section provides a very brief introduction to creating a custom GUI in R. The book *Programming Graphical User Interfaces with R* by John Verzani and Michael Lawrence provides a thorough treatment of this area.

A GUI toolkit consists of three main parts: there are functions to to create GUI components, such as buttons, menus, and dialog boxes, which are often collectively referred to as *widgets*; there are functions to arrange these widgets next to each other within a container, such as a window; and there is some way to link R functions, called *event handlers*, to the actions of those components.

Two examples of GUI toolkits that are conveniently available in R are provided by the **tcltk** package and the **RGtk2** package. Another major set of packages, that are still in development at the time of writing, is focused on the Qt toolkit library.[*] One advantage of the **tcltk** package is that it is distributed as part of the default R installation, along with the underlying tcltk software library.[†]

---

[*]http://qtinterfaces.r-forge.r-project.org/.
[†]http://www.tcl.tk/.

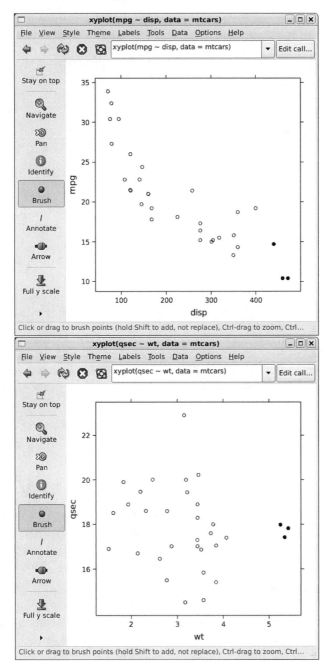

Figure 17.13
Two **playwith** scatterplots showing linked brushing of points. The top plot shows miles-per-gallon versus engine displacement and the bottom plot shows quarter-mile times versus weight.

```
 1 drawClock <- function(hour, minute) {
 2     t <- seq(0, 2*pi, length=13)[-13]
 3     x <- cos(t)
 4     y <- sin(t)

 6     grid.newpage()
 7     pushViewport(dataViewport(x, y, gp=gpar(lwd=4)))
 8     # Circle with ticks
 9     grid.circle(x=0, y=0, default="native",
10                 r=unit(1, "native"))
11     grid.segments(x, y, x*.9, y*.9, default="native")
12     # Hour hand
13     hourAngle <- pi/2 - (hour + minute/60)/12*2*pi
14     grid.segments(0, 0,
15                      .6*cos(hourAngle), .6*sin(hourAngle),
16                      default="native", gp=gpar(lex=4))
17     # Minute hand
18     minuteAngle <- pi/2 - (minute)/60*2*pi
19     grid.segments(0, 0,
20                      .8*cos(minuteAngle), .8*sin(minuteAngle),
21                      default="native", gp=gpar(lex=2))
22     grid.circle(0, 0, default="native", r=unit(1, "mm"),
23                 gp=gpar(fill="white"))
24 }
```

Figure 17.14

A function that draws an analog clock for a specified time.

Rather than address these packages that interface directly to GUI toolkit libraries, the focus in this section will be instead on the package **gWidgets**, which abstracts the problem to a slightly higher level.

An interactive clock

To provide an example of building a simple graphical GUI from scratch, this section describes a simple interactive clock that can be used to help teach children to read the time from an analog clock.

The core image in this example is a representation of a clock: a circle with 12 ticks plus two hands to show the hours and minutes. Figure 17.14 shows a simple R function that uses **grid** to draw this clock for a given time specification.

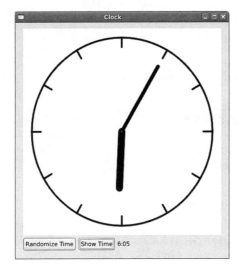

Figure 17.15

A clock image with a simple GUI, created using **gWidgets** and **gWidgetsRGtk2**.

The goal is to produce an interactive clock as shown in Figure 17.15. Besides the clock face image itself, there is a button to randomize the time shown on the clock and there is a button to display the time as text.

The gWidgets package

The idea of the **gWidgets** package is to provide a single consistent interface to creating GUIs in R. It must be used in combination with a "backend" package that implements the **gWidgets** functionality in a lower-level toolkit. For example, the **gWidgetsRGtk2** package provides a backend based on the **RGtk2** package, which in turn is based on the GTK+ toolkit. Other backends are also available.

```
> library(gWidgetsRGtk2)
```

The first step involved in creating a GUI is to create the widgets and lay them out. The first widget is usually a window to hold everything else. The following code creates a window with the gwindow() function.

```
> window <- gwindow("Clock")
```

The next widget is just a container to arrange the contents of the window. This is created using the ggroup() function. The resulting widget is placed within the window that was created in the previous step and, in this case, the widgets that are placed within this container will be arranged vertically.

```
> allContent <- ggroup(container=window, horizontal=FALSE)
```

The next widget is a graphics canvas, which is created by the `ggraphics()` function. The widget is a normal R graphics device, so it is possible to draw any R graphics output within it. This is placed within the `allContent` widget from the previous step.

```
> graphicTime <- ggraphics(container=allContent)
```

The remaining widgets are to be arranged horizontally, so a new container is created to hold these.

```
> timeContent <- ggroup(container=allContent)
```

The following code creates a label widget, with the `glabel()` function, but this widget is not yet placed in a container. That will be done later. The text label is blank initially because it will be set and changed by the user interacting with the buttons.

```
> textLabel <- glabel("")
```

The buttons require a little more code to create because it is necessary to provide an R function *event handler* that will be run when the button is clicked. The following code defines the function for the Randomize Time button. This function generates a new time, draws a clock representing that time, with the `drawClock()` function shown in Figure 17.14, makes the `textLabel` widget (created above) invisible, with the `visible()` function, and sets the text of that label to be the appropriate time, with the `svalue()` function.

```
> randomizeTime <- function(h, ...) {
      hour <- sample(1:12, 1)
      minute <- sample(seq(0, 55, 5), 1)
      drawClock(hour, minute)
      visible(textLabel) <- FALSE
      svalue(textLabel) <- paste(hour,
                                 sprintf("%02d", minute),
                                 sep=":")
  }
```

The button itself is created with the following call to the gbutton() function.

```
> reset <- gbutton("Randomize Time",
                   handler=randomizeTime)
```

The final button also needs an event handler, but this time it is quite simple because all it needs to do is make the textLabel widget visible.

```
> textButton <- gbutton("Show Time",
                        handler=function(h, ...) {
                            visible(textLabel) <- TRUE
                        })
```

The final step is to add these buttons and the label widget to the timeContent container, using the add() function.

```
> add(timeContent, reset)
> add(timeContent, textButton)
> add(timeContent, textLabel)
```

The result is the simple GUI shown in Figure 17.15. When the Randomize Time button is clicked, the clock is redrawn with a new time and the time label is removed. When the Show Time button is clicked, the time label is shown with the correct time for the current clock.

17.4 Interactive graphics for the web

Producing interactive graphics for web pages is an area of rapid development. This section briefly describes a few of the packages that are available.

The **gridSVG** package provides some limited interactivity by producing **grid**-based plots in an **SVG** format and allowing access to some of the more advanced SVG features, such as hyperlinks, animation, and the ability to embed scripts.

```
> library(gridSVG)
```

The following example produces an SVG version of the simple animation from Section 17.1. The first step is to draw a simple **grid** image. The important

detail about the following code is that the circle grob has been given the name
`"planet"`.

```
> grid.newpage()
> pushViewport(dataViewport(x, y))
> grid.lines(x, y, default.units="native")
> grid.circle(x[1], y[1], default.units="native",
               r=unit(2, "mm"), gp=gpar(fill="black"),
               name="planet")
```

The `grid.animate()` function is then used to add animation information to
the circle grob called `"planet"`. The x- and y-values used here were defined
back at the beginning of this chapter.

```
> grid.animate("planet",
               x=unit(x, "native"), y=unit(y, "native"))
```

The final step is to save the drawing to an SVG file, which can then be viewed
in a web browser.

```
> gridToSVG("animation.svg")
```

A more sophisticated alternative is the **SVGAnnotation** package, which
provides functions for augmenting R plots with a wide range of interactive
features. This package is part of the Omegahat Project and depends on a
Cairo-based graphics device (see Chapter 9).

```
> library(SVGAnnotation)
```

With this package, graphical output is produced in the SVG format, then
various functions are provided to modify the SVG content and possibly add
javascript code to add interactive features, such as hyperlinks, tool tips, animations, and even linked plots. For example, the following code creates an
initial image containing two scatterplots, with the `svgPlot()` function, then
augments the plot using the `linkPlots()` function so that moving the mouse
over a point in one plot causes corresponding points in the other plot to be
highlighted (see Figure 17.16). The plot is actually created and modified in
memory and the `saveXML()` function (from the **XML** package) is used to
write it to disk as an SVG file.

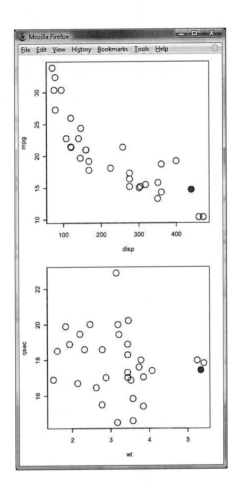

Figure 17.16
Two linked scatterplots being viewed in a browser. The plots are a combination of
SVG and javascript produced by the **SVGAnnotation** package.

```
> doc <- svgPlot({ par(mfrow=c(2, 1), cex=.7,
                     mar=c(5.1, 4.1, 1, 1))
              plot(mpg ~ disp, mtcars, cex=2)
              plot(qsec ~ wt, mtcars, cex=2) },
             width=4, height=8)
> linkPlots(doc)
> saveXML(doc, "linkedplots.svg")
```

One advantage of these approaches that produce SVG files is that the resulting file does not depend on R, so the plot can be used and shared by anyone with a compliant web browser. It is also possible to create even more complex interactivity by having R sitting in behind the web browser on the web server, but that sort of setup goes beyond the scope of this book.

Several other packages are being developed in this space, including packages that provide interfaces to other web graphics libraries such as protovis and Processing.js (for example, the **webvis** package). There are likely to be rapid developments in this area.

Chapter summary

The **animation** package provides a convenient interface for generating dynamic graphics from R plots. The **rggobi** and **iplots** packages provide access to interactive graphics systems from R. Several packages also implement less sophisticated interaction with native R plots. The **gWidgets** package provides a convenient interface to GUI toolkit packages for building interfaces that allow interactive control of R graphics via GUI elements such as buttons and menus. Other packages provide tools for generating interactive R plots for inclusion in web pages.

18

Importing Graphics

Chapter preview

This chapter describes packages and functions that import images
from external files and allow them to be included as part of R graphics
output. There are separate packages for importing raster images and
importing vector images.

Sections 3.4.1 and 6.2 describe the set of graphical primitives that are available in the traditional graphics system and the **grid** graphics system. These graphical primitives make it possible to draw basic shapes, text, and bitmap images and they form the basis for drawing more complex images with R.

By combining basic shapes, it is possible to produce an infinite variety of pictures, however, there are still some images that cannot be produced with R and R is not the best way to produce many kinds of images. For example, it is not possible to generate a photographic image with R and there are much better programs than R for producing artistic images such as logos.

Images like photographs and logos can be useful in plots or pictures, for example, to provide a background image for a plot, or to annotate a plot with the logo of a company or institution. In such cases, it may be necessary, or just more convenient, to source or create the image outside of R and *import* the image into R.

A number of packages provide tools for importing graphics into R and the choice of which one to use will depend on the format of the original image and what is to be done with the image once it has been imported. Image formats can be divided into *raster* formats and *vector* formats (see Section 9.2.1) and packages that import images into R typically address one of these options.

Figure 18.1
Two images of the Moon. On the left is a JPEG photograph of the North Pole of the Moon that has been assembled from images taken by the Galileo spacecraft, courtesy of NASA (`image #: PIA00130`). On the right is a cartoon image of the Moon from the Open Clip Art Library `http://openclipart.org/media/files/rg1024/10351`.

18.1 The Moon and the tides

To provide a concrete example of importing images into R, this section looks at producing a plot that shows the relationship between the timing of low tide and the phase of the Moon. The main plot shows the hour during the day at which low tide occurs as a function of the day of the month and the phases of the Moon and the plot is "dramatized" by adding an image of the Moon in the background.

Two versions of this plot are considered: one using a raster JPEG photograph of the North Pole of the Moon, taken by NASA's Galileo spacecraft and one using a vector SVG file from the Open Clip Art Library (see Figure 18.1).

The complete plots are shown in Figure 18.2 and code for these plots is available on the book web site. The focus of this chapter is on the two *conceptual* steps involved in producing the plots:

1. The external image has to be read into R. This involves having the ability to read the file format of the external image *and* being able to work with the data structure that is used to use to store the image in R.

2. The image has to be rendered by R. How this occurs will depend on the data structure that was created in the previous step. It may be important to be able to retain the original aspect ratio of the original image *and* it is usually important to be able to draw the image relative to the coordinate systems of the R plot.

Table 18.1

A selection of packages that can read external raster images into R. The **Dependencies** are third-party software applications or libraries that must also be installed for the package to work.

Package	Function	File Formats	Dependencies
pixmap	read.pnm()	PBM, PGM, PPM	-
png	readPNG()	PNG	libpng
rtiff	readTiff()	TIFF	**pixmap**, libtiff
ReadImages	read.jpeg()	JPEG	libjpeg
EBImage	readImage()	MANY	ImageMagick
RImageJ	IJ$openImage()	MANY	**rJava**, Java

18.2 Importing raster graphics

Examples of raster formats include JPEG, PNG, and GIF. A standard source of JPEG files are photographs from digital cameras, while GIF and PNG images are commonly encountered on the web.

The first step is to find a function that can read the file format of the external image. A large number of packages provide functions for reading image formats and Table 18.1 describes some of these. The important differences between these functions are the range of file formats that they can handle and how much they depend on other software (i.e., how much other software must also be installed).

The packages that are built on more sophisticated systems, such as **EBImage** and **RImageJ**, also provide numerous functions for manipulating images, from simple operations such as cropping an image to more complex tasks such as identifying features within an image.

There are also many application-specific packages that can read more specialized formats and perform image manipulations relative to the area of research. For example, there are packages such as **fmri** for reading and analyzing MRI medical images. Another example is the **raster** package for reading and working with geographical maps (see Section 14.5).

Having read an image into R, the next step is to view the image. Some packages provide functions for viewing an image by itself. For example, there is the `display()` function in the **EBImage** package. However, if the image is to be included as part of R graphics output, it is more useful to be able to draw the image with either the `rasterImage()` function or the `grid.raster()`

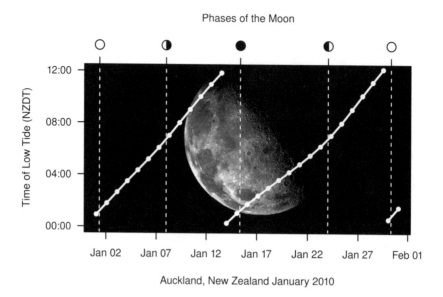

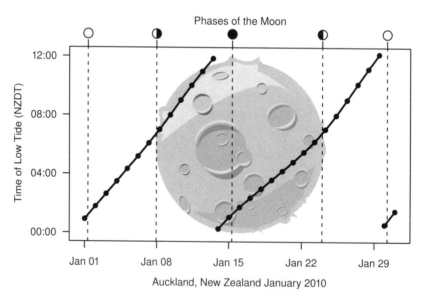

Figure 18.2

Two versions of a plot with a background image. In the top plot, a raster photograph of the Moon provides a backdrop and, in the bottom plot, the backdrop is a vector cartoon of the Moon. The data on low tides and phases of the Moon for Auckland in January 2010 were obtained from Land Information New Zealand (`http://hydro.linz.govt.nz`).

function because they allow the image to be drawn relative to the plot regions and coordinate systems of an R plot (for example, Figure 18.2).

The best solution is for the package to provide a method for the as.raster() function, to convert the image to a "raster" object. This allows the image to be used directly with rasterImage() or grid.raster(). At the time of writing few packages have provided this support, so another option is to convert the image to a matrix or array, which the functions will also accept and automatically convert.

As an example, the following code reads the Moon image (in a PGM format) into R.

```
> library(pixmap)
```

```
> moon <- read.pnm("Moon/GPN-2000-000473.pgm")
```

The result, moon, is a "pixmapGrey" object. With an as.raster() method defined for this type of object, drawing the image is as simple as the following code.

```
> grid.raster(moon)
```

If no such method exists, the information in the object can be used to create a matrix to draw, as in the following code.

```
> grid.raster(matrix(moon@grey, nrow=moon@size[1]))
```

In the case of **pixmap**, there is also a convenience function addlogo() for adding an image to a traditional graphics plot.

18.2.1 Manipulating raster images

Once an image has been read into R, the result is typically something like a matrix object. Because R has many facilities for working with matrix-like objects, it is relatively simple to manipulate an image in R. For example, cropping an image is simply a subsetting operation in R.

Figures 18.3 and 18.4 show a slightly more complex example. In this case, the grayscale values from the right-hand image in Figure 18.3 are used to set the alpha channel of the left-hand image in Figure 18.3; the result is the image in Figure 18.4. Wherever the right image is black, the left image

Figure 18.3

Two raster images that are used as the source material for producing the raster image in Figure 18.4. The right-hand image is the same as in Figure 18.1. The left-hand image is also courtesy of NASA (http://grin.hq.nasa.gov/ABSTRACTS/GPN-2001-000013.html).

becomes transparent; where the right is white, the left remains the same, and where the right is gray, the left becomes ghostly. This is achieved simply by manipulating matrices (the code is available on the book web site).

18.3 Importing vector graphics

Examples of vector image formats include PDF, PostScript, and SVG. There is only one package that is directly aimed at reading vector images into R, so the choice of which package to use is straightforward.

18.3.1 The grImport package

In the simplest case, where the original is a simple PostScript image and all that is needed is to draw the entire image somewhere on the page, the following code will suffice to read the image into R.

```
> library(grImport)
> PostScriptTrace(system.file("extra", "comic_moon.ps",
                              package="RGraphics"))
```

Figure 18.4
Manipulating raster images. The two images in Figure 18.3 have been combined to form this image.

```
> vectorMoon <- readPicture("comic_moon.ps.xml")
```

The `PostScriptTrace()` function converts a PostScript image into an XML format (using Ghostscript). This step only needs to be performed once for each image; it creates a new file with `.xml` attached to the file name. The `readPicture` function reads the XML file into R.

The object that is created, `vectorMoon`, can then be rendered using the `picture()` function, which will draw the image in the current traditional graphics plot region, or using the `grid.picture()` function, which will draw the image in the current **grid** viewport, as in the following code.

```
> grid.picture(vectorMoon)
```

Unfortunately, the situation is rarely this simple. For a start, the image may not be in the PostScript format. In that case, the only option is to use another software tool to convert the image to PostScript. Many tools exist to do this job; ImageMagick is one and, for converting from an SVG image, Inkscape*

---

*http://inkscape.org/.

produces good results. One danger is that, for complex vector images, some tools may convert the image, or parts of it, to a raster format. A number of other issues may also arise, mainly due to the fact that the content of a vector image can vary much more than the content of a raster image.

A raster image can be thought of as simply a two-dimensional array of pixels. There are many different ways that an array of pixels can be stored in a file, but the image structure is fundamentally always the same and very simple. This means that there are very few variations on how to read a raster image into R or how to draw a raster image as part of an R plot. The functions to read and draw raster images have relatively few arguments.

By contrast, a vector image is made up of a number of shapes or *paths*. There may be very few paths, or very many paths. The paths may overlap each other or even intersect with themselves. There may be text (letters are essentially quite detailed and complex paths) and, in more complex cases, one path may be used just to define a clipping region and not be drawn at all.

Sometimes, these complications mean that R will not be able to import an image or it may not render the original image properly. In any case, reading in a vector image and rendering the image may require more than a single step. In particular, it may be necessary to work with individual paths within a vector image and the **grImport** package provides several tools for doing so.

18.3.2 Manipulating vector images

One convenient feature is the ability to subset the object that is created by the `readPicture()` function. For example, the following code just draws the first four paths in the image (see Figure 18.6).

```
> grid.picture(vectorMoon[1:4])
```

There is also a `picturePaths()` function that allows each path to be inspected in isolation. The following code shows the first six paths within the cartoon Moon image (see Figure 18.5).

```
> picturePaths(vectorMoon[1:6], fill="white",
              freeScales=TRUE, nr=2, nc=3)
```

It is also useful to note that the imported image is essentially just a series of polygon outlines. The following code draws a "wireframe" version of the Moon image by ignoring the colors from the original image and just drawing the outline of each path (see Figure 18.6).

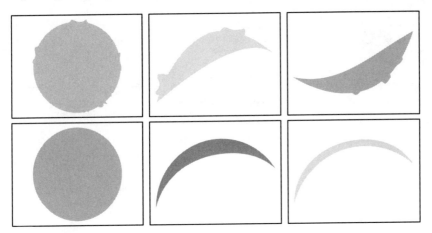

Figure 18.5

The first six paths (shapes) in the cartoon Moon image from Figure 18.1.

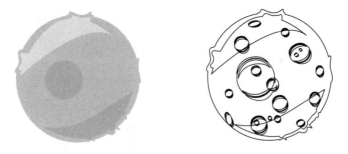

Figure 18.6

On the left, a "subset" of the Moon image (Figure 18.1), consisting of only the first four paths. On the right, the paths from the cartoon Moon image drawn as simple outlines, ignoring the fill colors from the original image.

```
> grid.picture(vectorMoon, use.gc=FALSE)
```

These facilities can be used, for example, to exclude certain parts of an image, or to render paths in a different order, which can sometimes be useful in reproducing the original image faithfully with R graphics.

Chapter summary

A number of packages provide functions for reading raster images into
R. The images can be drawn using traditional or **grid** graphical primitives. The **grImport** package provides functions for reading vector
images into R and drawing them. Both raster and vector images can
be manipulated in R using standard data manipulation tools, such as
subsetting.

19

Combining Graphics Systems

Chapter preview

This chapter describes the **gridBase** package, which makes it possible
to combine the output from the traditional graphics system with the
output from the **grid** graphics system.

The **grid** graphics system and the traditional graphics system work completely
independently of each other. This means that, while it is possible to produce
output from both systems on the same page, there should normally be no
expectation that the output from the two systems will correspond in any
sensible way.

This chapter describes the **gridBase** package, which provides functions that
can be used, in some situations, and with a little care, to overcome this in-
herent incompatibility and combine the output from the two systems in a
coherent manner.

19.1   The gridBase package

The **grid** graphics system offers more power and flexibility than the traditional
graphics system, and the **lattice** and **ggplot2** packages provide some facilities
not available in the traditional graphics system. However, it is often necessary
to use the traditional system because many plotting functions in extension
packages for R are built on the traditional system. Clearly, a combination of

the wide range of traditional plots and the power and flexibility of **grid** and **lattice** would be desirable and this is what the **gridBase** package provides.

19.1.1 Annotating traditional graphics using grid

The **gridBase** package has one function, `baseViewports()`, that supports adding **grid** output to a traditional graphics plot. This function creates a set of **grid** viewports (see Section 6.5) that correspond to the current traditional plot regions (see Section 3.1.1). By pushing these viewports, it is possible to do simple annotations to a traditional plot, such as adding lines and text using **grid**'s units to locate them relative to a wide variety of coordinate systems, or to attempt more complex annotations involving pushing further **grid** viewports.

The `baseViewports()` function returns a list of three grid viewports. The first corresponds to the traditional graphics inner region. This viewport is relative to the entire device and it only makes sense to push this viewport from the "top level" (i.e., only when no other **grid** viewports have been pushed). The second viewport corresponds to the traditional graphics figure region and is relative to the inner region, and it only makes sense to push it after the inner viewport has been pushed. The third viewport corresponds to the traditional graphics plot region and is relative to the figure region, and it only makes sense to push it after the other two viewports have been pushed in the correct order.

A simple application of this facility involves adding text to the margins of a traditional graphics plot at an arbitrary orientation. The traditional graphics function `mtext()` allows text to be located in terms of a number of lines away from the plot region, but only at rotations of 0 or 90 degrees. The traditional graphics `text()` function allows arbitrary rotations, but only locates text relative to the user coordinate system in effect in the plot region (which is inconvenient for locating text in the margins of the plot). By contrast, the **grid** function `grid.text()` allows arbitrary rotations and can be used in any **grid** viewport. In the following, a traditional graphics plot is created with the x-axis tick labels left off.

```
> midpts <- barplot(1:10, col="gray90", axes=FALSE)
> axis(2)
> axis(1, at=midpts, labels=FALSE)
```

In the next code, `baseViewports()` is used to create grid viewports that correspond to the traditional graphics plot and those viewports are pushed.

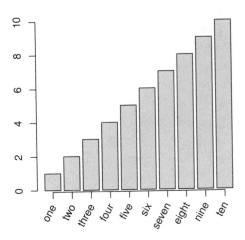

Figure 19.1
Annotating a traditional plot with **grid**. Most of the plot is drawn using the traditional `barplot()` function, but the x-axis labels are drawn using `grid.text()` to make use of both a convenient coordinate system (lines of text away from the x-axis) and the ability to rotate text to any angle.

```
> library(gridBase)
> vps <- baseViewports()
> pushViewport(vps$inner, vps$figure, vps$plot)
```

Finally, rotated labels are drawn using `grid.text()` (and the viewports are popped to clean up). The final output is shown in Figure 19.1.

```
> grid.text(c("one", "two", "three", "four", "five",
              "six", "seven", "eight", "nine", "ten"),
            x=unit(midpts, "native"), y=unit(-1, "lines"),
            just="right", rot=60)
> popViewport(3)
```

19.1.2 Traditional graphics in grid viewports

The **gridBase** package provides several functions for adding traditional graphics output to **grid** output. There are three functions that allow traditional graphics plotting regions to be aligned with the current **grid** viewport. These make it possible to draw one or more traditional graphics plots within a **grid** viewport. The fourth function, `gridPAR()`, provides a set of graphical parameter settings so that traditional graphics `par()` settings can be made to correspond to some of the current **grid** graphical parameter settings.

The first three functions are `gridOMI()`, `gridFIG()`, and `gridPLT()`. They return the appropriate `par()` values for setting the traditional graphics inner, figure, and plot regions, respectively.

The main usefulness of these functions is to allow the user to create a complex layout using **grid** and then draw a traditional graphics plot within relevant elements of that layout. The following example uses this idea to create a **lattice** plot where the panels contain dendrograms drawn using traditional graphics functions.

The first step just involves preparing some data to plot. A dendrogram object is created and cut it into four subtrees.*

```
> hc <- hclust(dist(USArrests), "ave")
> dend1 <- as.dendrogram(hc)
> dend2 <- cut(dend1, h=70)
```

Next, some dummy-variables are created that correspond to the four subtrees.

```
> x <- 1:4
> y <- 1:4
> height <- factor(round(sapply(dend2$lower,
                                 attr, "height")))
```

Now a **lattice** panel function is defined to draw the dendrograms. The first thing this panel function does is push a viewport that is smaller than the viewport **lattice** creates for the panel. The purpose of this is to ensure that there is enough room for the labels on the dendrogram. The `space` variable contains a measure of the length of the longest label. The panel function then calls `gridPLT()` and makes the traditional graphics plot region correspond to the viewport that has just been pushed. It also sets `new=TRUE` so that the following call to `plot()` does not start a new page. Finally, the traditional `plot()` function is used to draw the dendrogram (and then the viewport is popped).

---

*This example uses data on violent crimes in the United States, available as the `USAr-rests` data set in the **datasets** package.

```
> space <- 1.2 * max(stringWidth(rownames(USArrests)))
> dendpanel <- function(x, y, subscripts, ...) {
    pushViewport(viewport(gp=gpar(fontsize=8)),
                 viewport(y=unit(0.95, "npc"), width=0.9,
                          height=unit(0.95, "npc") - space,
                          just="top"))
    par(plt=gridPLT(), new=TRUE, ps=8)
    plot(dend2$lower[[subscripts]], axes=FALSE)
    popViewport(2)
  }
```

Now the main plot can be drawn, using **lattice** to set up the arrangement of panels and strips (**grid** viewports) and the panel function defined above to draw a traditional graphics dendrogram in each panel.

```
> library(lattice)
```

The final plot is produced by a call to the xyplot() function (see Figure 19.2).

```
> plot.new()
> print(xyplot(y ~ x | height, subscripts=TRUE,
               xlab="", ylab="",
               strip=strip.custom(style=4),
               scales=list(draw=FALSE),
               panel=dendpanel),
         newpage=FALSE)
```

The code above includes a call to plot.new() before the call to xyplot(). It is generally a good idea to start the new page with a call to plot.new() like this, rather than with grid.newpage(), or a high-level **lattice** or **ggplot2** function, because the **grid**-based functions tend to be more accepting of the fact that there may already be other drawing on the page.

This also explains the explicit call to print() around the xyplot() call, so that the newpage argument can be used to prevent xyplot() from starting its own new page.

19.1.3 Problems and limitations

The functions provided by the **gridBase** package allow the user to mix output from two quite different graphics systems and there are limits to how much the systems can be combined:

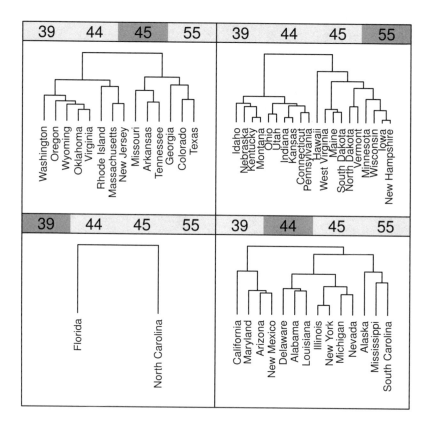

Figure 19.2

Embedding a traditional plot within **lattice** output. The arrangement of the panels and the drawing of axes and strips is all done by **lattice** using **grid**, but the contents of each panel is a dendrogram plot produced by the traditional graphics system.

- It is not possible to embed traditional graphics output within a **grid** viewport that is rotated.

- There are certain traditional graphics functions that modify settings like omi and fig themselves (e.g., coplot()). Output from these functions will not embed properly within **grid** viewports.

- The calculations used to match **grid** graphics settings with traditional graphics settings (and vice versa) are only valid if the device size does not change. If these functions are used to draw into a window, then the window is resized, the traditional graphics and **grid** settings will almost certainly no longer match and the graph may become nonsensical. This also applies to copying output between devices of different sizes.

The recordGraphics() function provides one way to avoid this problem, though proper use of the function requires expert knowledge. A very naive use is shown in the following code.

```
> plot.new()
> recordGraphics({ print(xyplot(y ~ x | height,
                                subscripts=TRUE,
                                xlab="", ylab="",
                                strip=strip.custom(style=4),
                                scales=list(draw=FALSE),
                                panel=dendpanel),
                   newpage=FALSE)
                 },
                 list(),
                 globalenv())
```

Some other solutions to this problem are discussed in Section 8.3.11.

Chapter summary

The **gridBase** package provides functions for aligning **grid** viewports with traditional graphics plot regions. This makes it possible to draw **grid**-based output within a traditional plot and traditional graphics output within **grid** viewports, including **lattice** and **ggplot2** plots.

Bibliography

Daniel Adler and Duncan Murdoch. *rgl: 3D Visualization Device System (OpenGL)*, 2010. R package version 0.91.

Adobe Systems Inc. *PostScript Language Reference Manual*. Addison-Wesley Longman, 2nd edition, 1990.

Felix Andrews. *latticist: A Graphical User Interface for Exploratory Visualisation*, 2010a. R package version 0.9-43.

Felix Andrews. *playwith: A GUI for Interactive Plots Using GTK+*, 2010b. R package version 0.9-52.

Baptiste Auguie. *gridExtra: Functions in grid Graphics*, 2010. R package version 0.7.

Richard A. Becker and John M. Chambers. *Extending the S System*. Chapman & Hall, 1985.

Richard A. Becker, William S. Cleveland, and Ming-Jen Shyu. The visual design and control of trellis display. *Journal of Computational and Graphical Statistics*, 5:123–155, 1996.

Richard A. Becker, Allan R. Wilks, and R version by Ray Brownrigg. *mapdata: Extra Map Databases*, 2010a. R package version 2.1-3.

Richard A. Becker, Allan R. Wilks, and R version by Ray Brownrigg and Thomas P Minka. *maps: Draw Geographical Maps*, 2010b. R package version 2.1-4.

Roger Bivand, Friedrich Leisch, and Martin Mächler. *pixmap: Bitmap Images ("Pixel Maps")*, 2009. R package version 0.4-10.

Roger S. Bivand, Edzer J. Pebesma, and Virgilio Gomez-Rubio. *Applied Spatial Data Analysis with R*. Springer, 2008.

Carole Blanc and Christophe Schlick. X-splines: A spline model designed for the end-user. In *SIGGRAPH '95: Proceedings of the 22nd Annual Conference on Computer Graphics and Interactive Techniques*, pages 377–386. ACM, 1995.

Carter T. Butts, Mark S. Handcock, and David R. Hunter. *network: Classes for Relational Data*, July 26, 2008. R package version 1.4-1.

Dan Carr and ported by Nicholas Lewin-Koh and Martin Mächler. *hexbin: Hexagonal Binning Routines*, 2010. R package version 1.22.0.

David Carslaw and Karl Ropkins. *openair: Open-Source Tools for the Analysis of Air Pollution Data*, 2010. R package version 0.3-10.

J. M. Chambers. Structured computational graphics for data analysis. *Proceedings of the International Statistical Institute*, 40:501–507, 1975.

Herman Chernoff. The use of faces to represent points in k-dimensional space graphically. *Journal of the American Statistical Association*, 68(342):361–368, 1973.

William S. Cleveland. *The Elements of Graphing Data*. Wadsworth Publ. Co., 1985.

William S. Cleveland. *Visualizing Data*. Hobart Press, 1993.

William S. Cleveland and Robert McGill. Graphical perception: The visual decoding of quantitative information on graphical displays of data (C/R: p210-229). *Journal of the Royal Statistical Society, Series A, General*, 150: 192–210, 1987.

Shane Conway. *webvis: Create graphics for the web from R.*, 2010. R package version 0.0.1.

D. Cook, A. Buja, J. Cabrera, and C. Hurley. Grand tour and projection pursuit. *Journal of Computational and Graphical Statistics*, 4:155–172, 1995.

Dianne Cook and Deborah F. Swayne. *Interactive and Dynamic Graphics for Data Analysis: With R and GGobi*. Springer, 2007.

Gabor Csardi and Tamas Nepusz. The igraph software package for complex network research. *InterJournal*, Complex Systems:1695, 2006.

Peter Dalgaard. *Introductory Statistics with R*. Springer, 2002.

Felipe de Mendiburu. *agricolae: Statistical Procedures for Agricultural Research*, 2010. R package version 1.0-9.

Seth Falcon and Robert Gentleman. *hypergraph: A Package Providing Hypergraph Data Structures*, 2009. R package version 1.17.0.

M.W. Felgate, S.H. Bickler, and P. Murrell. Quantifying broken objects: Estimating the death assemblage by integrating sample assemblage brokenness and completeness. *Journal of Archaeological Science*, In Press.

John Fox. *An R and S-Plus Companion to Applied Regression*. Sage Publications, 2002.

John Fox. *Rcmdr: R Commander GUI for R*, 2010. R package version 1.6-0.

John Fox. Effect displays in R for generalised linear models. *Journal of*

Statistical Software, 8(15):1–27, 2003.

Romain Francois and Philippe Grosjean. *RImageJ: R Bindings for ImageJ*, 2009. R package version 0.1-142.

Michael Friendly. A fourfold display for 2 by 2 by k tables. Technical Report 217, Psychology Department, York University, 1994a.

Michael Friendly. Mosaic displays for multi-way contingency tables. *Journal of the American Statistical Association*, 89:190–200, 1994b.

Michael Friendly. *Visualizing Categorical Data*. SAS Publishing, 2000.

Michael Friendly. Graphical methods for categorical data. *SAS User Group International Conference Proceedings*, 17:190–200, 1992.

Michael Friendly, Heather Turner, David Firth, Achim Zeileis, and Duncan Murdoch. *vcdExtra: vcd Additions*, 2010. R package version 0.5-0.

Emden R. Gansner and Stephen C. North. An open graph visualization system and its applications to software engineering. *Software — Practice and Experience*, 30(11):1203–1233, 2000.

R. Gentleman, Elizabeth Whalen, W. Huber, and S. Falcon. *graph: A Package to Handle Graph Data Structures*, 2010. R package version 1.26.0.

Jeff Gentry, Li Long, Robert Gentleman, Seth Falcon, Florian Hahne, Deepayan Sarkar, and Kasper Hansen. *Rgraphviz: Provides Plotting Capabilities for R Graph Objects*, 2009. R package version 1.23.6.

Enrico Glaab, Jonathan M. Garibaldi, and Natalio Krasnogor. vrmlgen: An R package for 3d data visualization on the web. *Journal of Statistical Software*, 36(8):1–18, 2010.

Mark S. Handcock, David R. Hunter, Carter T. Butts, Steven M. Goodreau, and Martina Morris. *statnet: Software Tools for the Statistical Modeling of Network Data*. Seattle, WA, 2003. Version 2.0.

Mark A. Harrower and Cynthia A. Brewer. Colorbrewer.org: An online tool for selecting color schemes for maps. *The Cartographic Journal*, 40:27–37, 2003.

J.A. Hartigan and B. Kleiner. A mosaic of television ratings. *The American Statistician*, 38:32–35, 1984.

Richard M. Heiberger. *HH: Statistical Analysis and Data Display*, 2009. R package version 2.1-32.

Richard M. Heiberger and Burt Holland. *Statistical Analysis and Data Display: An Intermediate Course with Examples in S-PLUS, R, and SAS*. Springer, 2004a.

Richard M. Heiberger and Burt Holland. *Statistical Analysis and Data Display: An Intermediate Course with Examples in S-PLUS, R, and SAS.* Springer, 2004b.

M. Helbig, M. Theus, and S. Urbanek. Jgr: Java gui for r. *Statistical Computing and Graphics Newsletter*, 16:9–11, 2005.

A.V. Hershey. A contribution to computer typesetting techniques: Tables of coordinates for Hershey's repertory of occidental type fonts and graphic symbols. *NBS Special Publication 424*, April 1976.

Robert J. Hijmans and Jacob van Etten. *raster: Geographic Analysis and Modeling with Raster Data*, 2010. R package version 1.4-10.

Toby Dylan Hocking. *directlabels: Direct Labels for Plots*, 2010. R package version 1.0/r378.

H. Hofmann and M. Theus. Interactive graphics for visualizing conditional distributions. Unpublished manuscript, 2005.

Torsten Hothorn, Kurt Hornik, and Achim Zeileis. Unbiased recursive partitioning: A conditional inference framework. Research Report Series 8, Department of Statistics and Mathematics, WU Wien, 2004.

Ross Ihaka, Paul Murrell, Kurt Hornik, and Achim Zeileis. *colorspace: Color Space Manipulation*, 2009. R package version 1.0-1.

Frank E Harrell Jr. *Hmisc: Harrell Miscellaneous*, 2010. R package version 3.8-3.

L. Kaufman and P.J. Rousseeuw. *Finding Groups in Data: An Introduction to Cluster Analysis*. Wiley, New York, 1990.

Timothy H. Keitt, Roger Bivand, Edzer Pebesma, and Barry Rowlingson. *rgdal: Bindings for the Geospatial Data Abstraction Library*, 2010. R package version 0.6-28.

Eric Kort. *rtiff: A tiff Reader for R*, 2010. R package version 1.4.1.

Duncan Temple Lang. *SVGAnnotation: Tools for Post-Processing SVG Plots Created in R*, 2010a. R package version 0.7-2.

Duncan Temple Lang. *XML: Tools for Parsing and Generating XML within R and S-Plus*, 2010b. R package version 3.1-1.

Michael Lawrence. *cairoDevice*, 2010. R package version 2.14.

Michael Lawrence and Duncan Temple Lang. *RGtk2: R Bindings for Gtk 2.8.0 and Above*, 2010. R package version 2.12.18.

Michael Lawrence and John Verzani. *gWidgetsRGtk2: Toolkit Implementation of gWidgets for RGtk2*, 2010. R package version 0.0-68.

Nicholas J. Lewin-Koh and Roger Bivand. *maptools: Tools for Reading and Handling Spatial Objects*, 2010. R package version 0.7-38.

H. W. Lie and B. Bos. *Cascading Style Sheets, Level 1*, 1996. W3C Recommendation.

Uwe Ligges and Martin Mächler. scatterplot3d: An R package for visualizing multivariate data. *Journal of Statistical Software*, 8(11):1–20, 2003.

Markus Loecher. *ReadImages: Image Reading Module for R*, 2009. R package version 0.1.3.1.

Markus Loecher and Sense Networks. *RgoogleMaps: Overlays on Google Map Tiles in R*, 2010. R package version 1.1.9.1.

Thomas Lumley. *dichromat: Color Schemes for Dichromats*, 2009. R package version 1.2-3.

Ulric Lund and Claudio Agostinelli. *circular: Circular Statistics*, 2010. R package version 0.4.

Martin Mächler, Peter Rousseeuw, Anja Struyf, and Mia Hubert. *Cluster Analysis Basics and Extensions*, 2010. R package version 1.13.1.

John Maindonald and John Braun. *Data Analysis and Graphics Using R: An Example-Based Approach*. Cambridge University Press, 2003.

Sam McClatchie, John F. Middleton, and Tim M. Ward. Water mass analysis and alongshore variation in upwelling intensity in the eastern Great Australian Bight. *Journal of Geophysical Research*, page 111, 2006.

Doug McIlroy and Packaged for R by Ray Brownrigg and Thomas P Minka and Roger Bivand. *mapproj: Map Projections*, 2009. R package version 1.1-8.2.

A.E. Miller. The analysis of unreplicated factorial experiments from a geometric perspective. *Canadian Journal of Statistics*, 31:311–327, 2003.

Julien Moeys. *soiltexture: Functions for Soil Texture Plot, Classification and Transformation*, 2010. R package version 1.01.

Albert H. Munsell. A pigment color system and notation. *The American Journal of Psychology*, 23:236–244, 1912.

P. R. Murrell. Layouts: A mechanism for arranging plots on a page. *Journal of Computational and Graphical Statistics*, 8:121–134, 1999.

Paul Murrell. Integrating grid graphics output with base graphics output. *R News*, 3(2):7–12, 2003.

Paul Murrell. Importing vector graphics: The grImport package for R. *Journal of Statistical Software*, 30(4):1–37, 2009.

Paul Murrell. *gridBase: Integration of Base and grid Graphics*, 2006. R package version 0.4-3.

Paul Murrell. *gridSVG: Export grid Graphics as SVG*, 2010a. R package version 0.5-10.

Paul Murrell. *hyperdraw: Visualizing Hypergaphs*, 2010b. R package version 0.1.1.

Jaroslav Myslivec. *symbols: Symbol plots*, 2009. R package version 1.1.

Kurt Nassau, editor. *Color for Science, Art and Technology*. Elsevier, 1998.

Erich Neuwirth. *RColorBrewer: ColorBrewer palettes*, 2007. R package version 1.0-2.

E. Paradis, J. Claude, and K. Strimmer. Ape: analyses of phylogenetics and evolution in R language. *Bioinformatics*, 20:289–290, 2004.

Tom Patterson and Nathaniel Vaughn Kelso. *Natural Earth project*, 2010.

Edzer J. Pebesma and Roger S. Bivand. Classes and methods for spatial data in R. *R News*, 5(2):9–13, 2005.

Tony Plate. *RSVGTipsDevice: An R SVG graphics device with dynamic tips and hyperlinks*, 2009. R package version 1.0-1.

Jörg Polzehl and Karsten Tabelow. fmri: A package for analyzing fmri data. *RNews*, 7(2):13–17, 2007.

R Development Core Team. *R: A Language and Environment for Statistical Computing*. R Foundation for Statistical Computing, Vienna, Austria, 2010.

Andrew Redd. *gmaps: functions for maps package to work with grid*, 2009. R package version 0.1.1.

Naomi Robbins. *Creating More Effective Graphs*. Wiley, 2005.

P.J. Rousseeuw. A visual display for hierarchical classification. In E. Diday, Y. Escoufier, L. Lebart, J. Pages, Y. Schektman, and R. Tomassone, editors, *Data Analysis and Informatics 4*, pages 743–748. North-Holland, Amsterdam, 1986.

Deepayan Sarkar. *Lattice: Multivariate Data Visualization with R*. Springer, New York, 2008.

Deepayan Sarkar and Felix Andrews. *latticeExtra: Extra Graphical Utilities Based on Lattice*, 2010. R package version 0.6-14.

Jon T. Schnute, Nicholas Boers, Rowan Haigh, and Alex Couture-Beil. *PBSmapping: Mapping Fisheries Data and Spatial Analysis Tools*, 2010. R package version 2.61.9.

Dave Schreiner. *OpenGL Reference Manual: The Official Reference Document to OpenGL, Version 1.2*. Addison-Wesley Longman, 1999.

David W. Scott. *Multivariate Density Estimation: Theory, Practice, and Visualization*. Wiley, 1992.

Charlie Sharpsteen and Cameron Bracken. *tikzDevice: A Device for R Graphics Output in PGF/TikZ Format*, 2010. R package version 0.5.2.

Oleg Sklyar, Gregoire Pau, Mike Smith, and Wolfgang Huber. *EBImage: Image Processing Toolbox for R*, 2009. R package version 3.3.1.

Greg Snow. *TeachingDemos: Demonstrations for Teaching and Learning*, 2010. R package version 2.7.

Karline Soetaert. *diagram: Functions for Visualising Simple Graphs and Flow Diagrams*, 2009a. R package version 1.5.

Karline Soetaert. *shape: Functions for Plotting Graphical Shapes, Colors*, 2009b. R package version 1.2.2.

Andy South. *rworldmap: For Mapping Global Data*, 2010. R package version 0.110.

A. Struyf, M. Hubert, and P.J. Rousseeuw. Integrating robust clustering techniques in S-PLUS. *Computational Statistics and Data Analysis*, 26: 17–37, 1997.

Deborah F. Swayne, Duncan Temple Lang, Andreas Buja, and Dianne Cook. GGobi: Evolving from XGobi into an extensible framework for interactive data visualization. *Computational Statistics and Data Analysis*, 43:423–444, 2003.

Duncan Temple Lang. *FlashMXML*, 2010a. R package version 0.2-0.

Duncan Temple Lang. *A framework for developing R graphics devices entirely in R*, 2010b. R package version 0.4-0.

Duncan Temple Lang, Debby Swayne, Hadley Wickham, and Michael Lawrence. *rggobi: Interface between R and GGobi*, 2009. R package version 2.1.14.

Martin Theus and Simon Urbanek. *Interactive Graphics for Data Analysis: Principles and Examples*. Chapman & Hall, 2008.

Luke Tierney. *tkrplot: TK Rplot*, 2010. R package version 0.0-19.

E. R. Tufte. *The Visual Display of Quantitative Information*. Graphics Press, 1989.

Edward R. Tufte. *Envisioning Information*. Graphics Press, 1990.

Simon Urbanek. *JavaGD: Java Graphics Device*, 2010a. R package version

0.5-2.

Simon Urbanek. *Acinonyx: Next-generation Interactive Graphics*, 2010b. R package version 3.0-0.

Simon Urbanek. *png: Read and Write PNG Images*, 2010c. R package version 0.1-1.

Simon Urbanek. *rJava: Low-level R to Java interface*, 2010d. R package version 0.8-2.

Simon Urbanek and Jeffrey Horner. *Cairo: R Graphics Device Using Cairo Graphics Library*, 2009. R package version 1.4-5.

K. Gerald van den Boogaart, Raimon Tolosana, and Matevz Bren. *compositions: Compositional Data Analysis*, 2008. R package version 1.01-1.

Bill Venables and Kurt Hornik. *oz: Plot the Australian Coastline and States*, 2009. R package version 1.0-18.

John Verzani. *Using R for Introductory Statistics*. Chapman & Hall/CRC, Boca Raton, FL, 2005.

John Verzani. *gWidgets: gWidgets API for Building Toolkit-independent, Interactive GUIs*, 2010a. R package version 0.0-41.

John Verzani. *pmg: Poor Man's GUI*, 2010b. R package version 0.9-42.

John Verzani and Michael Lawrence. *Programming Graphical User Interfaces with R*. Chapman and Hall/CRC, 2011.

Gregory R. Warnes. *gplots: Various R Programming Tools for Plotting Data*, 2010. R package version 2.8.0.

P. Wessel and W.H.F. Smith. A global, self-consistent, hierarchical, high-resolution shoreline database. *Journal of Geophysical Research*, 101:8741–8743, 1996.

Charlotte Wickham. *munsell: Munsell Colour System*, 2009a. R package version 0.1.

Hadley Wickham. *ggplot2: Elegant graphics for data analysis*. Springer, 2009b.

Hadley Wickham, Di Cook, Andreas Buja, and Barret Schloerke. *DescribeDisplay*, 2010. R package version 0.2.2.

Leland Wilkinson. *The Grammar of Graphics*. Springer, 2nd edition, 2005.

Peter Wolf and Uni Bielefeld. *aplpack: Another Plot PACKage*, 2010. R package version 1.2.3.

Yihui Xie. *animation: Demonstrate Animations in Statistics*, 2010. R package version 1.1-2.

Achim Zeileis, Torsten Hothorn, and Kurt Hornik. Model-based recursive partitioning. *Journal of Computational and Graphical Statistics*, 17(2):492–514, 2008.

Achim Zeileis, Kurt Hornik, and Paul Murrell. Escaping RGBland: Selecting colors for statistical graphics. *Computational Statistics & Data Analysis*, 53:3259–3270, 2009.

Index

3D plots, 3, 32, 37, 106, 124, 431–451

abline(), 84, 85
Acinonyx, 319, 467
add(), 475
addGrob(), 229
addlogo(), 483
addtable2plot(), 340, 342
adjustcolor(), 324
Adobe Flash, 319, 454
aes(), 148, 151, 155
Aesthetics, 150
agopen(), 422
agricolae, 365
ani.options(), 455
ani.start(), 455
ani.stop(), 455
animation, 454, 456, 478
Animations, 454, 475, 476
ape, 426
aplpack, 366
applyEdit(), 291
applyEdits(), 291
arctext(), 347, 349
Arranging plots, *see* Layouts
    in ggplot2, 223
    in lattice, 129–132
    in traditional plots, 72–78
arrow(), 181
Arrows, *see* Graphical primitives
arrows(), 20, 79, 83, 85, 350
as.raster(), 483
Aspect ratio, 74, 130, 317, 398, 433,
    480
assoc(), 388
Association plots, 35
assocplot(), 35

Axes, 357
    in ggplot2, 154
    in grid, 175
    in lattice, 132
    in traditional plots, 41, 65, 90
axis(), 65, 90–94, 97, 108, 111, 184,
    357, 375
axis.break(), 357, 358
axis.circular(), 376
axis.Date(), 94
axis.POSIXct(), 94
axTicks(), 94

Banner plots, 30
barchart(), 124
Barcharts, 3, 32, 35, 37, 124, 157,
    161
barplot(), 32, 34, 35, 38, 41, 59,
    102, 383, 385, 491
Barplots, *see* Barcharts
baseViewports(), 490
Bioconductor, xxvi
bitmap(), 308, 317, 318
Bitmaps, 84, 183, 481–484
BMP, 315
bmp(), 308
borders(), 412
box(), 68, 85, 108
boxed.labels(), 347, 349
boxplot(), 32, 34, 35, 38, 99, 102,
    104, 112
boxplot.stats(), 112
Boxplots, 3, 32, 35, 37, 124
brewer.pal(), 352
Brushing, 460
bwplot(), 124, 142
bxp(), 99

c(), 143, 188
Cairo, 318, 319
Cairo graphics, 311–313, 316, 318
Cairo(), 319
cairo_pdf(), 311
cairo_ps(), 312
cairoDevice, 318
canonical.theme(), 142
canvas, 319
car, xxiv
cbind(), 73
cd_plot(), 383
cdplot(), 34, 35, 383
chartSeries(), 13
Chernoff faces, 363
CIDFont(), 310
circleGrob(), 179, 236
Circles, see Graphical primitives
circular, 372, 374
circular(), 374
clip(), 71
Clipping
    in grid, 185, 205–206
    in traditional plots, 71
close.screen(), 78
cloud(), 124, 126, 439, 440, 451
cloud3d(), 449
cluster, 30
cm(), 96
cm.colors(), 325
co.intervals(), 112
col.whitebg(), 142
col2rgb(), 322, 324
Color spaces, 322, 351
ColorBrewer, 352
colorRamp(), 326
colorRampPalette(), 326
Colors, 58, 193, 321, 351
colors(), 322
colorscheme(), 463
colorspace, 351, 352
colours(), 322
compositions, 367–369
Conditional density plots, 34, 383
Conditioning plots, 38,

see Multipanel conditioning
contour(), 37, 99, 112, 313
contourLines(), 106, 112, 442
contourplot(), 124
convertColor(), 324, 351
convertHeight(), 188
convertUnit(), 188
convertWidth(), 188, 288
convertX(), 188
convertY(), 188
coord_polar(), 375
coord_trans(), 163
Coordinate systems, 354
    in ggplot2, 163
    in grid, 174, 176, 185–188,
        see Units
    in traditional plots, 49–51, 94–
        96
coordinates(), 428
coplot(), 37, 38, 72, 104, 112
corner.label(), 354, 355
cotabplot(), 392
crop(), 410
Cross-hatching, 59
CRS(), 407
current.vpTree(), 207, 298
curve(), 44
curveGrob(), 179
cylinder3d(), 448
cylindrect(), 347, 349

Data frames, 26, 126, 150
Data symbols, 64–65, 79, 179, 330
datasets, 99, 106, 149
dataViewport(), 175
Dendrograms, 30, 44, 492
density(), 42
Density plots, 124
densityplot(), 124
DescribeDisplay, 463
Design, xxiv
dev.control(), 318
dev.copy(), 317, 318
dev.copy2eps(), 318
dev.cur(), 307

dev.list(), 307
dev.new(), 306
dev.next(), 307
dev.off(), 306
dev.prev(), 307
dev.print(), 318
dev.set(), 307
dev.size(), 307
dev2bitmap(), 318
devAskNewPage(), 72, 174, 317
devSVGTips(), 319
dia, 420
diagram, 428–430
dichromat, 354
direct.label(), 347
directlabels, 347
display(), 462, 481
Display list
    in grid, 228–229, 287–290
    in graphics engine, 317–318
display.brewer.all(), 353
diverge_hcl(), 352, 353
dotchart(), 32, 34, 35, 383
dotplot(), 124, 142
Dotplots, 32, 35, 124
Double-decker plots, 388
doubledecker(), 388
downViewport(), 176, 203, 208, 210,
    219, 262, 270
draw.arc(), 347, 349
draw.circle(), 347, 349
drawDetails(), 189, 270–274, 276,
    278, 282, 289
Drawing context, 199
Dynamic graphics, 453,
    see Interactive graphics

EBImage, 481
ecdfplot(), 143
edgeRenderInfo(), 420
editDetails(), 275–278, 282
editGrob(), 229, 236, 247, 268, 275,
    291
effects, xxiv
ellipse3d(), 448

Ellipsis argument, 112, 135
embedFonts(), 310
emptyspace(), 341, 343, 344
Encoding, 333
engine.display.list(), 287
equal.count(), 129
erase.screen(), 78
errbar(), 20, 350
Error bars, 20, 350
Event handlers, 470
example(), 26
expression(), 81, 333
Expressions, 333

faces(), 365, 366
faces2(), 365, 366
facet_grid(), 164
facet_wrap(), 164
Facetting, 164,
    see Multipanel conditioning
ffmpeg, 454
Figure margins, 48
Figure region, 48
Fill patterns, 59, 347, 349
filled.contour(), 37, 90, 104
findFn(), 20
FlashMXML, 319
Flow charts, 426
Fluctuation diagrams, 388
fmri, 481
Fonts, 60, 195, 331
Formulae,
    see Mathematical formulae
    as data to plot, 26, 81, 126–129
fourfold(), 388
Fourfold displays, 35, 388
fourfoldplot(), 35
frame(), 107
frameGrob(), 240, 242
Frames, 239–242
    packing grobs, 240–241
    placing grobs, 241
fromGXL(), 415

gbutton(), 475

gEdit(), 291
gEditList(), 291
Generic functions
    in grid, 265–266, 282–283
    in lattice, 142
    in traditional plots, 27, 81
geom_abline(), 152
geom_bar(), 152, 155
geom_boxplot(), 152
geom_contour(), 152
geom_density(), 152
geom_errorbar(), 351
geom_hex(), 378
geom_histogram(), 152
geom_hline(), 169
geom_line(), 148, 152, 158
geom_path(), 152
geom_point(), 148, 150–152
geom_pointrange(), 351
geom_polygon(), 152
geom_rect(), 152
geom_ribbon(), 351
geom_segment(), 152
geom_smooth(), 152, 158
geom_text(), 151, 152
Geometric context, 199
Geoms, 150
get.gpar(), 194
getGraphicsEvent(), 44, 219
getGrob(), 229
GGobi, 461
ggobi(), 461
ggplot(), 148, 150, 160
ggplot2, 5, 11, 18, 145–171, 222, 251
ggraphics(), 474
ggroup(), 473
Ghostscript, 317, 485
GIF, 454
glabel(), 474
gList(), 232
gLists, 232
glyph_color(), 463
gmaps, 399
gpar(), 192
gPath(), 233

gPaths, 233
gplots, 20, 340, 350, 354, 362, 364
gradient.rect(), 347, 349
Grammar of Graphics, 5, 145
graph, 414, 416, 417, 422, 423, 430
graph(), 424
graph.atlas(), 424
graph.famous(), 424
graph.formula(), 424
graph.full(), 424
graph.tree(), 424
graphBPH(), 423
Graphical context, 194, 199, 211
Graphical Manual, 6
Graphical parameters,
    see Graphics state
    in 3D, 434
    in ggplot2, 150–154
    in grid, 192–199, 233–234
    in lattice, 139–142
    specifying, 321–335
    in traditional plots, 41, 57–72
Graphical primitives, 347
    in grid, 178–184
    in traditional plots, 79–84
graphics, 18, 19, 21, 25, 30
Graphics devices, 20, 306–307
Graphics engine, 18
Graphics formats, 308, 319
Graphics Gallery, 6
Graphics state, 51–57
Graphics systems, 18
    combining, 489
    comparison of, 21–22
graphics.off(), 307
Graphs, 414
graphviz, 417
gray(), 323
gray.colors(), 325, 326
grconvertX(), 88, 96, 354, 355
grconvertY(), 88, 96, 354, 355
grDevices, 18, 19, 305, 318, 351, 352
grid, 18, 173–301
grid(), 84

grid.add(), 229, 287
grid.animate(), 476
grid.barbed(), 349, 350
grid.circle(), 179, 228, 236
grid.clip(), 185
grid.curve(), 179–182
grid.display.list(), 229, 287
grid.draw(), 236, 249, 270, 271, 292
grid.edit(), 176, 228–231, 236, 247,
        268, 275, 287, 291
grid.ellipse(), 349, 350
grid.frame(), 240
grid.gedit(), 231
grid.get(), 228, 229, 287
grid.gget(), 231
grid.grab(), 238, 249, 251
grid.grabExpr(), 238
grid.layout(), 212–216
grid.line.to(), 178, 179, 182, 205,
        217, 218
grid.lines(), 179–182, 218, 260
grid.locator(), 217, 461
grid.ls(), 231, 232, 250, 253, 296,
        297
grid.move.to(), 178, 179, 205
grid.newpage(), 174, 493
grid.null(), 179, 205
grid.pack(), 240
grid.path(), 179, 182, 198
grid.pattern(), 349, 350
grid.picture(), 485
grid.place(), 241
grid.points(), 179, 183
grid.polygon(), 179, 182, 197, 217,
        218
grid.polyline(), 179, 180, 182, 198
grid.raster(), 179, 183, 481, 483
grid.record(), 189, 290
grid.rect(), 178, 179, 257, 266
grid.refresh(), 287
grid.remove(), 229, 287
grid.roundrect(), 179
grid.segments(), 179, 181, 182
grid.set(), 229
grid.table(), 341, 342

grid.text(), 178, 179, 200, 201, 220,
        223, 491
grid.xaxis(), 179, 183
grid.xspline(), 179, 181, 182, 198
grid.yaxis(), 179, 183
gridBase, 21, 489–491, 493, 495
gridExtra, 341, 349
gridFIG(), 492
gridlines(), 409
gridOMI(), 492
gridPAR(), 492
gridPLT(), 492
gridSVG, 475
grImport, 18, 484, 486, 488
grob(), 265, 266
Grob paths, 233
grobHeight(), 190, 242, 243
Grobs, 228–231
grobWidth(), 190, 242, 243
grobX(), 190, 205, 245, 246, 251
grobY(), 190, 205, 245, 251
GSHHS, 401
GTK+, 461
gTree(), 237, 265, 266
gTrees, 232–237
gWidgets, 472, 473, 478
gWidgetsRGtk2, 473
gwindow(), 473
GXL, 415

hcl(), 323, 324
heat.colors(), 325
heat_hcl(), 352
heightDetails(), 279–280, 283
help(), 25
Hershey outline fonts, 331, 332
Hexagonal binning, 378
hexbin, 20, 378
hexbinplot(), 378
hexplom(), 378
HH, xxiv
High-level functions, 20
hist(), 32, 59, 112
histogram(), 124, 126
Histograms, 3, 32, 35, 124

Hmisc, 20, 343, 345, 350, 351
hsv(), 322, 323
HTML, 319
hyperdraw, 422, 423
hypergraph, 422, 423
Hyperlinks, 475, 476

I(), 148, 151
iabline(), 464
ibar(), 464
ibox(), 464
identify(), 44, 461
idev(), 319
ievent.wait(), 465, 466
igraph, 413, 424–426, 430
igraph.to.graphNEL(), 425
IJ$openImage(), 481
ilines(), 464
image(), 37, 313, 411
ImageMagick, 312, 455, 485
imap(), 464
imosaic(), 464
Inkscape, 486
Inner region, 48
Interactive graphics, 453
    in 3D, 443
    in grid, 217
    in traditional plots, 44
iobj.rm(), 465
iplot(), 464
iplot.set(), 465
iplots, 20, 463–467, 478
is.finite(), 113
is.na(), 113
iset.select(), 465
iset.selected(), 465
itext(), 465

Java, 319, 466
JavaGD, 20, 319
JavaGD(), 319
javascript, 319, 476
Jittering data, 102
JPEG, 313
jpeg(), 308

kde3d(), 446
KML, 412
Krig(), 256

labcurve(), 345, 346
lapply(), 106
largest.empty(), 343, 344
LaTeX, 315
lattice, 4, 10, 18, 119–144, 219, 250,
    439
latticeExtra, 142–144, 412
latticist, 467, 469, 470
latticist(), 467
layer(), 143
Layers, 143, 148
layout(), 72–76, 78, 104
Layout algorithms, 417
layout.circle(), 424
layout.reingold.tilford(), 424
layout.show(), 74
layout.spring(), 424
layoutGraph(), 420
Layouts
    in grid, 212–216
    in traditional plots, 73–76
lcm(), 74
leaves(), 416
legend(), 59, 71, 89, 96, 340, 354
Legends
    in ggplot2, 154
    in lattice, 129
    in traditional plots, 89–90
Level plots, 37, 124
levelplot(), 124
libjpeg, 481
libpng, 481
libtiff, 481
light3d(), 448
Lighting, 434
Line ends, 327
Line joins, 327
Line style, 327
Lines, *see* Graphical primitives
lines(), 27, 79–81, 88, 99, 106, 111,
    354, 375, 438, 442

lines.circular(), 376
lines3d(), 446, 449
linesGrob(), 179
lineToGrob(), 179
Linked plots, 460
linkPlots(), 476
LiveGraphics3D, 449
lm(), 27
load(), 249
Locales, 333
locator(), 44, 466
Low-level functions, 20

map(), 399
map2SpatialPolygons(), 401
mapdata, 5, 12, 399, 400
mapplot(), 412
mapproj, 5, 12
Maps, 5, 259, 261, 397–412
maps, 5, 12, 18, 20, 399–401, 412
maptools, 345, 401, 412
marginal.plot(), 143
material3d(), 448
Mathematical formulae, 333–335
matlines(), 81
matplot(), 27, 37, 81
matpoints(), 81
max(), 187
Mesa 3D, 443
misc3d, 446
Missing values
    in grid, 216–217
    in traditional plots, 86
mnsl2hex(), 353
mosaic(), 385
Mosaic plots, 34, 385, 464
mosaic3d(), 395
mosaicplot(), 34, 35, 37, 38, 385
moveToGrob(), 179
mtext(), 58, 86–88, 104, 111
Multipanel conditioning, 126–127,
    see Facetting
munsell, 353
my.symbols(), 348, 349

n2mfrow(), 112
na.omit(), 113
Natural Earth Project, 401
nclass.FD(), 112
nclass.scott(), 112
nclass.Sturges(), 112
network, 425–427
network(), 425
nodeRenderInfo(), 420
Non-finite values
    in grid, 216–217
    in traditional plots, 62, 86
Normalized coordinates,
    see Coordinate systems
nullGrob(), 179

Omegahat Project, xxvi
on.exit(), 113
openair, 371–373, 375–377
OpenGL, 443
opts(), 166, 167
Outer margins, 48
oz, 259
ozRegion(), 259

packGrob(), 242
Painters model, 1, 174
pairs(), 37, 103–105, 388, 391
palette(), 323
Panel functions
    in lattice, 135
    in traditional plots, 104
panel.2dsmoother(), 144
panel.3dbars(), 144
panel.3dpolygon(), 144
panel.3dtext(), 144
panel.abline(), 135, 138
panel.arrows(), 138
panel.bwplot(), 137
panel.curve(), 138
panel.ellipse(), 144
panel.grid(), 138
panel.histogram(), 137
panel.identify(), 461
panel.key(), 144

panel.lines(), 138, 139
panel.lmline(), 136
panel.loess(), 138
panel.my.symbols(), 348
panel.points(), 138, 139, 219
panel.polygon(), 138
panel.rect(), 138
panel.rug(), 138
panel.segments(), 138
panel.smooth(), 104
panel.smoothScatter(), 138
panel.superpose(), 137
panel.text(), 135, 138, 219
panel.violin(), 138
panel.xyarea(), 144
panel.xyplot(), 135–137
par(), 48, 51–73, 94, 113, 140, 492
par3d(), 448
parallel(), 124
Parallel coordinate plots, 124
party, 5, 14
pathGrob(), 179
Paths, 83, 182, 486
PBM, 481
PBSmapping, 412
PDF, 309
pdf(), 306, 308, 310, 311
pdfFonts(), 310, 331
persp(), 37, 106, 435, 437–439, 443,
         445, 451
persp3d(), 443, 445, 446
PGM, 481
pictex(), 308, 315
picture(), 485
picturePaths(), 486
pie(), 32, 59
Pie charts, 3, 32, 164, 368
pixmap, 18, 481, 483
placeGrob(), 242
playwith, 469–471
playwith(), 469
plot(), 1, 26–42, 44, 58
Plot region, 48
plot.agnes(), 30
plot.circular(), 376

plot.dendrogram(), 44
plot.density.circular(), 376
plot.lm(), 27
plot.new(), 107, 108, 428, 493
plot.window(), 108
plot.xy(), 108
plot3d(), 446
plotCI(), 20, 350
plotmat(), 429
plotmeans(), 350
plotrix, 339–351, 354, 357, 367, 369,
         370, 372, 373
plotViewport(), 175
pmg, 467, 468
PNG, 313
png, 481
png(), 308, 313
pointLabel(), 345, 346
points(), 27, 58, 79, 88, 438, 442
points3d(), 446, 449
pointsGrob(), 179
Polar coordinates,
    see Coordinate systems
Polar plots, 368–375
polar.plot(), 370, 372, 373
polarFreq(), 371, 373
polygon(), 59, 79, 83, 87, 348, 442
polygon.shadow(), 348, 349
polygonGrob(), 179
Polygons, *see* Graphical primitives
polylineGrob(), 179
polypath(), 79, 83
popViewport(), 202, 203, 261
postDrawDetails(), 280–282
PostScript, 311
postscript(), 308, 311, 312
PostScriptTrace(), 485
PPM, 481
preDrawDetails(), 280–282
print(), 220, 239, 493
PROJ.4, 407
proj4string(), 407
projection(), 410
Projections
    in 3D, 432

in maps, 406
pushViewport(), 200, 206, 207

qplot(), 146–148, 151, 171
qq(), 124
qqmath(), 124
qqnorm(), 42
qqplot(), 42
Qt, 470
quantmod, 5, 13
Quartz, 313
quartz(), 308

R-Forge, xxvi
rainbow(), 325, 326
rainbow_hcl(), 352
raster, 410, 412, 481
raster(), 410
Raster graphics, 312, 409, 481
Raster images, *see* Bitmaps
rasterGrob(), 179
rasterImage(), 79, 84, 481, 483
rbind(), 73
Rcmdr, 467, 468, 470
RColorBrewer, 352
rcomp(), 367
read.graph(), 425
read.jpeg(), 481
read.pnm(), 481
readImage(), 481
ReadImages, 481
readPicture(), 486
readPNG(), 481
readShapeSpatial(), 401
readTiff(), 481
recordGraphics(), 96, 495
recordGrob(), 290
recordPlot(), 318
rect(), 58, 59, 79, 83, 96, 111
Rectangles, *see* Graphical primitives
rectFill(), 348, 349
rectGrob(), 179, 266
Recycling rule, 81
Regular expressions, 229
removeGrob(), 229

renderGraph(), 420
renderInfo(), 420
replayPlot(), 318
revaxis(), 357, 359
rgb(), 322, 323
rgb2hsv(), 322
rgdal, 408, 412
rggobi, 20, 319, 461, 464, 478
rgl, 20, 319, 395, 431, 443, 446–449, 451
rgl.postscript(), 446
RgoogleMaps, 412
RGraphics, xxvi
RGraphicsDevice, 319
Rgraphviz, 413, 415, 417–419, 421, 423, 425, 426, 430
Rgshhs(), 401
RGtk2, 470, 473
RImageJ, 481
rJava, 363, 466, 481
rootogram(), 143
rose.diag(), 374, 376
rotate3d(), 448
roundrectGrob(), 179
RSVGTipsDevice, 319
rtiff, 481
rug(), 85
rworldmap, 412

save(), 249
saveMovie(), 455
saveSWF(), 455
saveXML(), 476
scale_color_manual(), 155, 157
scale_fill_manual(), 157, 161
scale_linetype(), 157
scale_shape(), 157
scale_shape_manual(), 160
scale_size(), 157
scale_x_continuous(), 154, 157, 163
scale_x_date(), 157
scale_x_discrete(), 157
scale_y_continuous(), 154
Scales, 154
Scatterplot matrix, 37, 124

scatterplot3d, 20, 440, 451
scatterplot3d(), 440, 442
Scatterplots, 3, 32, 35, 37, 124, 151
screen(), 78
seekViewport(), 204
segments(), 79, 81, 83, 96, 102, 111
segmentsGrob(), 179
segplot(), 143
sequential_hcl(), 352, 353
set.edge.attribute(), 425
set.vertex.attribute(), 425
setEPS(), 311
setGrob(), 229
shade3d(), 448
shadowtext(), 348, 349
shape, 350, 428
Shapefiles, 400
shingle(), 129
Shingles, 129
show.settings(), 140
showViewport(), 298
sinkplot(), 340
Small multiples, 37, 164
smartlegend(), 354, 355
smoothScatter(), 34, 35
snapshot3d(), 446
soiltexture, 368, 369
sos, 20
source(), 249
sp, 401–403, 406, 407, 412
sp.lines(), 405
sp.points(), 405
sp.text(), 405
spine(), 383
Spine plots, 34, 161, 383
spineplot(), 34, 35, 383
Spinograms, 34, 383
Splines, *see* X-Splines
split.screen(), 76, 78
splom(), 124
spplot(), 403, 404
spread.labels(), 343, 344, 346
spread.labs(), 344, 346
spTransform(), 408
stars(), 37, 365

stat_bin(), 159
stat_binhex(), 378
stat_boxplot(), 159
stat_contour(), 159
stat_identity(), 159
stat_smooth(), 159
staxlab(), 357, 358
stem(), 32
Stem-and-leaf plots, 32
straightarrow(), 428
strheight(), 95
stringHeight(), 190
stringWidth(), 190
stripchart(), 32, 34, 35
Stripcharts, 32, 35, 124
stripplot(), 124
structable(), 388
strwidth(), 95, 111
subGraph(), 416
subplot(), 356
substitute(), 335
sunflowerplot(), 34, 35
surface3d(), 446
svalue(), 474
SVG, 312, 318, 475, 476
svg(), 308, 312
SVGAnnotation, 476, 477
svgPlot(), 476
Sweave, xxvii
SWFTools, 455
symbol(), 365, 366
symbols, 365, 366
symbols(), 37, 98, 99, 348

tcltk, 319, 470
tcltk, 470
TeachingDemos, 344, 348, 356, 358,
          365, 366
Ternary plots, 365
ternaryplot(), 367, 369
terrain.colors(), 325
terrain_hcl(), 352
Text, *see* graphical primitives
text(), 62, 79–81, 85, 87, 96, 106,
          340, 347, 354, 438, 442

text3d(), 446, 449
textbox(), 347, 349
textGrob(), 179, 236
textplot(), 340, 341
textrect(), 428
theEconomist.theme(), 143
theme_blank(), 167
theme_bw(), 166
theme_text(), 166, 167
Themes
    in ggplot2, 166
    in lattice, 142
thigmophobe.labels(), 344–346
Three-dimensional plots,
    see 3D plots
TIFF, 313
tiff(), 308
tikz(), 319
tikzDevice, 20, 318, 319
tile(), 388, 394
Tile displays, 388
tileplot(), 143
title(), 87
tkplot(), 425
tkrplot, 319
toFile(), 422
Tool tips, 460, 476
topo.colors(), 325
Traditional graphics, 18
trans3d(), 106, 438
translate3d(), 448
Trellis graphics, 4, 119
trellis.focus(), 139, 219
trellis.panelArgs(), 139
trellis.par.get(), 140
trellis.unfocus(), 139
triax.plot(), 367, 369
TT.plot(), 368, 369
Type1Font(), 310, 331

unit(), 185–192
unit.c(), 188
Units, 185–188, 213–214
updateusr(), 358, 359

upViewport(), 176, 203, 210, 219,
    261, 270
User coordinates, 49,
    see Coordinate systems

validDetails(), 268–270, 282
vcd, 367, 369, 381, 383, 385, 388–
    390, 392–396
vcdExtra, 395
Vector graphics, 307, 484
venn(), 362
Venn diagrams, 362
venneuler, 363, 364
venneuler(), 363
Viewpoints, 432
viewport(), 199–216
Viewport lists, 207
Viewport paths, 208
Viewport stacks, 207
Viewport trees, 207–208
Viewports, 199–212
    in gTrees, 234–235
    navigating between, 200–204
vignette(), 173
visible(), 474
vpList(), 207
vpPath(), 208
vpStack(), 207
vpTree(), 207
VRML, 449
vrmlgen, 448, 449

webvis, 478
Widgets, 470
widthDetails(), 279–280, 283
Wiki, 6
win.metafile(), 308
Wind rose, 375
windows(), 308
windowsFont(), 331
windowsFonts(), 331
windRose(), 375, 377
wireframe(), 124, 126, 144, 439, 451
WMF, 312
write.graph(), 425

Writing graphics functions
    in grid, 257–264
    in traditional plots, 111–114

X-splines, 83, 181
X11(), 308
x11(), 308
xaxisGrob(), 179
xDetails(), 279, 282
xfig, 308, 420
xfig(), 308
xinch(), 96
XML, 476
XML, 485
xspline(), 79, 83
xsplineGrob(), 179
xy.coords(), 112
xyinch(), 96
xYplot(), 351
xyplot(), 124, 126, 135–137, 142,
        146, 493
xyz.coords(), 112, 440

yaxisGrob(), 179
yDetails(), 279, 282
yinch(), 96

zoomplot(), 358, 359